T0143384

INTELLIGENT INTERACTIVE MULTIMEDIA SYSTEMS FOR E-HEALTHCARE APPLICATIONS

INTELLIGENT INTERACTIVE MULTIMEDIA SYSTEMS FOR E-HEALTHCARE APPLICATIONS

Edited by

Shaveta Malik, PhD

Amit Kumar Tyagi, PhD

First edition published 2023

Apple Academic Press Inc.
1265 Goldenrod Circle, NE,
Palm Bay, FL 32905 USA

760 Laurentian Drive, Unit 19,
Burlington, ON L7N 0A4, CANADA

CRC Press
6000 Broken Sound Parkway NW,
Suite 300, Boca Raton, FL 33487-2742 USA

4 Park Square, Milton Park,
Abingdon, Oxon, OX14 4RN UK

Library and Archives Canada Cataloguing in Publication

Title: Intelligent interactive multimedia systems for e-healthcare applications / edited by Shaveta Malik, PhD, Amit Kumar Tyagi, PhD.
Names: Malik, Shaveta, 1987- editor. | Tyagi, Amit Kumar, 1988- editor.
Description: First edition. | Includes bibliographical references and index.
Identifiers: Canadiana (print) 20220244413 | Canadiana (ebook) 20220244421 | ISBN 9781774910221 (hardcover) | ISBN 9781774910238 (softcover) | ISBN 9781003282112 (ebook)
Subjects: LCSH: Artificial intelligence—Medical applications. | LCSH: Machine learning. | LCSH: Interactive multimedia. | LCSH: Multimedia systems.
Classification: LCC R859.7.A78 I58 2023 | DDC 610.285/63—dc23

Library of Congress Cataloging-in-Publication Data

CIP data on file with US Library of Congress

ISBN: 978-1-77491-022-1 (hbk)
ISBN: 978-1-77491-023-8 (pbk)
ISBN: 978-1-00328-211-2 (ebk)

About the Editors

Shaveta Malik, PhD

Shaveta Malik, PhD, is Associate Professor in the Computer Engineering Department (NBA accredited) at Terna Engineering College, University of Mumbai, Nerul, India. She has more than 11 years of teaching and research experience. She is a member of numerous editorial boards and scientific and advisory committees of international conferences and journals. She has been a co-chair at international conferences also. Her research area focuses on image processing, machine learning, deep learning, and artificial intelligence. She received an international best paper award at the 20th ICCSA–2020, University of Caglari, Italy. She is a member of the Association for Computing Machinery and IEEE. She earned her PhD at Lingaya's Vidyapeeth, Faridabad, India.

Amit Kumar Tyagi, PhD

Amit Kumar Tyagi, PhD, is Assistant Professor (Senior Grade) and Senior Researcher at Vellore Institute of Technology, Chennai Campus, India. His current research focuses on machine learning with big data, blockchain technology, data science, cyber physical systems, smart computing, and security and privacy. He has contributed to several projects such as AARIN and P3-Block to address some of the open issues related to privacy breaches in vehicular applications (such as parking) and medical cyber physical systems. He received his PhD from Pondicherry Central University, India. He is a member of IEEE.

Contents

Contributors .... *ix*
Abbreviations .... *xiii*
Preface .... *xix*
Introduction .... *xxi*

PART 1: INTRODUCTION OF MULTIMEDIA IN HEALTHCARE APPLICATIONS **1**

1. **Data Visualization for Healthcare** **3**
S. Usharani, P. Manju Bala, R. Rajmohan, T. Ananth Kumar, and M. Pavithra

2. **Health Monitoring and Management System Using Machine Learning Techniques** **33**
Bharati Patil and Vydeki D.

3. **A Study on Fusion Methods and Their Applications in Medical Image Processing** **53**
P. Nischitha and Chandra Singh

4. **A Study on an Elderly Caretaker Bot Using Computer Vision and Artificial Intelligence** **69**
Chandra Singh, P. Nischitha, Ashwath Rao, Shailesh S. Shetty, and Pavanalaxmi

5. **Knowledge Shift for Candidate Categorization in Lung Nodule Detection Using 3D Convolutional Neural Network** **85**
P. Manju Bala, S. Usharani, R. Rajmohan, M. Pavithra, and T. Ananth Kumar

PART 2: METHODS AND TECHNIQUES FOR SUPPORTING MULTIMEDIA SYSTEMS FOR E-HEALTHCARE **103**

6. **Data Mining Techniques: Risk-Wise Classification of Countries toward Prediction and Analysis of Worldwide COVID-19 Dataset** **105**
Sachin Kamley

7. **Mathematical Model of COVID-19 Diagnosis Prediction Using Machine Learning Techniques** **131**
V. Kakulapati and M. Nagaraju

8. **Detection and Classification of Knee Osteoarthritis Using Texture Descriptor Algorithms** **151**
Anjani Hegde, Rishma Mary George, and H. D. Ranjith

9. **Sensor Cloud-Based Theoretical Machine Learning Models for Predicting Pandemic Diseases** 167
Prashant Sangulagi and Ashok V Sutagundar

PART 3: APPLICATIONS FOR INTELLIGENT AND AUTOMATED HEALTHCARE SYSTEMS 197

10. **Applications of Machine Learning Algorithms in Fetal ECG Enhancement for E- Healthcare** 199
Yojana Sharma, Shashwati Ray, and Om Prakash Yadav

11. **Blockchain for Wearable Internet of Things in Healthcare** 223
Shubhangi Kharche and Rizwana Shaikh

12. **AI-Based Robotics in E-Healthcare Applications** 249
P. Praveen Kumar, T. Ananth Kumar, R. Rajmohan, and M. Pavithra

13. **Automated Health Monitoring System Using the Internet of Things for Improving Healthcare** 271
Vergin Raja Sarobin M., Sherly Alphonse, and Jani Anbarasi L.

14. **Biomedical Data Analysis: Current Status and Future Trends** 297
Amit Kumar Tyagi, S. U. Aswathy, and Shaveta Malik

PART 4: FUTURE RESEARCH DIRECTIONS FOR INTELLIGENT AND AUTOMATED HEALTHCARE ENVIRONMENT 323

15. **Healthcare 4.0 in Prospective of Respiratory Support System and Artificial Lung** 325
Moupali Roy, Arpan Das, Biswarup Neogi, and Prabir Saha

16. **Lifestyle Revolution: The Way to Healthcare, Case of India** 349
Raju K. Kurian, Jyotsna Haran, and Saurabh Ojha

PART 5: SHIFTING OF HEALTHCARE SYSTEMS TOWARD EMERGING TECHNOLOGIES 371

17. **Internet of Things-Based Cloud Applications: Open Issues, Challenges, and Future Research Directions** 373
Siddharth M. Nair, R. Varsha, Amit Kumar Tyagi, and S. U. Aswathy

18. **Genomics and Genetic Data: A Third Eye for Doctors** 395
M. Shamila, Amit Kumar Tyagi, and S. U. Aswathy

Index *425*

Contributors

Sherly Alphonse
Assistant Professor, Senior Grade 2 School of Computer Science and Engineering, VIT University, Chennai, India

S. U. Aswathy
Department of Computer Science and Engineering, Jyothi Engineering College, Thrissur, Kerala, India

P. Manju Bala
Department of Computer Science and Engineering, IFET College of Engineering, Villupuram, Tamil Nadu, India; E-mail: pkmanju26@gmail.com

Vydeki D.
School of Electronics Engineering, VIT, Chennai, Tamil Nadu, India

Arpan Das
Narula Institute of Technology, Kolkata, India

Rishma Mary George
Department of Electronics and Communication Engineering, Mangalore Institute of Technology and Engineering, Moodabidri, India

Jyotsna Haran
Royal College of Arts Science and Commerce, Mira Road, Mumbai, India

Anjani Hegde
Department of Electronics and Communication Engineering, Mangalore Institute of Technology and Engineering, Moodabidri, India

V. Kakulapati
Sreenidhi Institute of Science and Technology, Yamnampet, Ghatkesar, Hyderabad, Telangana 501301, India

Sachin Kamley
Department of Computer Applications, S.A.T.I., Vidisha (M.P.), India; E-mail:skamley@gmail.com

Shubhangi Kharche
Electronics and Telecommunication Department, SIES Graduate School of Technology, Nerul, Navi Mumbai, India; E-mail: Shubhangi.kharche@siesgst.ac.in

T. Ananth Kumar
Department of Computer Science and Engineering, IFET College of Engineering, Villupuram, Tamil Nadu, India; E-mail: ananth.eec@gmail.com

P. Praveen Kumar
Department of Computer Science and Engineering, IFET College of Engineering, Tamil Nadu, India

Raju K. Kurian
Royal College of Arts Science and Commerce, Mira Road, Mumbai, India

Jani Anbarasi L.
School of Computer Science and Engineering, VIT Chennai, India

Vergin Raja Sarobin M.
School of Computer Science and Engineering, VIT Chennai, India; E-mail: verginraja.m@vit.ac.in

Shaveta Malik
Department of Computer Engineering, Terna Engineering College, Navi Mumbai, India

M. Nagaraju
Sreenidhi Institute of Science and Technology, Yamnampet, Ghatkesar, Hyderabad, Telangana 501301, India

Siddharth M. Nair
School of Computer Science and Engineering, Vellore Institute of Technology, Chennai Campus, Chennai 600127, Tamil Nadu, India.

Biswarup Neogi
JIS college of Engineering, West Bengal, India

P. Nischitha
Department of ECE, Mangalore Institute of Technology & Engineering, Mangalore, India

Saurabh Ojha
Royal College of Arts Science and Commerce, Mira Road, Mumbai, India

Pavanalaxmi
Department of Electronics & Communication Engineering, Sahyadri College of Engineering & Management, India

Bharati Patil
VIT, Chennai, Tamil Nadu, India
GHRCEM, Pune, Maharashtra, India
School of Electronics Engineering, VIT, Chennai, Tamil Nadu, India; E-mail: bharti.patil@raisoni.net

M. Pavithra
Department of Computer Science and Engineering, IFET College of Engineering, Villupuram, Tamil Nadu, India

R. Rajmohan
Department of Computer Science and Engineering, IFET College of Engineering, Villupuram, Tamil Nadu, India

H. D. Ranjith
Department of Electronics and Communication Engineering, Mangalore Institute of Technology and Engineering, Moodabidri, India

Ashwath Rao
Department of Electronics & Communication Engineering, Sahyadri College of Engineering & Management, India

Shashwati Ray
Department Electrical Engineering, Bhilai Institute of Technology, Durg 491001, Chhattisgarh, India

Moupali Roy
Narula Institute of Technology, Kolkata, India

Prabir Saha
National Institute of Technology, Meghalaya, India

Rizwana Shaikh
Computer Engineering Department, SIES Graduate School of Technology, Nerul, Navi Mumbai, India;
E-mail: rizwana.shaikh@siesgst.ac.in

Yojana Sharma
Department Electronics and Communication Engineering, Bhilai Institute of Technology,
Durg 491001, Chhattisgarh, India

M. Shamila
Gokaraju Rangaraju Institute of Engineering and Technology, India

Shailesh S. Shetty
Department of Computer Science &Engineering,Sahyadri College of Engineering & Management, India

Chandra Singh
Department of ECE, Sahyadri College of Engineering & Management, Mangalore, India;
E-mail: Chandrasingh146@gmail.com

Prashant Sangulagi
Bheemanna Khandre Institute of Technology, Bhalki 585328, Karnataka, India and Affiliated to
Visvesvaraya Technological University, Belagavi Karnataka, India

Ashok V. Sutagundar
Basaveshwar Engineering College, Bagalkot 587102, Karnataka, India

Amit Kumar Tyagi
School of Computer Science and Engineering, Vellore Institute of Technology, Chennai Campus,
Chennai 600127, Tamil Nadu, India
Centre for Advanced Data Science, Vellore Institute of Technology, Chennai 600127,
Tamil Nadu, India

S. Usharani
Department of Computer Science and Engineering, IFET College of Engineering, Villupuram,
Tamil Nadu, India; E-mail: ushasanchu@gmail.com

R. Varsha
School of Computer Science and Engineering, Vellore Institute of Technology, Chennai Campus,
Chennai 600127, Tamil Nadu, India

Om Prakash Yadav
Department Electrical and Electronics Engineering, PES Institute of Technology, Shivamogga 577204,
Karnataka, India; E-mail: omprakashelex@gmail.com

Abbreviations

AAL	ambient assisted living
AHMS	automated health monitoring system
AI	artificial intelligence
ANFIS	adaptive neuro fuzzy inference system
ANN	artificial neural network
API	application programming interface
ARIMA	autoregressive integrated moving average
ARM	regression, association rule mining
ATM	all time money
AUC	arca under the curve
BBN	Bayesian belief network
BDL	Bayesian deep learning
BERT	bidirectional encoder representations for the transformers
BIDS	brain imaging data structure
BIG	bioinformatics, imaging, and genetics
BLE	Bluetooth low energy
BVP	blood volume pulse
CAD	computer-based diagnostics
CADe	computer-aided diagnosis PC supported location
CART	classification and regression trees
CC	cloud computing
CFR	case fertility rate
CNN	convolutional neural network
COVID-19	Coronavirus disease 2019
CPMP	cases per million population
CPS	cyber and physical systems
CR	correlation
CSV	comma-separated values
CT	computed tomography
CXR	chest radiography
DAS	data acquisition system
DDOS	distributed denial of service
DL	deep learning
DLV	discrete localized variations

DNN	deep neural networks
DOI	degree of impact
DR	death rate
DVH	directed vector histograms
DWT	discrete wavelet transform
ECG	electro cardiogram
EE	end effector
EEG	electroencephalography
EMG	electromyogram
ESN	echo state recurrent neural network
EVD	Ebola virus disease
FDA	The Food and Drug Administration
FECG	fetal ECG
FF	fusion factor
FL	fuzzy logic
fMRI	functional magnetic resonance imaging
FN	false negative
FP	false positive
GIS	geographic information systems
GLCIL	gray level co-occurrence indexed list
GPS	global positioning system
GWAS	genome-wide association studies
HBDF	hospital beds facility
HIS	intensity-hue saturation
HN	head node
HWF	hand washing facility
IHDPS	Intelligent Heart Disease Prediction System
ICarMa	inexpensive cardiac arrhythmia management
ICT	information and communication technology
ICU	intensive care unit
IFR	International Robotics Federation
IoHT	internet of healthcare things
IoT	Internet of Things
KNN	K-nearest neighbor
LDCT	low-parcel figured tomography
LIDC	Lung Image Database Consortium
LOGISMOS	layered optimal graph image segmentation of multiple objects and surfaces
LPT	Laplacian pyramid transform

LiDAR	light detection and ranging
LINAC	linear accelerator
LR	logistic regression
LSTMNN	long short term memory neural network
LTE	4G-Long Term Evolution
MAE	mean absolute error
MALLET	ML for language toolkit
MECG	Mother ECG
MEG	magneto encephalography
MEMS	microelectromechanical systems
MIEE	multisource information exchange encoding
ML	machine learning
MLP	multilayer perceptron
MLR	multinomial logistic regression
MQTT	message queue telemetry transport
MRI	magnetic resonance imaging
MSVD	multi-resolution singular value decomposition
NB	Naïve Bays
NFC	near-field communication
NGS	next-generation sequencing
NLP	natural language processing
NLST	National Lung Screening Trial
NN	neural network
NOI	no. of infection
OA	osteoarthritis
OAP	old age population
OctConv	octave convolution
OSA	obstructive sleep apnea
PA	prophet algorithm
PCA	principal component analysis
PCNN	pulse coupled neural network
PDA	personal digital assistant
PET	positron emission tomography
PHCs	primary health centers
PHEIC	Public Health Emergency of International Concern
PPPs	point-to-point protocols
PRD	percentage root mean square difference
PRDN	normalized PRD
PSNR	peak signal to noise ratio

RBF	radial bias function
RCED-Net	residual convolutional encoder–decoder network
ResNet	residual network
RF	random forest
RFID	radio frequency identification
RL	reinforcement learning
RMSE	root mean square error
RPC	remote procedure calls
RR	recovery rate
RT-PCR	reverse transcription-polymerase chain reaction
SARC	severe acute respiratory syndrome
SC	sensor cloud
SCG	scaled conjugate gradient
SCS	sensor cloud server
SE	spin echo
SEIR	sensitive exposed infectious recovered
sEMG	surface electromyogram
SHS	Smart Hospital System
SDAE	stacked denoising autoencoder
SIR	susceptible–infectious–recovered
SISTA	signals, identification, system theory and automation
SLA	service-level-agreement
SN	sink node
SNR	signal-to-noise ratio
SSIM	structural similarity index measurement
SSO	single sign-on
SVM	support vector machine
SVR	support vector regression
TE	echo time
TN	true negative
TP	true positive
THI	temporal health index
TLS	transport layer security
TR	repetition time
VAR	vector auto regressive
VOI	volume of interest
VM	virtual machine
USG	ultrasonography
UWB	ultra wideband

WBANs	wireless body area networks
WIoT	wearable internet of things
WHO	World Health Organization
WMC	wearable music creator
WSN	wireless sensor network
WSR	wireless sensor network
WT	wavelet transform

Preface

A tremendous amount of multimedia data or big data is produced every day. These multiple data include audio, video, image, text and are available in various applications in the healthcare sector, education field, environment field, and automobile industry. Eventually, it is difficult to store, collect, and manage such big data analyses. It is a demanding research topic and has attracted significant attention in the community of multimedia. Multimedia data are large unstructured and heterogeneous data. It is the field integrated with text, audio, video, graphics, animation, and much more. All the information in various forms can be stored, handled, and communicated digitally. It can be accessed, played, recorded, information displayed, and also presented live.

We have seen various multimedia applications such as natural language processing, speech recognition, identifying diseases types in healthcare applications (using images), biomedical imaging, etc. in the recent years. Deep learning is also a growing field of artificial intelligence that is focused on neural networks. At present, we use some existing deep learning models such as CNN, ANN, etc. CNN is a convolutional neural network that is well used for analyzing images such as MRIs and X-rays. Also, these models can be used during the COVID-19 pandemic.

The spread of COVID-19 coronavirus has been infecting people around the world, but artificial intelligence may help in detecting the virus. For example, in a Singapore hospital, public health facilities are performing real-time temperature checks, and in Wuhan (China), they are using drones and robots to deliver medicines and other necessary things to patients. They sprayed disinfectant and sanitized the surrounding vicinity.

Apart from that, this technology is connected with the Internet of Things (IoT), Internet of Nano Things (IoNT), and Internet of Medical Things (IoMT). IoMT is the internetworking of physical (smart) devices connecting through sensors; the idea behind this is to connect or interface any smart devices or gadget to the Internet. Numerous technologies are converged to produce novel dimension of facilities toward e-healthcare services (for curing diseases in pre-stages) that expand the quality of life of patients and health officials.

The purpose of this book is to guide you in directing the concepts of interactive multimedia with the help of artificial intelligence in healthcare and to explain how the latest technologies, that is, artificial intelligence and deep learning, will help and manage the data in every field and how IoT will connect or help with artificial intelligence. Artificial intelligence, including modern techniques such as deep learning, is being used for solving complex pattern recognition problems.

This book also links artificial intelligence, deep learning, big data, and IoT smart devices) together and suggests how IoT will help in many applications such as healthcare with machine intelligence/artificial intelligence.

This book can be used by undergraduate or graduate students who plan their careers in either industry or research, and by software engineers who want to begin using multimedia with artificial intelligence, deep learning, big data, and IoT to develop healthcare applications.

Introduction

ABSTRACT

Today, we can see major development in many sectors due to recent developments in technology; for example, post-COVID-19 intelligent analytics is used in healthcare. Many countries have used artificial intelligence (AI) and the Internet of Things (IoT) together to track, monitor, and diagnose patients. Intelligent analytics (or advanced analytics) is a part of intelligent automation. Also, it analyzes a large amount of data without any interruption, that is, it makes computer systems more decisions supported. In other words, computers are able to provide intelligence in doing tasks and do every task efficiently and perfectly (without any delay/error). With this feature, many organizations have built several vaccines to overcome COVID-19 virus. If we think about the previous centuries, we took several years to develop vaccines for curing disease such as TB, polio, H1N1 virus, etc. Hence, this book tries to include some good attempts from several researchers/scientists from around the world. Also, this work provides a basic introduction to this topic and its related terms in brief.

INTRODUCTION: INTELLIGENT INTERACTIVE MULTIMEDIA

The spread of COVID-19 coronavirus is increasing day by day and it impacts lives and economy but artificial intelligence (AI) may help in finding the virus. For example, in China, Unified Video Dissemination System (UVDs) robots help in disinfecting rooms that are contaminated with virus or bacteria by spreading UV light and detect fever in public places. In a Singapore hospital, public health facility perform real-time temperature check and n Wuhan (China) they detected CT scan images of 51 patients with laboratory confirmed COVID-19 pneumonia and more than 45,000 anonymized CT scan images through AI and they claim 95% accuracy of model. Here, intelligent automation systems for healthcare come into picture to help patients and healthcare officials. Interactive multimedia enhances the user's experience as compared to the traditional media. The interactive medium needs one of the subsequent elements in order to do so:

- Moving images and graphics
- Animation
- Digital text
- Video
- Audio

By manipulating one or more of these elements, a user can participate in this experience, somewhat what traditional media does not propose. The interactive multimedia integrate computer, memory storage, data, telephone, television, and other information technologies and interacts and allows the user to combine, control, and manipulate different types of media, that is, text, sound, video, graphics, and animation. The importance and the size of multimedia grew rapidly over the last few years. In today's world, people are surrounded by interactive multimedia and you can find number of examples everywhere, that is, in social networking websites such as Facebook, Instagram, Twitter, etc. These social networking sites use text, video, and permit users to share the photos and information related to chat. Even through applications in smart phones, interactive multimedia can predict the weather, location, etc. Applications also help you to shop, for example, through number of shopping sites such as Amazon, Flipkart, etc.

MULTIMEDIA SYSTEM FOR HEALTHCARE

The most important rapidly growing part of multimedia is the medical field or the healthcare field. In the 21st century, we cannot even think without computers. There are number of multimedia applications in the healthcare domain, that is, teleconsultation, telediagnosis, etc., to store medical data and forward later. It is available in many more areas, for example, radiology, dermatology, and pathology. Medical images are used for monitoring of diseases, diagnosis, and treatment purpose.

INTELLIGENT INTERACTIVE MULTIMEDIA SYSTEM FOR HEALTHCARE APPLICATIONS

Intelligent interactive multimedia technology is developed to assist people in the healthcare domain, that is, to monitor remotely or on site medical visit. There is a wide range of wearable devices to monitor the lifestyle of a person, that is, sleep quality, activities, and to measure the health parameters, that is, blood pressure, body temperature, heart rate, etc. M-Health is the mobile health used to monitor the health of the patient using the mobile. Artificial intelligent, big data, and IOT have been applied in number of applications in healthcare. Moreover, various sensors are used in applications of healthcare, that is, microphone sensors, camera sensors, etc. Electronic data or dataset that is related to health is complex to maintain or manage but it can be managed by big data analytics,. Big data in healthcare includes healthcare payer provider, that is, EMR prescription of pharmacy, records of insurance, etc, and along with genotyping, data of gene expression and other data are fetched from the smart web of Internet of Things (IOT). There are lots of IoT devices that are used in healthcare including biosensors, wearable devices, Fitbit, health tracking devices, etc. AI and machine learning also plays an important role in healthcare. It helps the patient in data-driven decision-making. Through AI and machine learning algorithms, we can do the retina scan, record skin color changes, predict diseases, etc. With AI technology, now doctors are focusing more on patients rather than entering the data and administrative work. Doctors can see and monitor patients remotely. In COVID-19, AI technology reduced the burden of the processes by accelerating monitoring and diagnoses. IoT in scanning can be done which helps the algorithms to learn and improve accuracy. Note that in many countries, robots/AI have been used to track, monitor, and diagnose COVID patients.[25] In fact, in future, robots can be used in rural areas to help doctors in examining and treating patients by disinfecting the hospital rooms, automating the labs, transporting medicines, etc.

MOTIVATION AND SCOPE

The technologies related to AI, blockchain, cloud computing, etc are powerful and big, AI uses algorithms to learn large amount of healthcare data with the help of features extraction. It reduces the errors related to diagnosis and therapies. The AI system gives better accuracy as AI can be prepared with self-correcting and learning abilities. Moreover, AI helps in analyzing the data of the patients and predicts the result or health outcome in real time. It also reduces the healthcare risk. The tracking of data at real time helps in recovering the patient health and reduces the number of death cases.

Scope: AI, big data, blockchain, cloud computing, these technologies are very effective. It is having a massive influence on the healthcare industry. The number of facts are listed below:

- The leading cause of the patient death can be hospital error but that can be reduced or prevented by AI.
- In the healthcare industry, natural language processing (NLP) algorithms help in finding the patterns among thousand patients it leads to finding or helping to find the active treatment.

ARTIFICIAL INTELLIGENCE APPLICATION AND MACHINE LEARNING-BASED SOLUTIONS FOR HEALTHCARE

IoT can help in preventing the number of diseases, that is, COVID-19. Internet of Medical Thing (IoMT) has made health officials' or doctors' lives easier as they can use smart devices to identify diseases. The benefits of intelligent and interactive multimedia systems for healthcare sector are:

- **Early diagnosis and detection of infection:** It provides faster diagnosis of COVID cases by scanning parts of the body, by computed tomography (CT), and by many more different types of imaging technologies.
- **Monitoring treatment:** Through AI and machine learning algorithms, one can automatically monitor diseases and predict the spread of the virus. Neural network gives better results in predicting or detecting of the disease and also provide the updates of patient to day-to-day life related to COVID-19.
- **Development of drugs and vaccine:** AI helps in analyzing data related to COVID-19. This technology helps in the development of drugs and vaccines. It helps in testing the drug in real time. It is also helpful for clinical trials. It is processed faster rather than testing the drug and vaccine through standard manual process.
- **Prevention of the disease:** Through real-time data analysis, AI provides updates and analyzes the data and provides prediction for the prevention of disease. It predicts the need of beds, virus-infected sites, etc. It also provides the information related to the future diseases or virus. After that blockchain technology also plays an important role in healthcare. It helps in recording of pharma, clinical trials, data storage, and a way that it will be unambiguous and relevant. Handing data in a correct way eventually helps in reducing the cost and improves the health results for everyone.

There are number of applications of AI, machine learning, and big data, which can be depicted as shown in Figure 1.

- **AI in astronomy:** AI is used to solve problems of the universe and also helps in understanding how it works and its origin.
- **AI in healthcare:** AI in healthcare helps in diagnoses of any kind of disease in a faster way. It also helps the doctors to monitor the patients remotely.
- **AI in gaming:** AI is useful in gaming as machine gives faster response than human in tactical games, for example, chess.
- **AI in finance:** AI is also very helpful in finance. Machine learning algorithms are used in implementing chatbot, automation, etc.
- **AI in data safety:** The safety of data is a very important part in cyber security. AI helps in making the data safe and protected, for example, AEG bot, etc.
- **AI in social media:** Nowadays, AI is very common in the social media sites as in Facebook, Twitter, etc. These sites maintain millions of user profiles and AI helps in managing the enormous amount of data.
- **AI in robotics:** AI has an amazing role in robotics. Robots nowadays play a major role in healthcare also. AI creates intelligent robot; these robots can carry out tasks without being explicitly programmed, for example, humanoid robots.
- **AI in entertainment:** AI in the entertainment services is very popular such as in Netflix or Amazon. Machine learning algorithms are used in it through algorithms and it analyzes the activities of customer and recommends shows for viewers.
- **AI in agriculture:** AI in agriculture helps in monitoring crops. Automatic water supply to plants, detection of AI disease, etc. are some of the features.
- **AI in education:** The concept of automated grading came through AI. In education, chatbots are used as teaching assistants to help students.

Industries impacted by the AI revolution: A major revolution of AI in the previous decade is being used almost in all sectors of society.

Today, AI has reduced the workload of many people and has kept tracks of diagnoses, etc., on behalf of human beings; for example, now IoT can be used as IoMT (a variant of IoT) for monitoring patients and AI can recommend appropriate solutions to patients at the required time or in any emergency. Note that several challenges and issues with IoT and AI/machine/deep learning have been discussed by many authors in Refs [14–24].

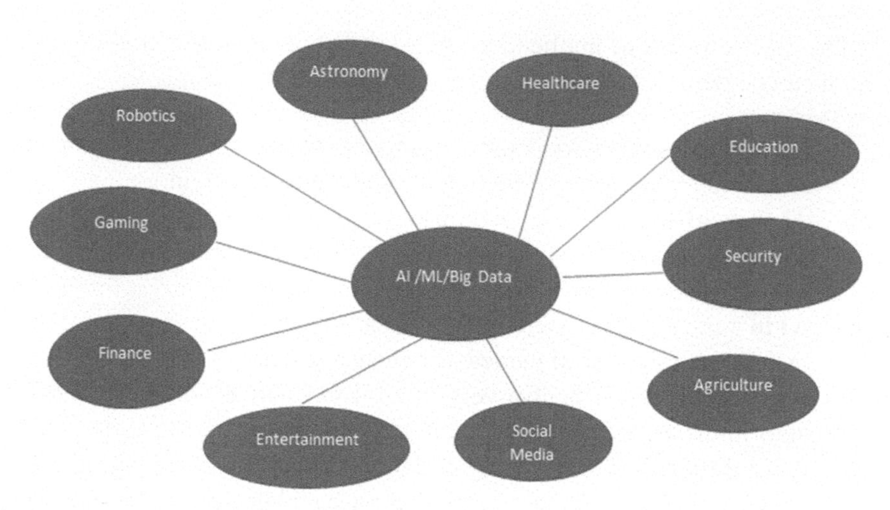

FIGURE 1 Artificial intelligence revolution.

RELATED WORK AND BACKGROUND

AI and machine learning have evolved intensely over the past five decades. Since the advent of machine learning and deep learning, applications of AI and machine learning have expanded from 1950s till 2020. AI is being used to help in many areas such as agriculture, medicine, manufacturing, healthcare, etc. It is increasing day by day to improve the efficiency and accuracy in all these sectors.

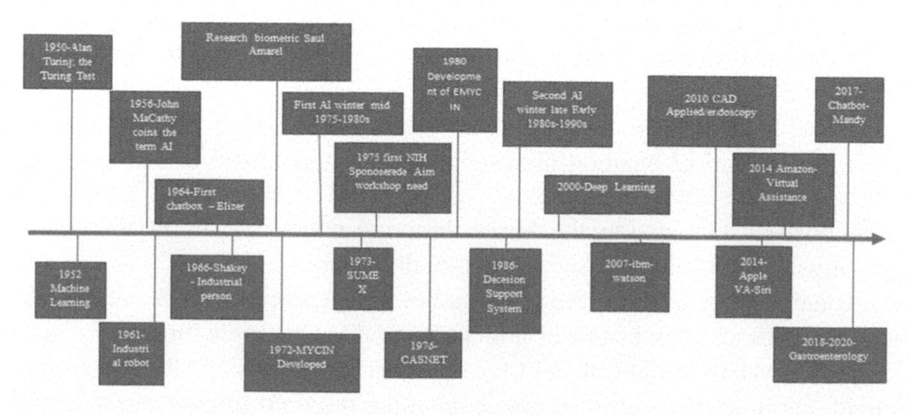

FIGURE 2 Artificial intelligence (AI) development year wise.

Figure 2 provides a chart of major works done in healthcare by AI or other sectors by researchers/scientists from around the world.

- 1950–1970s: AI is focused on the machine or the development of the machine that behaves like human, that is, machine that thinks or mimic humans.
- After a few years, in 1964, Eliza was introduced by Joseph Weizenbaum. Eliza was able to do the pattern matching methodology or concept using NLP.[1,2]
- In 1966: Shakey, developed by the first electronic person, was created at the Stanford Research Institute to interpret instructions for mobile robot.[3]
- In 1970s–2000s: In 1970s, the MYCIN was developed which is a "backward chaining" of AI system.[4]
- In 1986, DXplain that was released by the University of Massachusetts as a decision support system.[5]
- From 2000 to 2020:
- In 2007, Watson was created by IBM which is an open-domain question–answering system. It won the first place on the television game show *Jeopardy!* in 2011 while competing with a human.[6]
- In 2011, Siri was introduced and in 2014 Alexa, Amazon's virtual assistant, was introduced.[7]
- In 2015, Pharma bot was developed as a chatbot to assist pediatric patients in medication purpose.
- In 2017 for primary care practice, automated patient intake process, Mandy was created.[8]
- In 2000 with the deep learning algorithms, the limitations of large data set were overcome.[9–10]

CRITICAL CHALLENGES AND FUTURE OPPORTUNITIES TOWARD ARTIFICIAL INTELLIGENCE—INTERNET OF THINGS-BASED SYSTEMS

From the last two to three decades, there are number of advanced information technologies such as, mobile communication, IoT, wearable commuting, big data, AI, and machine learning that are used widely in the healthcare sector. There are few challenges for the storage of data, data management, and moreover in data processing:

- **Large Scale**: The advancement of electromedical and wearable devices, the volume of data has been widely increasing.[11]
- **High Throughput**: Data is released promptly by electromedical and wearable devices.[8]
- **Various Forms**: Data come from various sources related to healthcare, that is, hospital record, medical record, medical images. These data are collected from the different sources and in different standards.
- **Secrecy of Data and Ethical Concerns:** As the impact of AI is increasing, parallelly, the threats are also increasing. So, it also affects the privacy of the healthcare data.

FUTURE OPPORTUNITIES

AI in the healthcare sector is growing rapidly. In future, AI, machine learning, and eventually cloud can also be used in many applications such as robot-assisted surgery, assistants of virtual nursing, and automated image diagnosis. There are a few examples of latest tools or technology that influence AI and its subsets to expand several areas of medicine and healthcare. These are:

- **Virtual Assistant**: AI technology helps the patients to give reminder for their daily routine, for example, Alexa (from Amazon Echo Dot) gives reminders to Alzheimer's disease patients to eat, bathe, and take medication.[12]
- **Robotic Assistant Therapy**: Robotic arm helps patients in stoke recovery. Robotic arm or hand helps the patient to detect motion.[12]
- **Caption Guidance**: Images or echocardiographic images of a patient's heart can be taken through machine learning or AI technology. Machine learning trains the software to take high-quality 2D ultrasound images of the heart and record video clip of it. This technology has changed the way how heart diseases are diagnosed.

Hence, readers (including future researchers) are suggested to read various applications of AI in the current era[13] and also about open issues and challenges in IoT-based applications.[17–23] They can learn about issues and critical challenges in AI-based solutions (in current and for next decade).[14–16, and 24–27]

CONCLUSION

Intelligent automation and intelligent analytics are the need of many businesses today for expanding the value of their products or increase productivity and satisfy users/consumer all around the world. Intelligent multimedia systems, which are intelligent and automated, have the capability to take decision in minimum time and can help the society to live for a longer time. Today, technology is saving billions of lives and many scientists are conducting research on several problems which are related to human beings which impact human beings directly or indirectly such as caring of driver during accident, curing patient during highly critical operations, etc. We can see major development in the near future with respect to the combination of AI, IoT, blockchain, cloud computing, and big data. Readers (researchers/ scientists) are suggested to read this book and find out innovative solution/ for conducting research on the same problems in the near future.

Disclaimer: The papers cited on Machine Learning/ Deep learning, etc., by the Editors in this introduction section are only given as examples for future reference (for readers/ researchers). To leave any citation or link is not intentional.

KEYWORDS

- **emerging technologies**
- **e-healthcare**
- **intelligent automation**
- **interactive multimedia systems**

REFERENCES

1. Moran, M. E. Evolution of Robotic Arms. *J. Robot Surg.* **2007,** *1*, 103–111.
2. Weizenbaum, J. ELIZA da Computer Program for the Study of Natural Language Communication Between Man and Machine. *Commun. ACM* **1966,** *9*, 36–45.
3. Kuipers, B. F.; Hart, P. E.; Nilsson, N. J. Shakey: from Conception to History. *AI Magazine* **2017,** *38*, 88–103.
4. Weiss, S.; Kulikowski, C. A.; Safir, A. Glaucoma Consultation by Computer. *Comput. Biol. Med.* **1978,** *8*, 25–40.

5. The Massachusetts General Hospital Laboratory of Computer Science. Using Decision Support to Help Explain Clinical Manifestations of Disease. http://www.mghlcs.org/projects/dxplain (accessed April 30, 2020).
6. Ferrucci, D L.; Bagchi, S.; Gondek, D.; et al. Watson: Beyond Jeopardy! *Artif. Intell.* **2013,** *199–200*, 93–105.
7. Comendador, B.; Francisco, B.; Medenilla, J.; et al. Pharmabot: A Pediatric Generic Medicine Consultant Chatbot. *J. Automat. Control Eng.* **2015,** *3*, 137–140.
8. Ni, L.; Lu, C.; Liu, N.; et al. MANDY: Towards a Smart Primary Care Chatbot Application. In *Knowledge and Systems Sciences, KSS 2017. Communications in Computer and Information Science;* Chen, J., Theeramunkong, T., Supnithi, T., Tang, X., Eds.; Springer, Singapore, 2017; Vol. 780.
9. Yang, Y. J.; Bang, C. S. Application of Artificial Intelligence in Gastroenterology. *World J. Gastroenterol.* **2019,** *25*, 1666–1683.
10. Vinsard, D. G.; Mori, Y.; Misawa, M.; et al. Quality Assurance of Computer Aided Detection and Diagnosis in Colonoscopy. *Gastrointest. Endosc.* **2019,** *90*, 55–63.
11. Diagnostic Errors in the Intensive Care Unit: A Systematic Review of Autopsy Studies. *BMJ Qual. Saf.* **2012,** *21*, 894–902. DOI:10.1136/BMJQS-2012-000803.
12. https://healthtechmagazine.net
13. avatpoint.com/application-of-ai
14. Nair, M. M.; Kumari, S.; Tyagi, A. K.; Sravanthi, K. In *Deep Learning for Medical Image Recognition: Open Issues and a Way to Forward*, Proceedings of the Second International Conference on Information Management and Machine Intelligence. Lecture Notes in Networks and Systems; Goyal, D., Gupta, A. K., Piuri, V., Ganzha, M., Paprzycki, M., Eds.; Springer, Singapore, 2021; Vol. 166. https://doi.org/10.1007/978-981-15-9689-6_38.
15. Tyagi, A. K.; Chahal, P. Artificial Intelligence and Machine Learning Algorithms. In *Challenges and Applications for Implementing Machine Learning in Computer Vision*, IGI Global, 2020. DOI: 10.4018/978-1-7998-0182-5.ch008
16. Tyagi, A. K.; Rekha, G. Challenges of Applying Deep Learning in Real-World Applications. In *Challenges and Applications for Implementing Machine Learning in Computer Vision*, IGI Global, 2020, pp 92–118. DOI: 10.4018/978-1-7998-0182-5.ch004.
17. Reddy, K. S.; Agarwal, K.; Tyagi, A. K. Beyond Things: A Systematic Study of Internet of Everything. In *Innovations in Bio-Inspired Computing and Applications. IBICA 2019. Advances in Intelligent Systems and Computing*; Abraham, A., Panda, M., Pradhan, S., Garcia-Hernandez, L., Ma, K., Eds.; Springer, Cham, 2021; Vol. 1180. https://doi.org/10.1007/978-3-030-49339-4_23
18. Tyagi, A. K.; Rekha, G.; Sreenath, N. Beyond the Hype: Internet of Things Concepts, Security and Privacy Concerns. In *Advances in Decision Sciences, Image Processing, Security and Computer Vision. ICETE 2019. Learning and Analytics in Intelligent Systems*; Satapathy, S., Raju, K., Shyamala, K., Krishna, D., Favorskaya, M., Eds.; Springer, Cham, 2020; Vol. 3. https://doi.org/10.1007/978-3-030-24322-7_50
19. Tyagi, A. K.; Goyal, D. A Survey of Privacy Leakage and Security Vulnerabilities in the Internet of Things, In *2020 5th International Conference on Communication and Electronics Systems (ICCES)*, Coimbatore, India, 2020; pp 386–394. DOI: 10.1109/ICCES48766.2020.9137886.

20. Shamila, M.; Vinuthna, K.; Tyagi, A. A Review on Several Critical Issues and Challenges in IoT Based e-Healthcare System, 2019; pp 1036–1043. 10.1109/ICCS45141.2019.9065831.
21. Tyagi, A. K.; Nair, M. M.; Niladhuri, S.; Abraham, A. Security, Privacy Research Issues in Various Computing Platforms: A Survey and the Road Ahead. *J. Inform. Assur. Secur.* **2020,** *15*(1), 1–16.
22. Tyagi, A. K.; Nair, M. M. Internet of Everything (IoE) and Internet of Things (IoTs): Threat Analyses, Possible Opportunities for Future, **2020,** *15*(4).
23. Nair, S. M.; Ramesh, V.; Tyagi, A. K. Issues and Challenges (Privacy, Security, and Trust) in Blockchain-Based Applications, Book: Opportunities and Challenges for Blockchain Technology in Autonomous Vehicles, 2021; p 14. DOI: 10.4018/978-1-7998-3295-9.ch012.
24. Pramod, A.; Naicker, H. S.; Tyagi, A. K. Machine Learning and Deep Learning: Open Issues and Future Research Directions for Next Ten Years. In *Computational Analysis and Understanding of Deep Learning for Medical Care: Principles, Methods, and Applications*; Wiley Scrivener, 2020.
25. Tyagi, A. K.; Rekha, G.; Aswathy, S. U. Role of Emerging Technologies in COVID 19: Analyses, Predictions, and Future Countermeasures. SSRN: https://ssrn.com/abstract=3749782 or http://dx.doi.org/10.2139/ssrn.3749782 (accessed Dec 16, 2020).
26. Goyal, D.; Malik, S.; Goyal, R.; Tyagi, A. K.; Rekha, G. In *Emerging Trends and Challenges in Data Science and Big Data Analytics*, Proceeding of IEEE/International Conference on Emerging Trends in Information Technology and Engineering, VIT Vellore, Tamil Nadu, India, Feb 24–25, 2020.
27. Gillala, R.; Malik, S.; Nair, M. M.; Tyagi, A. K. Intrusion Detection in Cyber Security: Role of Machine Learning and Data Mining in Cyber Security, ASTESJ, May 2020.

PART 1

Introduction of Multimedia in Healthcare Applications

CHAPTER 1

Data Visualization for Healthcare

S. USHARANI*, P. MANJU BALA, R. RAJMOHAN, T. ANANTH KUMAR, and M. PAVITHRA

Department of Computer Science and Engineering, IFET College of Engineering, Villupuram, Tamil Nadu, India

**Corresponding author. E-mail: ushasanchu@gmail.com*

ABSTRACT

Data visualization is the pictorial representation of the key indicators and patterns in a data. Compared with a large dataset or numbers, humans are evolved in a way they can understand and analyze a pattern much faster. Our eyes are drawn to colors and shapes. We can easily recognize a circle from a rectangle. Data visualization is a process of representing the data using shapes, patterns, or graphs, such as bar chart, trend/line chart and heat map, etc. It helps people to interpret the data easily and help them identify certain patterns and trends. When we talk about data visualization, an old quote "a picture is worth a thousand words" will help us understand it, that is because a picture will make us understand our current position from the data more quickly, effectively, and sometime more efficiently. That enables the Key persons to understand what is going on at a glance and derive effective decisions based on it.

Day by day, as our world is getting digitalized, the amount of data it creates is growing more and more, and thanks to the new technologies like IoT (Internet of Things), the amount of data gets created day-to-day is growing on a large scale and all the Data science technologies and corporates

Intelligent Interactive Multimedia Systems for e-Healthcare Applications. Shaveta Malik, PhD Amit Kumar Tyagi, PhD (Eds.)

are now focusing on how to manage this huge data volume effectively to benefit them.

The term big data visualization is a process of collecting, processing, and modeling a huge volume of data in an effective manner and representing it in a visualization to get conclusion and deliver the key information in an efficient way. Data visualization does not always mean colorful pictures or simply converting data to graphs or patterns. Sometimes a plain bar chart can be too bland, it fails to get any attention even though it represents a key indicator, and on the other hand a graphic visualization may look nice but completely fails to represent the key point or gives volumes of information. Data visualization on big data is a delicate balance between the visual and the data, as important as how we show it, it is also more important that what we show.

Data visualization is the final target of all the data science technologies, no matter what whether it is big data consolidation, prediction, planning, or using machine learning algorithms. At last the result should be represented in a format which is easy to understand. When it comes to big data, if we are not getting any meaningful insights, it is simply useless to store the volumes and volumes of data. The main objective of Data science is to get needed insights from the collected raw data, to get the valuable information, as we cannot go through the millions and trillion rows of data, hence, we have to represent the huge amounts of data in a format which is easy to understand. A good data visualization always narrates a story, helps to remove data noises, and identify trends and correlations in data.

From a small-scale business to big industries, there is not a single industry or business which will not get benefited from making data more readable and easy to understand. All industries will benefit from understanding their current position, predictions, or analysis, and based on the data point, they take better decision for the future. All the fields, such as Government, Schools, Industries, and IT services, all gets benefited from the better data visuals and more understandable data.

Big data analytics in health care and medicine focus on the analysis and integration of large volume of complex heterogeneous data, such as diseases, symptoms, biomedical data, and electronic health records. To provide a best medical care, many Health Institutes use various big data analytics techniques and proposed numerous health care information models.[3] Figure 1.1 shows the data collection in health industries such as patient history, present situation data, such as, demographic data, CRM, etc. Data visualization on these models helps to improve the patient care, and numerous studies reveal that implementing data visualization techniques in health industries helps the doctors to focus more on the patient care and treatments instead of analyzing the raw data. Data visualization also reduced the human error in interpreting medical test results and helps to focus on the abnormal result values and readings.

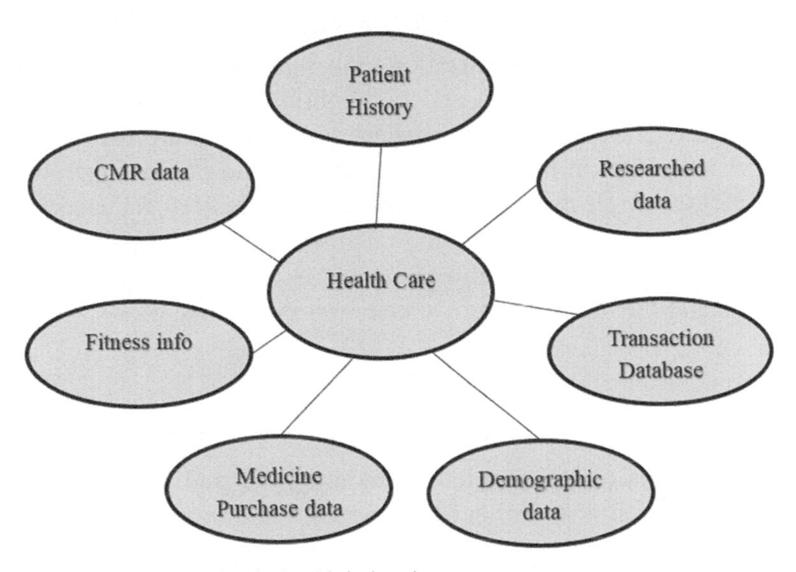

FIGURE 1.1 Data collection in health industries.

As an example, now, most of the countries were affected by COVID-19, in this pandemic, big data analysis and techniques can be used to collect the patient's record all over the world and process it and use data visualization to provide some of the most important information, such as areas which are most affected by the virus, number of patients affected, died, recovered, provides a trend, provides a heat map representation to show the most affected cities and states. The data visualizations can also help the health care professionals to understand the symptoms, test results of the patients, and how they are responding to the drugs in a simple graphical representation instead of comparing each result manually. This is just a tip of a Iceberg, and we discuss the benefits of the Data visualization in the health care industries much deeper in this book.

1.1 IMPORTANCE OF DATA VISUALIZATION IN HEALTH CARE

Data visualization is the process of collecting raw data and providing them in a visual representation like a graph or map. Data visualization enables the human brain to interpret the data irrespective of its size, and visualization also makes them easy to understand and identify outlines, styles, and clusters in big data.

A good visualization must provide insight into huge complex datasets so that the information it provides is clear and concise.

According to Curtis,[4] visualization tools help to reduce human error by providing understandable data and highlighting key focusing points. The tools also set a protective coat for patient safety. There are various benefits of data visualization in health care industries, some of the most important benefits include single dashboard for entire patient history, reduced response time, analyzing trends and patterns, maintaining the centralized narcotics history of the patient, cost of admission, admission and length of stay in ER and diagnosing the disease based on common symptoms, etc.

1.1.1 DIFFERENT FORMS OF VISUALS

In the initial days of information technology, the traditional approach to create visualization is to collect the raw data and structure the data in a relational format and transform the structured data into a visualization, such as bar graph, pie chart, or table. We are still using this traditional way which is the easiest, however, in the past 20 years, information technology developed tremendously in every aspect and vice versa, and the data volume increased along with it. So the customer needs more insights on data for better market advantage to compete in the market. They needed much complex data analysis which paved the way for intricate data visualization options, such as

- Bubble chart
- Cluster graphs
- Scattered plot
- Heat maps
- Gauges
- Radial trees
- Info graphics

When we plan our visualization, we have a lot of choices to choose. So, first we need to make sure that the dataset which we have is cleansed and transformed according to the requirement. If there is any mismatch in the dataset, it will be clearly represented in the visualization and will affect our goal, so the first thing to consider is to create a dataset which has all the information which we need and especially in the correct format. Usually a lot of the visualization tools also offer data quality transformation and cleansing.

Once the dataset is cleansed and is ready to be used, we need to focus on the type of data we are working with, what we want to convey to the audience, and who is the actual stakeholders for the particular data. The data visualization should always focus on the message we want to convey to the audience so depends on the data and the situation, a simple bubble chart may be enough to rely on the information or else we may need to use a complex visualization to convey the message. Figure 1.2 shows the simple bubble visualization for GDP per Capita versus life expectancy in 2015. Any objects or aspect of the visualization which we choose should never distract the user from the main area of focus. Both data and the visualization should work in harmony to identify the information in a connected manner.

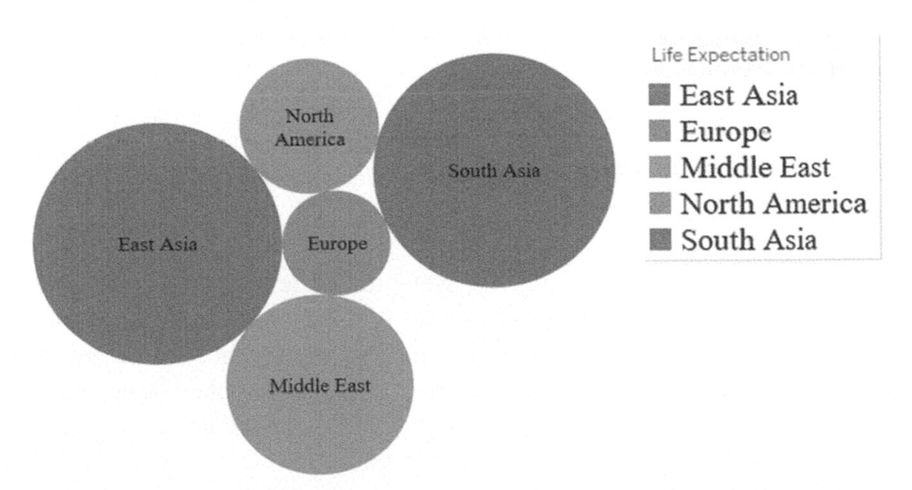

FIGURE 1.2 Bubble visualization.

The world we currently live is a digital world, the amount of data transferred around the world is too big to manage in the traditional way. The Department of Health Care and Science such as FDA, ISDH came up with the regulations that require all the Health Care and Health-related manufactures to record all the logs to maintain the integrity of the data for periodic auditing from the Health Department. Health care industries also started using machine learning to make use of the statistical data to understand the market trend and requirements.

While using machine learning, a great way to analyze the raw data quickly and effectively, it also brings us the question of how we are going to present the end result in a way which is understandable by the business users and stakeholders. This issue is also applicable to the advanced data analysis projects. While working on a complex project, it is essential to monitor and understand the collected dataset to make sure that they are on the right track, and dataset performs as per the expectation. The results obtained from the advanced algorithms are easier to comprehend in a visualization as opposed to detailed text lines or numbers.

Data visualization plays a vital role to any career, for example, by normal citizens to understand their day-to-day expense and also by data scientists trying to mark their presence in artificial intelligence. It would be difficult to imagine a career field where users will not need to better understand the data. Data visualization plays a key role in presenting the result to the key user for decision-making, forecasting, and understanding our current stand in the financial year. Tableau states that, "Data visualization is one of the hot skill to develop, the better we represent our data visually, the better we can use it for our advantage."

1.1.2 DATA VISUALIZATION AND SEO (SEARCH ENGINE OPTIMIZATION)

SEO is a method of improving the website traffic's quantity and quality by organic traffic results, whereas the quality defines the one who is accessing your website is actually the one looking for the service provided by you. For example, if a user wants to buy an apple phone and the search engine suggests your website, whereas you are actually a vendor selling organic apples, then it is not a quality traffic to your website. Usually, people pay a lot of amount to advertisement agents and most of the Search Engine Result Pages are filled with ads, so the organic traffic is the one which is free of cost, that is, the traffic which you get without paying any money.

According to Stewart,[13] People who have worked in Search Engine Optimization knows very well that sometimes it can be cumbersome to explain to the users who do not have any experience in that field, even some well-experienced persons in the field face some difficulties to explain and interpret the data from Keyword Planner, Google Ads, Google Analytics, and more.

This is where data visualization provides a better solution. SEO visualization will help to make the data more accessible to the stakeholders and business users. It also reduces the required time and understands the styles in the optimization method.

1.1.3 SISENSE

According to Hardwick,[10] "There are many elements that impact organic traffic, [and] it is sometimes tough to connect all the points during the optimization process, however, SEO visualization helps us to connect the points and identify whether a new link netted had an effect on the traffic and rankings. It also helps us to compare whether organic traffic is performing better than the PPC traffic, and which keywords brought in leads that become the actual sales." According to Himanshu Sharma,[7] the following SEO fields use data visualization most effectively:

- Back link analysis
- Competitive analysis

When we discuss about competitive analysis, we understand that most of the times we are collecting data on our competitors strategies. The competitive analysis helps us to understand the various aspects of competitors, such as

- Their reach in the social media
- The successful content produced by them
- Keyword techniques they use, etc.

But how we use this information is more important. One simple solution is to visualize the summarized data to better understand the trend. By transforming the data into table, graph, or other visualizations, we can easily look at a huge amount of data at once and hopefully identify the strategies followed by the competitor, that we can leverage it for our own advantage.

Data visualization can also aid us to improve a connection structure movement. We are able to use it to analyze:

- Geographic location of our links
- The quality of our links

The more the information we have, the more the power we have to handle our business in a rightful way, the data visualization will help us to understand the huge volume of data precisely.

Incorporating data visualization techniques in our business plan will ultimately be helpful to us.

No matter whether we use a simple graph or a complex graphic visualization, it will help us to visualize and understand the huge volume of data and observe the trends. Data visualization improves our decision-making more precisely. When we try to analyze strategies to come up with a budget or target, analyzing with clear graphs and linear algorithms can help us to derive the numbers considering current variables which may impact the result. Data visualization enables the business users to come up with right questions and to focus on the areas which are not performing well.

Data visualization will help to identify areas of business which may be affected in the future. It enables us to convey a story of the business in a precise and engaging way.

1.2 DATA VISUALIZATION TRANSFORMING THE HEATH CARE

In health care information system, data are collected from the electronic medical records, data of various departments, patient records, financial data, and administrative data. Nowadays, the patient test results are obtained as records and are collected in a systematic manner, and hence, it can be readily available for review, which reduces the response time by 60%. Since the patient's records are available in a dashboard, and physicians could understand the results and start the treatment and track the results in a single tool or dashboard. Visualization tools also help them to understand any epidemic or pandemic in the world so that they can diagnose the disease and compare the symptoms easily.

According to McCoy,[6] effective, meaningful, and efficient data investigation is at the mind of some most important health care medicinal studies. Throughout their everyday training, all the nurses, physicians, and other medical professionals need to work with health care interfaces and databases to view patient information, reports, and trends. They also generate their health reports using the same interface; shortly, this will result in a huge pile of data collected daily.

So, without a proper visualization, collecting this huge data is not useful to us, a good visualization will help us focus on most important information and identify correlations and patterns on the data.

Many physicians or medical professionals do not need the information of how a study was conducted in a detailed manner, they just need the end result of the study and the key discoveries, they only need the actionable takeaways where data visualization comes into play.

With all these information, it is safe to say that the health care industries will struggle to survive without data visualization, and to be precise, it is not only health care industries, all the industries could barely endure without data visualization.

1.2.1 INFOGRAPHICS AND MINI-INFOGRAPHICS

The common misunderstanding is that when conducting medical studies, it is better to have huge datasets, because bigger the dataset highly precise will be the result. That is true when we conduct medical studies,[8] however, when we finish our studies and post the results, we cannot publish the entire dataset for the sake of transparency, and to initiate discussions, we can include the important dataset in appendices.

But, it would be better to include perfect infographic, an effective form of visual representation, which represents the key takeaways. In this, representation, we should not try to picture everything altogether because that would distract the reader from the key information they were looking for. Hence, we have to highlight few key areas which provide the most compelling data and make sure that it is not grouped with some irrelevant information, for example, we can check out the visualizations published by Organization for Healthcare Research and Quality.

We can choose the same infographic which can be used to share important information to patients and physicians. The patients do not need detailed information on how the study was conducted or other technical details which impact the study results; hence we can use a mini dataset or small infographic which provides them the key takeaways they need. We can also use the mini dataset and share the infographic in the social media for information at the same time securing the key details. Best example of the mini-infographic is shown in Figure 1.3, which illustrates the six main symptoms of COVID-19.

Even people who do not know the language can easily understand by looking at the images. The main objective of infographic is to communicate the complex information in an easy way so that the audience will understand the content readily by looking at the image, even if the audiences do not know the language of the content.

Extended procedure, scrolling infographics are perfect for allocation on our blog or websites. Data visualization which helps medical professionals to analyze a huge datasets can reduce the time taken to understand the data and in turn helps to save the lives of people. A case study conducted with the medical professionals states that using the dashboards and Infographic to visualize electronic health records rescued 65% of the time consumed on data investigation in the initial year alone, with higher gains forecasted for the future.

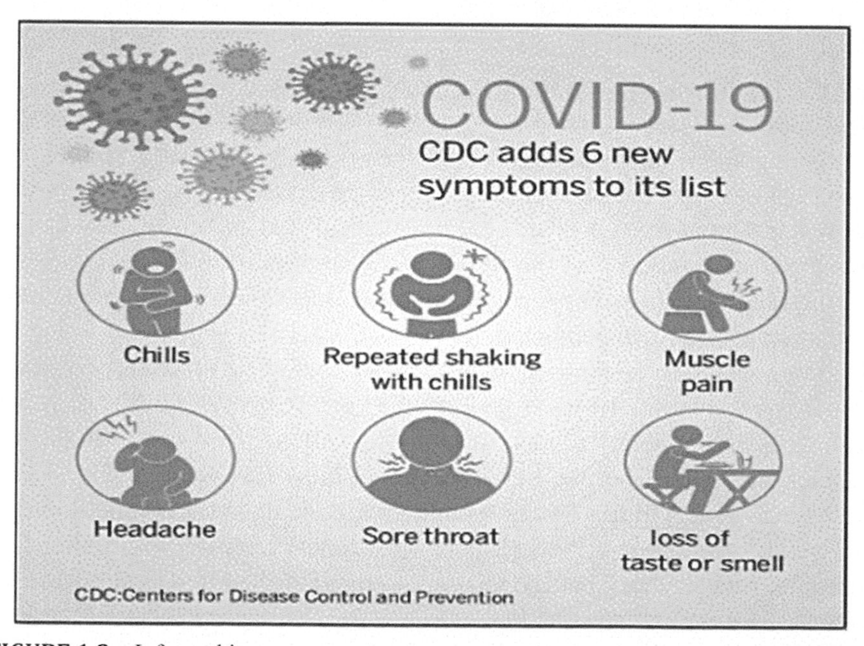

FIGURE 1.3 Infographics.

1.2.2 HEALTH INDUSTRY LEADERS ARE INTERESTED IN DATA VISUALIZATION

In a conference held with Health IT Analytics, Sanket Shah states that, "We do have a lot of data coming from various sources and business leaders who want to understand how they can use this data to figure out why something is happening and what we can do to leverage it to our advantage, the only way to get answer to those questions is to analyze that huge data, so if we are not an extremely accomplished data analyst, it would be difficult to present the data properly. So the only way to understand the data is to present the data in a human readable way. Hence, visualization is the best tool that can be adopted to make the data understandable to the business users."

1.2.3 CUSTOM DATA VISUALIZATION

Nowadays most of the health industries started using their own custom graphs and charts which are designed specific for the data they are presenting. Even better, they are also made cooperative and customizable rendering the data presented. Institute of Health Metrics and Evaluation is an organization that shares such custom graphs and visualizations. The National Center for Health Statistics also provides customizable graphs which help to visualize huge dataset into meaningful visualization.

1.2.4 INTERACTIVE WIDGETS

When we are working with a huge dataset and designing, a data dashboard which is used by everyone including all technical users, patients, and doctors cannot display a static graph which represents the overview. Different users will have different requirements. For example, a CEO may want to look at the overall sales, whereas the Plant Manager wants to visualize the sales of his own plant in detail. After all, different data points are more important for different members of your audience.

If this is the case, then interactive visualization widgets will solve the problem. Interactivity allows the user to filter or customize the graph or the content of the dataset according to their needs and to find the key takeaways they are looking for instead of working with the content which they are not interested in.

1.2.5 MOTION GRAPHICS

When we are working with a series of visualization that work together to help us understand the big picture, it would be better to represent them in the form of a motion picture or graphic, whereas a live video can also be used, but most of the times, the live video also requires some animation overlay to make it more informational and engaging. If you are not a trained data analyst, using simple visualizations most of the time, are the best way to convey the result to the end user efficiently.

For example, a fully animated video, "The History of Vaccines" a motion graphic of Carrington College, provides some necessary data of how the vaccines work and how we are using the vaccines over time. It was organized in a way that it provides a narrative, clear information to the user. It is one of the good examples to show how health industry can use the data visualization to break down difficult thoughts so that it is comprehensible even to a novice people.

1.3 VARIOUS HEALTH CARE METRICS

As always all industries focus on the financial reports mainly, some of the important health care metrics include admission cost, cost of admission by department, number of patients admitted, discharged, length of stay, diagnosis, stock of medicines, blood bank stock, waiting donor list, hospital donations, patients admitted by departments, death rate per department and doctor, number of operations by department, category and doctors, etc.

Health care performance indicators and metrics is a clear definition for the performance measurement which helps to monitor, investigate, and enhance all pertinent health care procedures to rise and improve patient satisfaction. Some of the measurements represented here are essentially the precise key performance indications for health care monitoring and hospitals.[14]

Average Infirmary Stay: Calculate the average time your patients have spent in the hospital.

Total Treatment Costs: Calculate the total cost of treatment to the hospital.

Hospital Readmission Rates: Track Patients repeat rate.

Average Wait Time: Calculate average wait time for treatment to improve customer satisfaction.

Customer Satisfaction: Calculate the patient satisfaction rate.

Patient Safety: Monitor and prevent incidents in our hospital.

Average ER Wait Time: Understand the rush hours in the ER.

Total Cost by Payer: Identify the total cost paid by the Insurance policy.

1.3.1 *AVERAGE INFIRMARY STAY*

Calculate the average time a patient spent on your facilities after admission.

The number of days stayed in hospital is a simple KPI which helps us to identify after admission. What is the total time a patient spent in the hospital? This KPI entirely depends on the type of treatment and rehabilitation time, for example, the average time a patient spent for heart surgery versus teeth removal will have vast difference. So in order to have an accurate result, we need to interpret this KPI using various categories of stays like procedures, treatment undertaken. We can also use the different units of measures that we use in our facility to calculate this KPI. Figure 1.4 shows the average length of stay in the hospital to be 5 days for any person being admitted for several reasons, such as fever, heart problems, diabetes, etc.

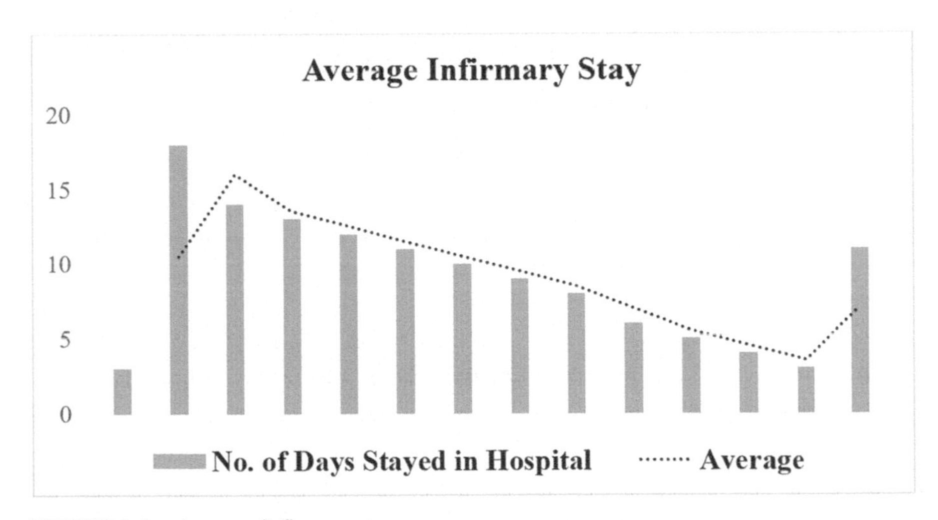

FIGURE 1.4 Average infirmary stay.

1.3.2 *TREATMENT COSTS*

Evaluate what is the total cost of patient to your facility.

Cost to the organization is an important KPI, which directly affects our financial statements, the profit margin, and the ability of our facility to withstand itself. Reducing the treatment cost to the patient helps us earn more profit, so it is important to identify the cost incurred for different classes per units, per procedure, per treatment, or as we have shown in our example, per patient oldness group. By identifying the cost of a procedure, we can also create a better budget and focus the money on areas which need it the most. Usually, the average of a younger patient's cost to the hospital will fall less than the senior patient's cost. Figure 1.5 shows the cost of treatment for different age persons. The graph shows 7350 is the average cost for all aged persons for treatment.

1.3.3 HOSPITAL READMISSION RATES

Calculate the number of patients coming back after discharge for the same problem.

This KPI helps us to understand the number of patients coming back to the hospital within a small duration of time after discharge. Figure 1.6 shows that out of 1700 people, around 830 people were readmitted once again for the same reason or some other reasons. As the chart shows 33% of patients were readmitted for the same reason. Figure 1.7 shows that more patients with the age 40–84 were readmitted in the hospital when compared with patients of other age. Readmission rate is an important KPI which provides a clear picture on the excellence of care provided in the hospital. However, it cannot be used as a stand-alone display for quality. It can also provide information on other defects, such as lack of staff, low stock of material, units which need special care, overloaded staff.[12] It may help us to focus on the areas which are affected and may reduce the cost of expensive and unnecessary readmission.

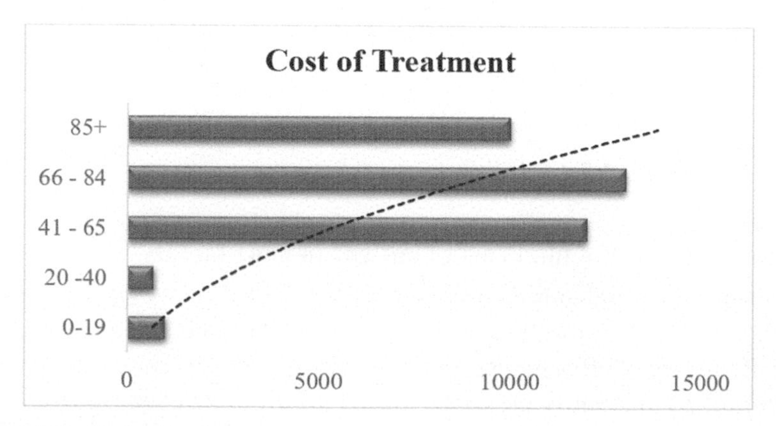

FIGURE 1.5 Cost of treatment.

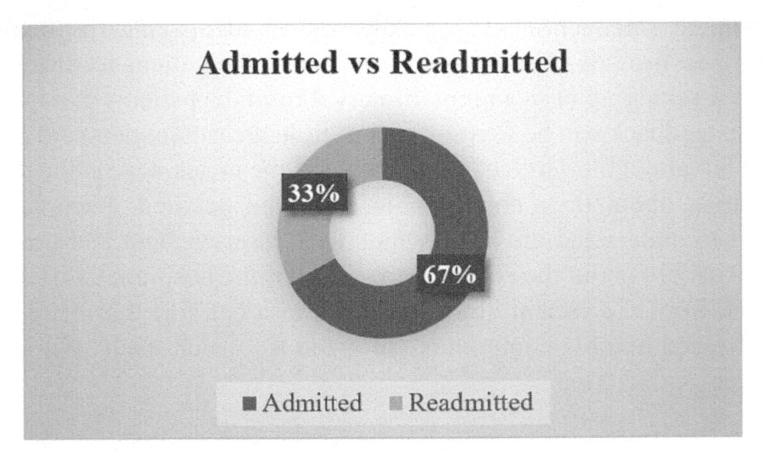

FIGURE 1.6 Admitted versus readmitted rate.

1.3.4 PATIENT WAIT TIME

Evaluate your patient wait time to increase the patient satisfaction.

As people of remote world were running fast, The waiting time in hospital is one of the major issue, and because of the waiting for too long, people do not like to access the health care to get their treatment done. So, reducing the waiting time is one of the important measurements to be taken by the health care system. Patient waiting period in the hospital is one of the greatest factor which impacts patient satisfaction, it calculates the average time taken for the patient visiting the hospital from the time of registration to the time they actually consult a physician and get treated. When we measure this KPI in emergency room, we can evaluate how fast the hospital is providing emergency services to its patients. The average patient wait time is directly related to the patient satisfaction score, as no one wants to spend their valuable time in hospital by waiting for the treatment. Figure 1.8 helps us to identify other factors in the hospital which need focus. Also, Figure 1.8 shows that the average waiting time for people to get treatment in health care is 1 h.

1.3.5 PATIENT SATISFACTION

Evaluate the Patient Satisfaction with the service.

Customer satisfaction is an important KPI for all the industries and business areas, especially in health care industries which work with patient care. Patient satisfaction plays a key role in identifying the quality of service they provide. The hospital can collect the patient feedback when they are discharging or in a regular interval from the patients in a long time care, the feedback can be focused on multiple areas to understand how the patient felt about the services like how good the meals was, is the doctor's explanation about their condition is clear and detailed. Such feedback helps us to understand the areas which need improvement and provides a detailed insight about the patient perception of the hospital. Collecting the feedback from the patient also provides him a comfort that his concerns are addressed and his complaints are listed to, which itself will improve the patient satisfaction.

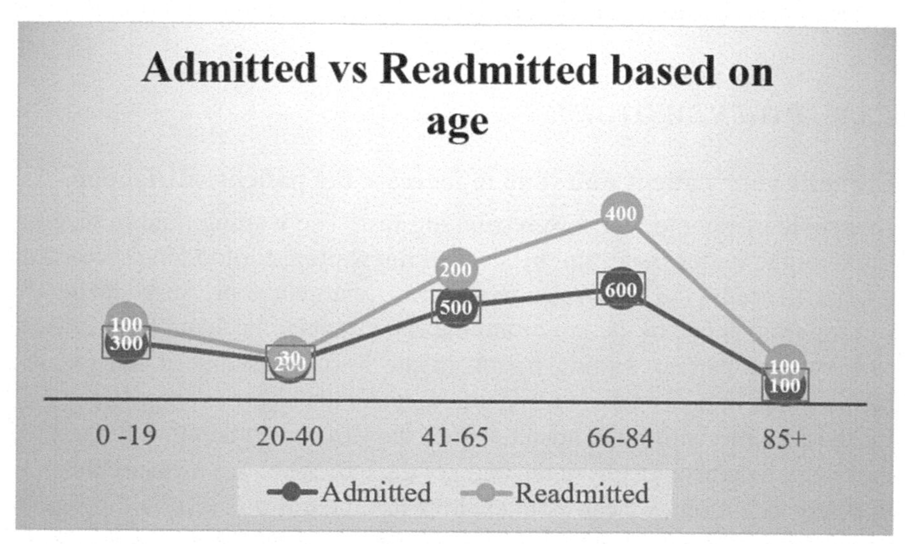

FIGURE 1.7 Readmission of patients based on age.

As Figure 1.9 shows, the good overall patient satisfaction score will help the hospital to attract future patients or on the other side, a low score can scare the patients away, so the higher the patient satisfaction, better it is for the hospital.

The lower the patient wait time, better the patient satisfaction. Patient wait time measured over a period of time helps us to identify the rush hours and trends, which in turn help in organizing the staffing needs and treatment process.

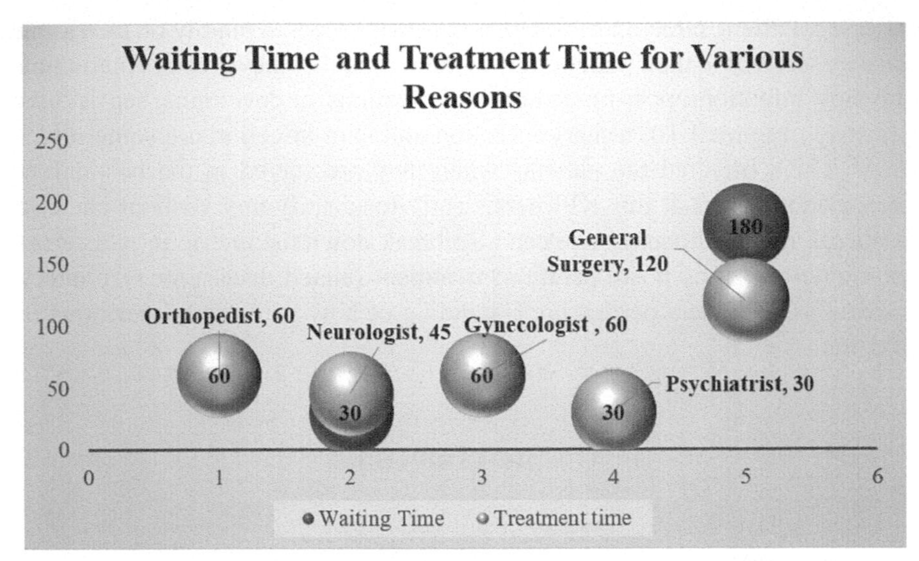

FIGURE 1.8 Treatment time and waiting time.

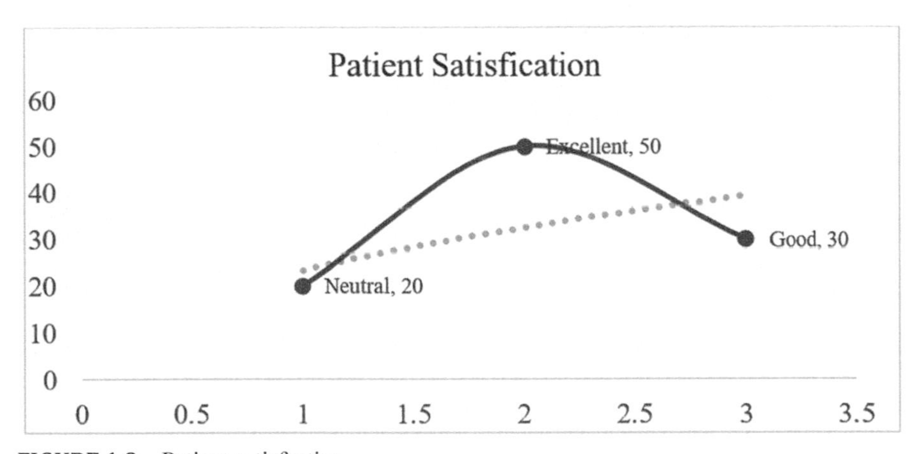

FIGURE 1.9 Patient satisfaction.

1.3.6 PATIENT SAFETY

Measure the number of incidents happening in the facility and take appropriate measures to avoid further risk.

Safety measures are important in all the industries which work with the hazardous elements, heavy machineries, and areas where people may get affected. Patient safety in health care industries focuses mainly on providing quality services to their patients and to make sure that they are not contracting any new infections, post-procedure complications, or developing sepsis.[11] As shown in Figure 1.10, urinary infection and skin infection are some infections which infected the patients when they are stayed in the hospital. It is important to track this KPI persistently to identify any virus or bacteria outbreak in your hospital. We can also break down the metric into different categories, such as post-operation, treatment-related disease or respiratory infections, to have a perfect understanding of how the hospital performs in the matter.

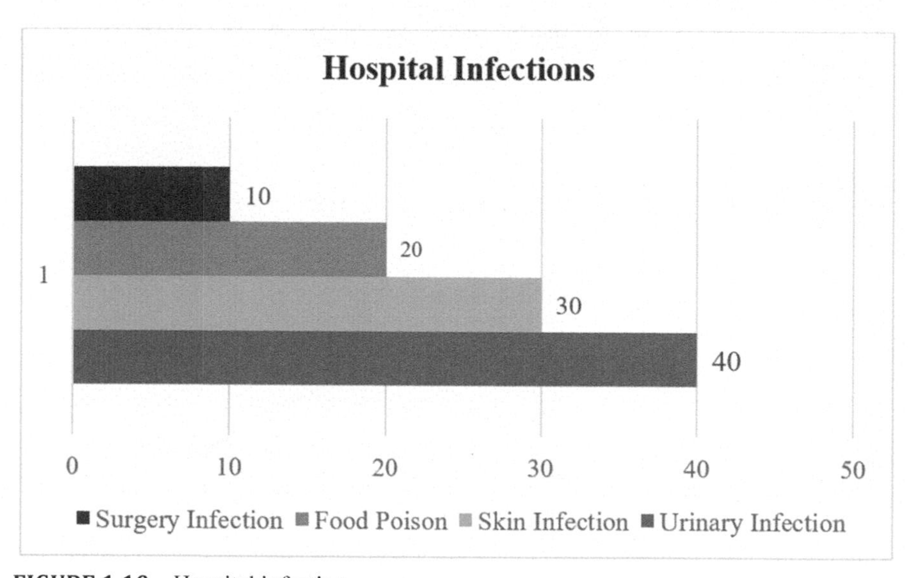

FIGURE 1.10 Hospital infection.

1.3.7 EMERGENCY WAIT TIME

Evaluate the average time patient wait in the ER to see a doctor and get treatment

Emergency situation patient wait time measures the average period a patient waits from the time of his arrival to the hospital to the time he or she actually consulted a doctor and gets treatment. Unlike general Out Patient, patients visit the ER during the emergency time and their condition needs to be asserted as quickly as possible, so by calculating this KPI, especially for the ER, we can understand the trend and identify[5] the rush hours, can set a target for the average wait period. It also shed graceful on other issues like change in current admission process, understaffed or staff-overloaded and take appropriate measures to mitigate it.

As Figure 1.11 shows that around 1 hour is the average emergency waiting time of the person. Once we started evaluating the ER wait time metric, we need to analyze it over time to understand any issues popping up and resolving it on time.

1.4 DIFFERENT TYPES OF DATA VISUALIZATION TOOL

For data visualization, there are different types of tools are available.[1] Some of them are as follows:

1.4.1 TABLEAU

Tableau is a data visualization software founded by Christian Chabot, Pat Hanrahan, and Chris Stolte, in Mountain View, California in January 2003. It helps anyone to understand their data by converting the data into pictorial representation. Any type of database can be easily connected, just by dragging and dropping the data to create their visualization and also the created visual data can easily be shared by a single click.

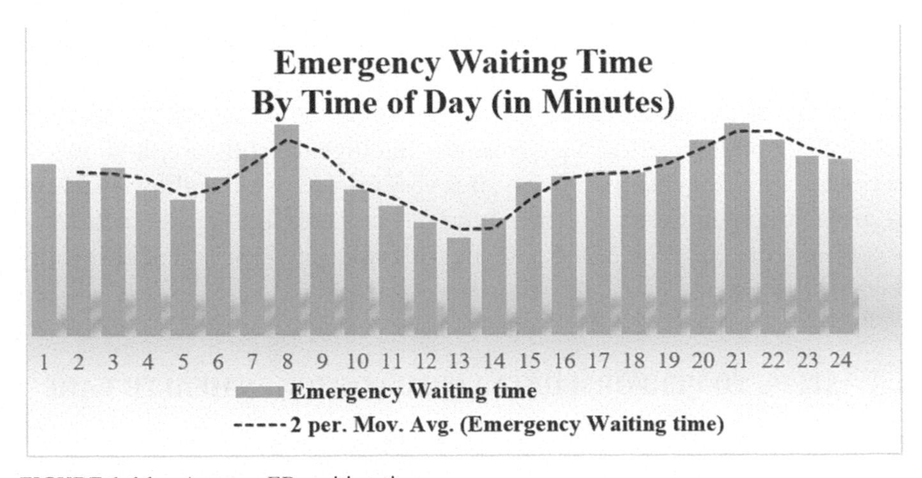

FIGURE 1.11 Average ER waiting time.

1.4.2 FUSION CHARTS

Fusion Charts is a part of InfoSoft Global Private Ltd. Fusion charts software is used to create different types of charts. Fusion Charts is a JavaScript Charting library. It contains almost over of 97+ charts and 1412+ maps to choose from, with incorporations available for all standard JavaScript contexts and back-end coding languages.

1.4.3 QLIKVIEW

QlikView is used to pursuing our data by analyzing with a slightly configurable dashboard, most of the data are static.

1.4.4 HIGH CHARTS

High charts is also a reliable and secure visualization tool, it is pure JavaScript-based projecting library which is intended to supplement web applications by accumulating interactive charting capability.

1.4.5 DATA WRAPPER

Data wrapper is an open source tool to create and publish charts, maps and tables for free.

1.4.6 SISENSE

Sisense is a built in self-service Business Intelligence software, it allows us to visualize the data in an easy way. It is very easy to use and allow everyone to generate a significant representation of graphs easily. In this method of graphical representation of the data, huge datasets are clearly and efficiently done by Sisense.

1.5 STEPS TO PERFORM DATA VISUALIZATION IN HEALTH CARE

1.5.1 COLLECTING THE DATA AS A DATA SOURCE

Data are facts collected together for reference or analysis. There are two kinds of data, they are Primary data Source and Secondary data Source.

Primary data sources are the data collected by the researchers, such as survey reports, experimental observed results, consulted, and collected data from different sets of groups. Secondary data sources are the data collected from different resources, such as internet, research articles, books, libraries, etc. Collection of all such data is used as a data source.

1.5.2 IMPORT THE DATA IN THE DATA VISUALIZATION TOOL

Data may be in any form, such as excel, CSV, database access file, etc., once the data are collected, import the data in to the visualization tool to perform calculations and analyses.

1.5.3 ANALYZE THE DIFFERENT DIMENSION OF DATA COLLECTED

For analyzing the data,[2] in visualization, different dimensions or measures are used for calculations which depend on the data sources.

1.5.4 PREPARE THE DATA FOR VISUALIZE AND MONITOR IN THE VISUALIZATION DASHBOARD

After performing analyses, project or present the data in visualization for better understandability of the results. As mentioned earlier, there are many visualization tools available for different types of visuals to represent the result in better understandability.

1.5.5 TEST AND DEPLOY THE VISUALIZED OUTPUT

Test the result with different visualization tools with different visuals for deployment.

1.6 CASE STUDY: DATA VISUALIZATION OF HEALTH CARE DATABASE

Nowadays, COVID-19 is the major problem in Health industry, so consider a COVID-19 dataset has a case study. In that, the number of cases founded in each region and number of death cases in each region can be visualized which will help to make aware the people to avoid the traveling to that place. Also finding target region using visualization makes that place to quarantine to avoid spreading of virus to other regions.[15]

With the help of visualization curve, people will easily know whether the virus spread is increased or decreased. It is also possible to know the number of people suffering and number of people died also. Also with the help of that information, heath care monitoring[9] system should allocate more resources of medical equipment's, doctors, and care takers for founded places.

1.7 HEALTH CARE DATABASE ANALYSIS USING TABLEAU DATA VISUALIZATION TOOL

As mentioned earlier, here we used a COVID-19 dataset to know the number of infected cases and number of death cases using different charts in tableau data visualization tool. Tableau has various charts, such as stacked bar chart, line graph, doughnut pie chart, side by side scatter plot, bullet graph, tree map, box plot, Gantt chart, etc. We discuss below how each chart and graph is used for visualizing the COVID-19 cases.

1.7.1 BAR CHART

A bar chart is used to epitomize data in rectangular bars with the distance of the bar relative to the value of the adjustable. In Tableau, horizontal bar chart is used to display the number of cases found in different countries. As shown in Figure 1.12, from the collected data of Brazil, so many cases are compared with other countries and territories. So it is easier to identify the countries with more cases of COVID-19 by seeing horizontal bar chart. With the help of this visualization, health care system will supply more doctors and care takers to the necessary places. Also to make awareness for other place people to avoid moving or traveling to the places where more cases are found to break the chain of COVID-19.

1.7.2 CIRCLE CHART

The *circle view* is another powerful visualization for proportional analysis. The number of cases versus different continents are visualized in Figure 1.13

in circle view visualization. From the chart, it is easier to identify which continent in Europe has COVID-19 cases and death. Also, America is having more number of cases and death. From Figure 1.12, the number of cases is easily identified by the bar chart. But the number of continents in the countries can be found from the circle chart. The number of cases in each continents are also easily viewed by the upcoming charts.

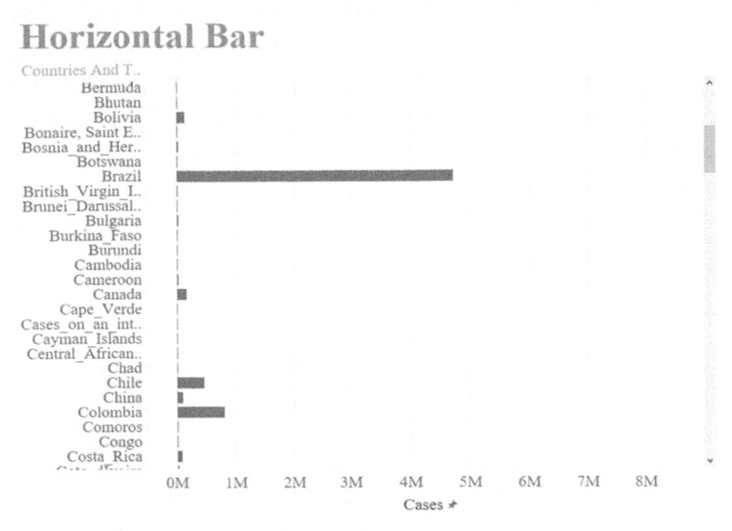

FIGURE 1.12 Horizontal bar chart.

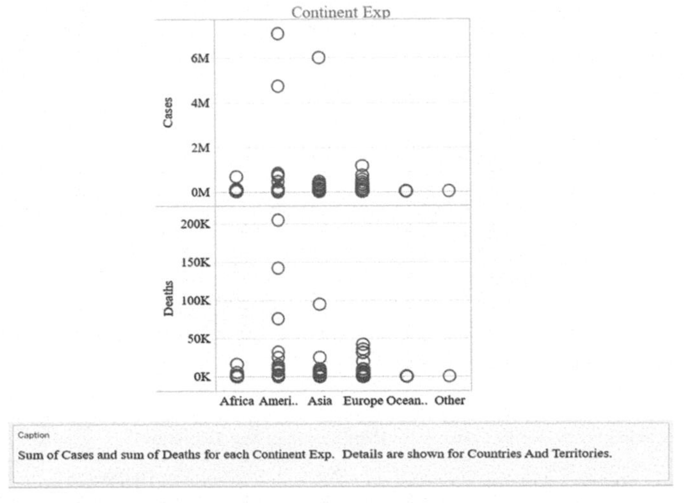

FIGURE 1.13 Circle view.

1.7.3 PACKED BUBBLES

We have learned from the bar chart and circle view chart that they both show some specific visualization of COVID-19 cases. But packed bubbles is a bubble chart which displays the data as a cluster of bubbles. The number of cases in different countries and continents is visualized in Figure 1.14. As the chart displays, in America, United States of America is having more number of cases rather than other countries in the same continent. In the same way for Asia, India is having more cases compared with other countries in the same continent. The packed bubbles show the clear picture of places where more cases of COVID are seen in the same continent. This helps the health industries to make necessary steps to reduce the cases and also to supply more medical equipment's to cure the patients. Also the analysis value of the continent can be easily visualized in Tableau. For easy understandability, the caption of the chart shows the data used for analysis.

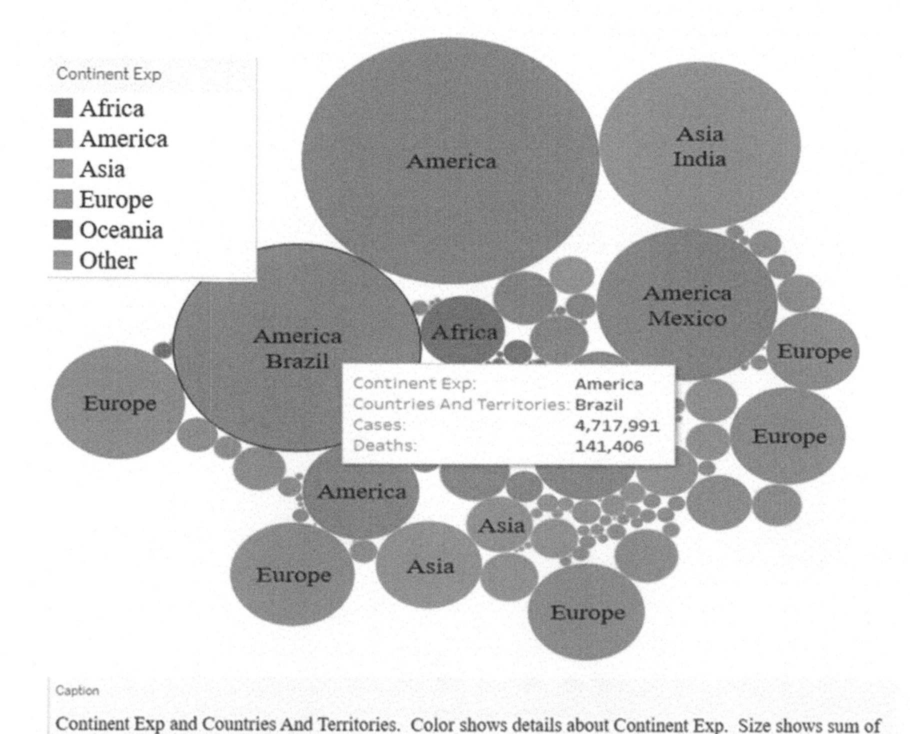

FIGURE 1.14 Packed bubbles.

1.7.4 TREE MAP

The tree map visualize the data in box of rectangles such as the nested one. The dimensions define the organization of the tree map and help to evaluate the individual rectangle based on the size and color. Size and color of the rectangle define the value and the data. If the measured value of the data is more, then the size of the rectangle is large and the color of the rectangle has more density. If the value of the data is less, then the size of the rectangle is small and the color density of the rectangle is less. The number of cases in different countries and continents are visualized as shown in Figure 1.15 in Tree map pattern. From the Tree map visualization, it is clear that India and America (USA and Brazil) have more cases compared with other places in the continents. As mentioned earlier, Tree map also helps to know the places where the cases are more, which helps the government and health care monitoring system to caution the people. By comparing the bar chart with tree map, the visualization of the number of infected cases and death count is obtained in detailed manner.

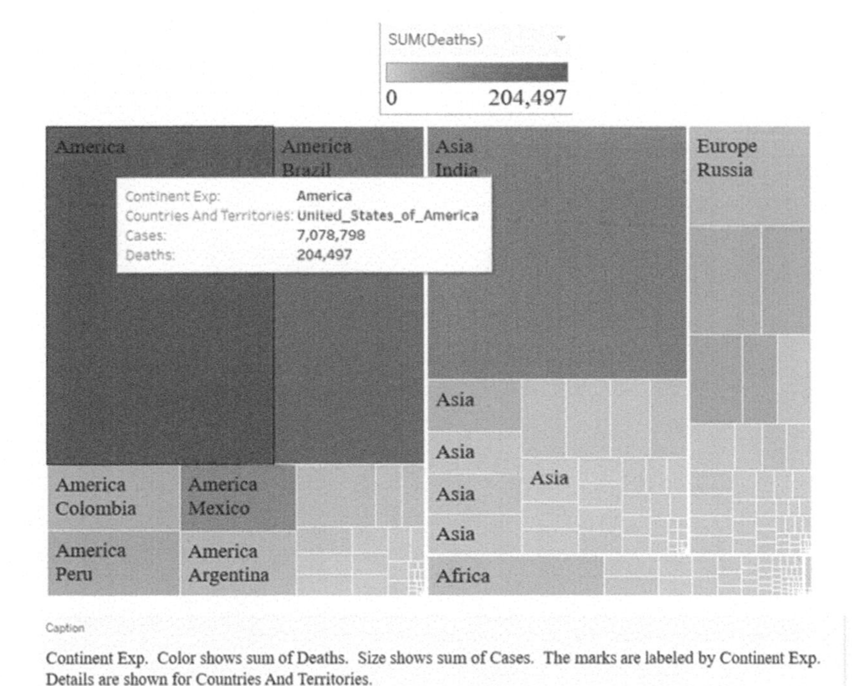

FIGURE 1.15 Tree map.

1.7.5 SCATTER PLOT

A scatter plot displays many points scattered in the XY plane. It is formed by plotting values of mathematical variables as X and Y coordinates in the XY plane. In this case, we used side by side circle to plot the number of cases in different countries and continents to obtain a scatter plot as shown in Figure 1.16. Side by side circle is one kind of scatter plot used in Tableau. We can also compare the visualization of side by side circle chart with circle view chart. In circle view, either the number of COVID-19 cases or number of death cases can be visualized. But in side by side chart, both the number of COVID-19 cases and the number of death cases in the continent can be easily visualized in a same chart. For example, the number of cases in America is more when compared with other places, but the number of death cases is more or less common as in other countries. Such analyses are easily performed by visualization which helps the health care industries to monitor and produce the needed equipment's to the necessary places.

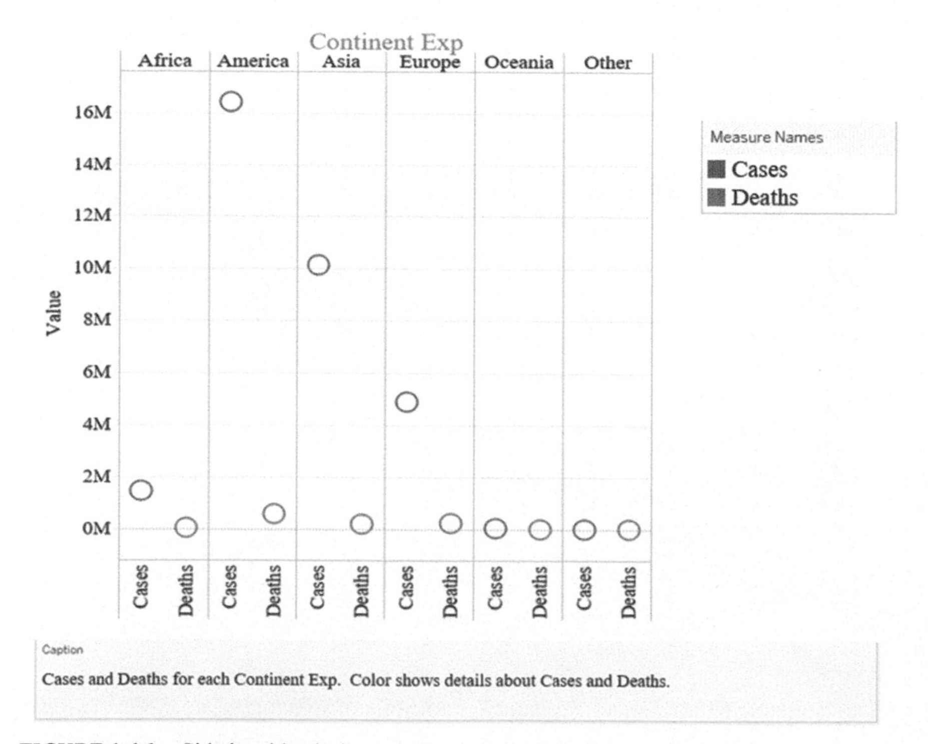

FIGURE 1.16 Side by side circle.

1.7.6 *LINE CHART*

A line chart is used to measure the value of the given data and to represent the chart in the XY dimension that is engaged beside the two axes of the graph area. Here, we used continuous line to plot the number of cases and number of death cases in different continents based on the day and month order as shown in Figure 1.17. The side by side circle chart will show the number of death cases. But the line chart will show the graph to understand the number of COVID-19 cases and number of death cases in increasing or decreasing manner. With the help of this graph, health care industries can reduce the resources and allot the resources to the facility needed places.

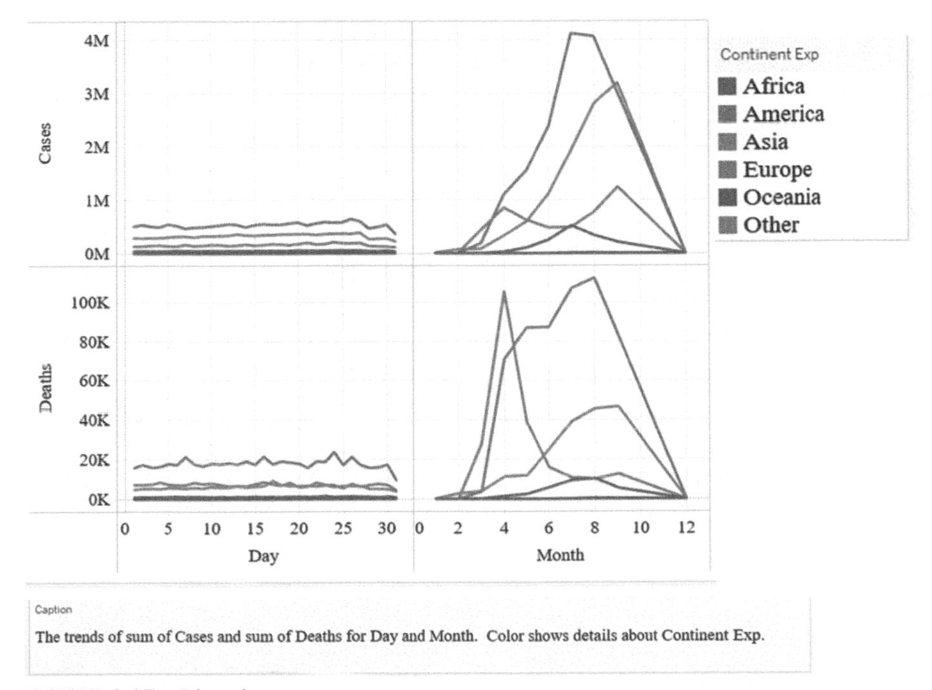

FIGURE 1.17 Line chart.

1.7.7 *AREA CHART*

An area chart epitomizes the adjustment in one or more measures over time. It is same as the line chart, but the measured value should be scheming with a series of data points over time, connecting those data points with line divisions, and then filling in the area between the line and the x-axis with color. Similar to the line chart, we used area chart to plot the number of cases and the number of death cases in different continents based on the days and month as shown in Figure 1.18. The area chart and line chart are more or less same. But in Figure 1.17, the curve shows either increasing or decreasing of cases to identify the number of cases or deaths. In area chart visualization, not only increasing or decreasing of cases is found but also which country is having more number of cases and more number of death cases.

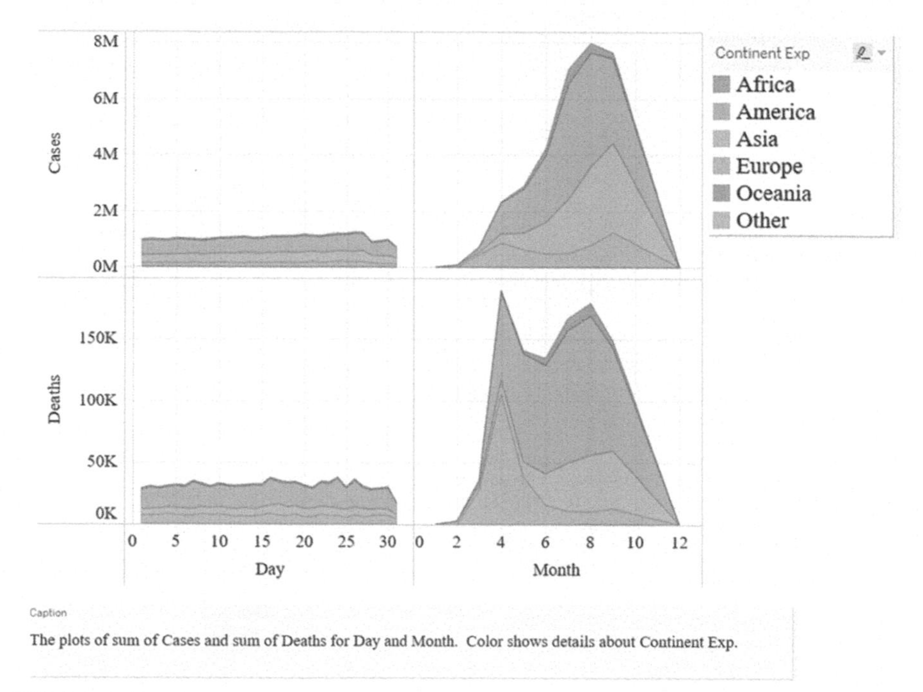

Caption

The plots of sum of Cases and sum of Deaths for Day and Month. Color shows details about Continent Exp.

FIGURE 1.18 Area chart.

1.8 CONCLUSION

Data Visualization helps the health care industries to monitor the patients' health. In this chapter, we analyzed the case study on COVID-19 dataset in Tableau data Visualization tool to understand the number of cases and death cases in different countries and continents. Also from the above observations, more number of charts are available to visualize the data. Each chart should be different and each has its own way of explaining the information. With the help of these charts, health care monitoring system should work more effectively. From the above comparison of each chart, visualization is more important and the advantages of visualization help the health care industries in many ways to perform their work in more simple and effective manner.

KEYWORDS

- **tableau**
- **data analysis**
- **visualization**
- **COVID-19 datasets**
- **patient monitoring**
- **performance indicator**

REFERENCES

1. Heitzman, A. *Data Visualization*; 2019. Available at https://www.searchenginejournal.com/what-is-data-visualization-why-important-seo/288127/.
2. Adimoolam, M.; John, A.; Balamurugan N. M.; Ananth Kumar, T. Green ICT Communication, Networking and Data Processing. In *Green Computing in Smart Cities: Simulation and Techniques*; Springer: Cham, 2020; pp 95–124.
3. Anwary, A. R.; et al. Gait Quantification and Visualization for Digital Healthcare. *Health Policy Technol.* **2020,** *9*, 204–212.
4. Curtis, B. *Impact of Data Visualization in Health Care*; 2019. Available at https://www.yourtechdiet.com/blogs/impact-data-visualization-healthcare/.
5. Deiva Ragavi, M.; Usharani, S. Social Data Analysis for Predicting Next Event. In *International Conference on Information Communication and Embedded Systems (ICICES2014)*, Chennai, 2014; pp 1–5. doi:10.1109/ICICES.2014.7033935.
6. McCoy, E. *How Data Visualization Is Transforming the Health Care Industry*; 2019. Available at https://modus.medium.com/how-data-visualization-is-transforming-the-healthcare-industry-6761d7293dd2#:~:text=interactive%20content%20below.)-,Data%20visualization%20brings%20the%20most%20important%20takeaways%20in%20the%20health,many%20organizations%20sharing%20such%20visualizations.
7. Sharma, H. *Data Visualization for Analysis of SEO Campaigns*; 2018. Available at https://www.optimizesmart.com/ultimate-data-visualization-guide-seos/.
8. Ko, I.; Chang, H. Interactive Data Visualization Based on Conventional Statistical Findings for Antihypertensive Prescriptions Using National Health Insurance Claims Data. *Int. J. Med. Inform.* **2018,** *116*, 1–8. doi:10.1016/j.ijmedinf.2018.05.003.

9. John, A.; Ananth Kumar, T.; Adimoolam, M.; Blessy, A. Energy Management and Monitoring Using IoT with Cup-Carbon Platform. In *Green Computing in Smart Cities: Simulation and Techniques*; Springer: Cham, 2020; pp 189–206.
10. Hardwick, J. *How to do an SEO Competitor Analyses*; 2019. Available at https://ahrefs.com/blog/competitive-analysis/.
11. Luo, M.; Yu, J.; Zhu, S.; Huang, L.; Chen, Y.; Wei, S. Detoxification Therapy of Traditional Chinese Medicine for Genital Tract High-Risk Human Papillomavirus Infection: A Systematic Review and Meta-analysis. *PLoS One* **2019,** *14* (3), e0213062. doi:10.1371/journal.pone.0213062. PMID: 30822331; PMCID: PMC6396931.
12. Narmadha, S.; Gokulan, S.; Pavithra, M.; Rajmohan R.; Ananthkumar, T. Determination of Various Deep Learning Parameters to Predict Heart Disease for Diabetes Patients. In *2020 International Conference on System, Computation, Automation and Networking (ICSCAN),* Pondicherry, India, 2020; pp 1–6. doi:10.1109/ICSCAN49426.2020.9262317.
13. Stewart, R. *Data Visualization Charts for SEO*; 2019. Available at https://webris.org/seo-data-visualization.
14. Sisense. *Hospital Readmissions*; 2020. Available at https://www.sisense.com/kpi-library/healthcare-kpis/hospital-readmissions/.
15. Yu, C.; Zhai, J.; Xun, J.; He, X. Clinical Observation of rhIL-2 Combined with Zhenqi Fuzheng and Baofu Kang Suppository in the Treatment of Cervical Intraepithelial Neoplasia II with HPV Infection. *Open J. Obstet. Gynecol.* **2020,** *10* (8), 1045–1055.

CHAPTER 2

Health Monitoring and Management System Using Machine Learning Techniques

BHARATI PATIL[1,2*] and VYDEKI D.[2]

[1]*GHRCEM, Pune, Maharashtra, India*

[2]*School of Electronics Engineering, VIT, Chennai, Tamil Nadu, India*

[*]*Corresponding author. E-mail: bharti.patil@raisoni.net*

ABSTRACT

Machine learning has ample number of applications in health care field. Right from administration to treating the patients, machine learning can be implemented. In healthcare field data mining strategies, such as classification, clustering, regression, etc. are widely used. Nowadays, due to hectic life schedule, major cause of deaths is HEART DISEASE. The most proficient and economical method to detect the risk of this disease is the use of machine learning algorithms. As per the data analysis available from the Centers for Disease Control, in last 10 years, nearly 80% of the people died due to heart attack. Heart diseases are also mentioned as cardiovascular disease in medical terms. Cardiovascular disease is very common in the age group of 40 and above which needs to be discovered and treated at the initial stage. The traditional method to deal with heart disease is not found to be efficient.

Machine learning algorithms which come as noninvasive methods play a vital role in forecasting the presence or absence of heart disease. These algorithms are used for mathematics bioinformatics to their deployment in clinical diagnosis, prognosis, and drug development. This noninvasive

Intelligent Interactive Multimedia Systems for e-Healthcare Applications. Shaveta Malik, PhD
Amit Kumar Tyagi, PhD (Eds.)

software-based system helps the doctors to detect and treat the heart disease more efficiently. Considering the most important parameters, such as accuracy and execution time, this proposed system will have a great impact on the classifiers. This chapter proposes machine learning-based heart attack prediction by using heart disease dataset. Here, the system will make use of SVM and logistic regression, Naïve Bayes, decision tree, and SVM algorithm to process the large-scale data to predict heart-related diseases.

2.1 INTRODUCTION

Machine learning has an incredible effect in the era of innovation, including clinical exploration and life science. Information mining procedures are most likely used to separate comprehension and decide exciting and helpful examples. Information mining is the examination of enormous measurement sets to discover styles and utilize those styles to conjecture or expect the chance of future events. The aims of data mining is to provide the various methodology for social insurance associations concerning assess treatment viability, spare existences of patients utilizing the prescient machine, oversee medicinal services at different levels, to deal with the client relationship.

As information mining or data mining got from the name of looking for important data in gigantic databases, and it is called information revelation in databases. For example, it can also be carried out in expectation.[1] Of these, most well-known grouping techniques may incorporate neural systems and choice tree. Numerous methods of information mining just as calculations have been utilized for anticipating cardiovascular maladies. The determination utilizing AI methods is one of the predominant strategies which have straightforward analytic information which gives increasingly exact outcomes.

Coronary illness or CVD is a sort of disease that includes the heart as well as veins of individuals all through the world. It is the fundamental purpose behind passing's and insufficiency on the planet principle. . Heart issue is a reason for death toll for individuals around the globe. According to WHO report with respect to coronary illness control and counteraction of the considerable number of sicknesses, heart ailments are the significant reasons for inability and demise around the world.

AI helps in information-driven dynamic, recognizable proof of key patterns and driving exploration effectiveness. With regards to social insurance, there are various manners by which AI strategies can be applied for compelling sicknesses forecast, analysis, and medicines, improving the general tasks of human services. AI built models that rapidly investigate information and convey results, utilizing recorded and ongoing information. With AI, social insurance specialist organizations can settle on better choices on patient's judgments and treatment alternatives, which lead to overall upgrading of medicinal services administrations.

2.2 RELATED WORK

Enormous work has already been done related to detection of the presence or absence of heart disease using different technologies. The main purpose of the study is to understand the performance of the classifier to predict heart disease in time.

Sushmita Manikandan,[6] in her work on "Heart Attack Prediction System," has reported that an attempt has been made for a system that is used to decrease the efforts and time taken by the doctor with the help of binary classifiers. A prototype invention of such a system is easy to use user interface. This graphic user interface is based on the web and Naïve Bayes. The resulting system gave an accuracy score of 81.25%.

C. Kalaiselvi,[7] in her work on "Diagnosing of Heart Disease using Average K-Nearest Algorithm of Data Mining," reports that heart disease is the most common disease in today`s world which causes 80% of the people to die. According to the author reports, around 25 million people are expected to die due to heart disease until the year 2030. By using less number of attributes that are relevant to heart disease, author tries to diagnose the heart disease by using Average KNN Algorithm, Naïve Bayes Algorithm, and Decision Tree Algorithm with the accuracy of 96.5–97%, 94.43–90.72%, and 96.1–96.62%, respectively.

Sellappan Palaniappan and Rafiah Awang,[8] in their work on "Intelligent Heart Disease Prediction System using Data Mining Techniques," suggested a prototype of IHDPS (intelligent heart disease prediction system) using data mining techniques. They have designed an expert medical diagnosing heart disease device and applied the gadget learning strategies inclusive of Naïve Bayes, selection tree, and ANN in the system.

2.3 METHODOLOGY

We present coronary illness forecast framework bolstered dependent on guileless Bayes' algorithmic rule. This framework is helpful, powerful and offers savvy forecast of illnesses to clients. This framework displays the investigation of shifted information handling strategies which can be useful for clinical examiners or professionals for the correct heart condition distinguishing proof.

Past exploration contemplates has inspected the use of machine learning procedures for the expectation and grouping of heart infection. To remove the noise data from the framework, dataset is going to utilize pre-preparing method of machine learning and to foresee the yield at its best level. The framework will make use of logistic regression, Naïve Bayes, decision tree and SVM algorithm. The fast reception of electronic well-being records has made an abundance of latest information about patients, which could be a goldmine for improving the comprehension of human well-being. The above method and algorithms are utilized to predict the disease by utilizing persistent details.

SYSTEM ARCHITECTURE

For the prediction of coronary illness different characteristics are viewed such that sex, age, cholesterol level, shade of blood, sort of chest torment, glucose level, most extreme pulse, resting of circulatory strain.[9]

The proposed framework incorporates a few stages that can be indicated utilizing engineering. The above figure shows the proposed framework architecture. The significant advances associated with the proposed framework are as follows:

a. *Rescaling data:* For information with characteristics of changing scales, the re-scale credits to have a similar scale. The re-scale properties fall into the range 0–1 and are call it standardization. The framework utilizes the MinMax Scalar class from scikit and gains proficiency. This gives us values somewhere in the range of 0 and 1. Rescaling information demonstrates of utilization with neural systems, optimization calculations, and those that utilization separation estimates like k-closest neighbors and weight inputs like relapse.
b. *Standardizing data:* With normalizing, the framework takes qualities with a Gaussian circulation and various methods and standard deviations change them into a standard Gaussian appropriation with a mean of 0 and a standard deviation of 1. For this, the framework utilized Standard Scalar class.

The proposed system (Fig. 2.1) works in five steps that included: (1) preprocessing, (2) attribute selection, (3) cross-endorsement methodology, (4) machine learning classifiers, and (5) classifiers' introduction appraisal systems. The structure of the proposed system is shown in Figure 2.1.

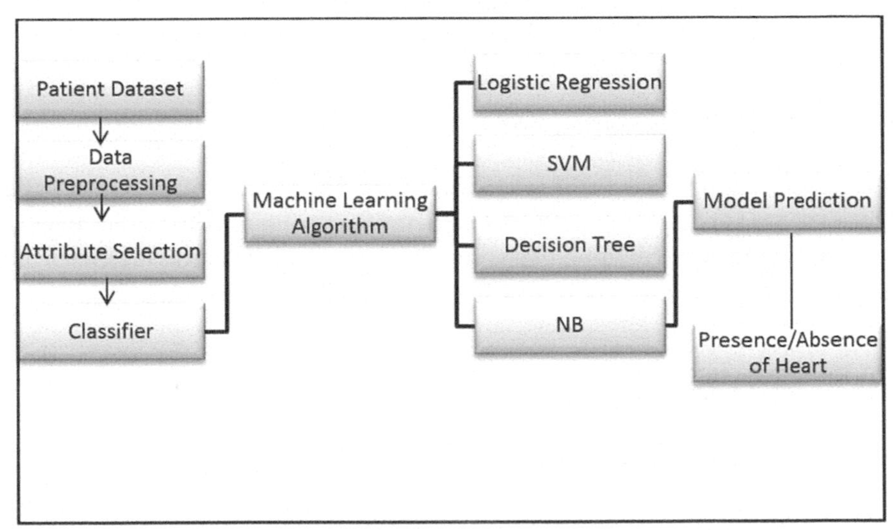

FIGURE 2.1 Proposed system architecture.

2.3.1 DATASET

The "Cleveland coronary sickness dataset 2016" is used by various experts.[10] The Cleveland coronary sickness dataset has a size of 303 patients, 76 features. 0 is the dataset used in this assessment concentrate for organizing a machine learning-based structure for coronary sickness end. During the examination, 6 models were removed as a result of missing characteristics in incorporated areas and the additional model size is 297 with 13 dynamically legitimate free information features, and the target yield mark was isolated and was used for diagnosing the coronary sickness.

2.3.2 PREPROCESSING OF DATASET

This is required for useful delineation of information and machine learning classifier which ought to be orchestrated and endeavored in an earth shattering way. Preprocessing framework, for instance, clearing missing characteristics, standard scalar, and MinMax Scalar have been applied to the dataset for convincing use in the classifiers. The standard scalar ensures that every segment has the mean 0 and distinction 1, carrying all features to a comparative coefficient. Correspondingly, the MinMax Scalar moves the data with the ultimate objective that all features are in some place in the scope of 0 and 1. The missing characteristics incorporated push is basically deleted from the dataset.

2.3.3 ATTRIBUTE SELECTION

Quality choice is vital for the Machine learning procedure in light of the fact that occasionally superfluous characteristic influences the order execution of the machine learning classifier. Property determination increases the arrangement exactness and diminishes the model execution time.

2.3.4 ML CLASSIFIERS

To detect cardiovascular patient, machine learning grouping calculations are utilized. Some well-known arrangement calculations and their hypothetical background are examined quickly.

2.3.5 DISEASE PREDICTION

The above techniques are utilized to get the better result from the proposed system. According to the output, it is understood that the SVM algorithm is the better one to predict heart disease than the other algorithms[1,2,3]. By using number of attributes with support vector machine algorithm, this system is able to predict the heart disease accurately. Expectation utilizing the conventional infection chance model typically includes machine learning and managed learning calculations which uses preparing information with the marks for the preparation of the models.

2.3.6 ACCURACY MEASURE

Precision is one measurement for assessing order models. Casually, precision is the division of expectations our model got right. Officially, precision has the accompanying definition: Accuracy = (Number of right Prediction)/ (Total Number of Prediction).

2.4 DESIGN AND IMPLEMENTATION

Tools and Technologies Used: HTML, CSS, Python, SQLite3

2.4.1 ALGORITHMS DETAILS

2.4.1.1 DECISION TREE

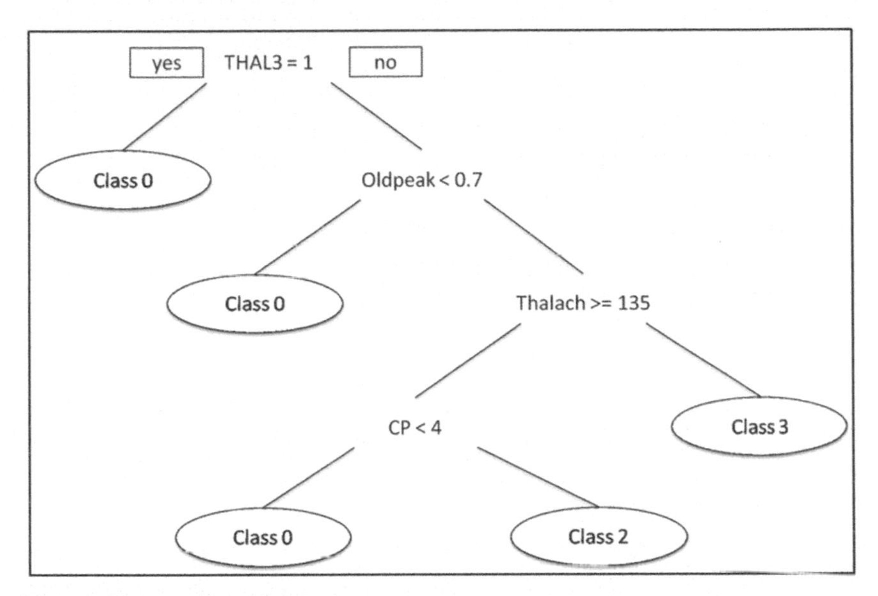

FIGURE 2.2 Sample tree.[1]

2.4.1.2 NAÏVE BAYES

In NB calculation (Fig. 2.2), the dataset utilized for preparing assists in discovering the restrictive likelihood estimation of vectors for a given class. It thinks that all the factors are free of one another. It is exceptionally adaptable.[11] Guileless Bayes depends on Bayesian x = Hypothesis,[12] which decides of how likely an occasion, can happen given that another occasion has just happened.

$$P(A/B) = \frac{P\left(\frac{B}{A}\right) * P(A)}{P(B)}$$

2.4.1.3 LOGISTIC REGRESSION

Logistic relapse is a one of the machine learning grouping calculation for dissecting a dataset in which there are at least one autonomous factor (IVs) that decides a result and furthermore clear cut ward variable (DV).[13] By utilizing condition, the strategic relapse calculation is spoken to in the diagrams indicating the contrast between the characteristics. From the preparation information, we need to evaluate the best and inexact coefficient and speak to it. Straight backslide uses yield in consistent numeric, while vital backslide changes its yield using the determined sigmoid ability to reestablish a probability regard which would then have the option to be mapped to in any event two discrete classes.[14]

The coordination relapse structures are of three sorts as given below.

a. Binary coordination relapse (two potential results in a DV).
b. Multinomial coordination relapse (at least three classifications in DV without requesting).
c. Ordinal coordination relapse (at least three classes in DV with requesting).[15]

Furthermore, strategic relapse model uses progressively complex cost function (known as sigmoid function or logistic function) rather than direct function.[12] Calculated relapse restrains the cost work between 0 and 1.

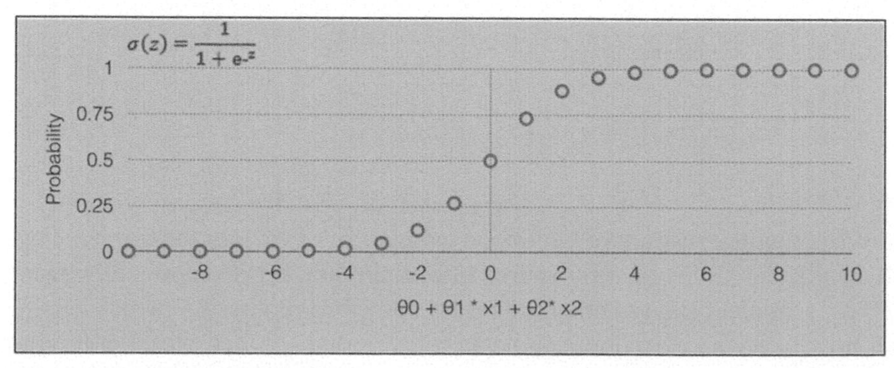

FIGURE 2.3 Sigmoid function.

As per the given informational collection (Fig. 2.3), 1 demonstrates the high danger of 10-year future coronary illness and 0 shows no heart dangers. The autonomous factor n in the strategic model is given as $x1$, $x2$, $x3$, …., xn.

$$\log\left(\frac{p}{1-p}\right) = \beta o + \beta 1x1 + \beta 2x2 + \beta 3x3 \ldots\ldots + \beta nxn$$

Logistic regression achieves this by taking the log odds of the occasion $\text{In}\left(\frac{p}{1-p}\right)$, where, P is the likelihood of occasion which is danger of coronary heart disease (CHD). Subsequently, P consistently lies between 0 and 1.

2.4.1.4 SVM ALGORITHM

Support vector machine (SVM) uses sound judgment for information focuses that are outside the preparation set. There are two classes of information in SVM. The information focuses are disengaged, so that they could draw a straight line on the figure. The line is made such that it isolates all the focuses on one side of one class and all the focuses on the opposite side of the different class. At the point when such circumstance happens, at that point, the information is directly distinct. The line used to isolate the dataset is known as an isolating hyperplane. The focuses nearest to the isolating hyperplane are known as help vectors. Portions are utilized to stretch out SVMs to a bigger number of datasets. Mapping of one element space to another is finished by portion.[16,17] This mapping starting with one element space then onto the next is finished by a part. Expect part as a wrapper or interface for the information to change over it from an extreme designing to a straightforward organizing. The radial bias function (RBF) is a part that is commonly required with the Support vector machines (SVM). The radial bias function (RBF) is an unmistakable part that quantifies the separation among two vectors. The execution of a SVM is additionally receptive to streamlining parameters and determinations of the portion utilized. Bolster vector machines (SVM) are a parallel classifier and different strategies can be proceeded to order of classes more prominent than two.

2.4.1.5 THE MATHEMATICS OF THE SUPPORT VECTOR MACHINE

We have *k* subspaces, so that there are *k* arrangement consequences of subspace, called CL_SS1, CL_SS2,..., CL_SSk. Consequently, the issue is the manner by which we incorporate those outcomes. The basic incorporating path is to compute the mean worth:

$$\text{CL} = \frac{1}{k}\sum_{i=1}^{k} CL\_SS_i$$

Or weighted mean value:

$$\text{CL} = \frac{1}{k}\sum_{i=1}^{k} w_i CL\_SS_i$$

Where w_i is the weight of classification result of subspace SSi, and satisfies

$$\sum_{i=1}^{R} w_i = 1$$

The centroid of a hand is calculated as follows:

$$\overline{\text{X}} = \frac{\sum_{i=0}^{k} xi}{k}, \quad \overline{\text{Y}} = \frac{\sum_{i=0}^{k} yi}{k}$$

Where (X¯,Y¯) speaks to the centroid of the hand, xi and yi are x and y directions of the *i*th pixel in the hand locale and k indicates the quantity of traits that speak to just the hand parcel. In the subsequent stage, the separation between the centroid and the fingertip was determined. For separation, the accompanying Euclidean separation was utilized:

$$\text{Distance} = \sqrt{(x2 - x1)^2 - (y2 - y1)^2}$$

Where $(x1, x2)$ and $(y1, y2)$ represent the two coordinate values.

TABLE 2.1 Description Database of Cleveland Data Set.

Sr. No.	Field	Description	Range and values
1	Age	Age of the patient	0–100 in years
2	Sex	Gender of the patient	1: male, 0: female
3	Chest pain	Type of chest pain	1: typical angina; 2: atypical angina; 3: non-anginal· 4: asymptotic
4	Resting blood pressure	Blood pressure during rest	mm Hg
5	Cholesterol	Serum cholesterol	mg/dl
6	Fasting blood sugar	Blood sugar content before food intake if >120 mg/dl	0: false; 1: true
7	ECG	Resting electrocardiographic results	0: normal; 1: having ST-T wave
8	Max heart rate	Maximum heart beat rate	Beats/min
9	Exercise induced angina	Has pain been induced by exercise	0: no; 1: yes
10	Old peak	ST depression induced by exercise relative to rest	0–4
11	Slope of peak exercise	Slope of the peak exercise ST segment	1: up sloping; 2: flat; 3: down sloping
12	Ca	Number of vessels colored by fluoroscopy	0–3
13	Thal	Defect type	3: normal 6: fixed defect 7: reversible defect
14	Num	Diagnostics of heart disease	0: <50% narrowing; 1: >50% narrowing

2.4.2 DESIGN: DATA FLOW DIAGRAM

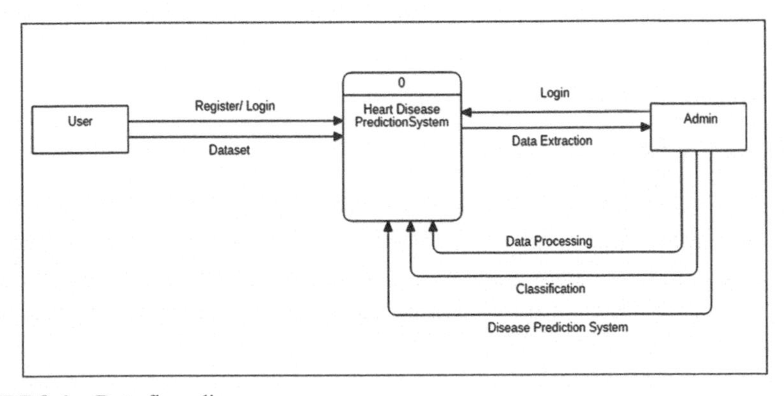

FIGURE 2.4 Data flow diagram.

Figure 2.4 shows the flow of the process.

UML DIAGRAM

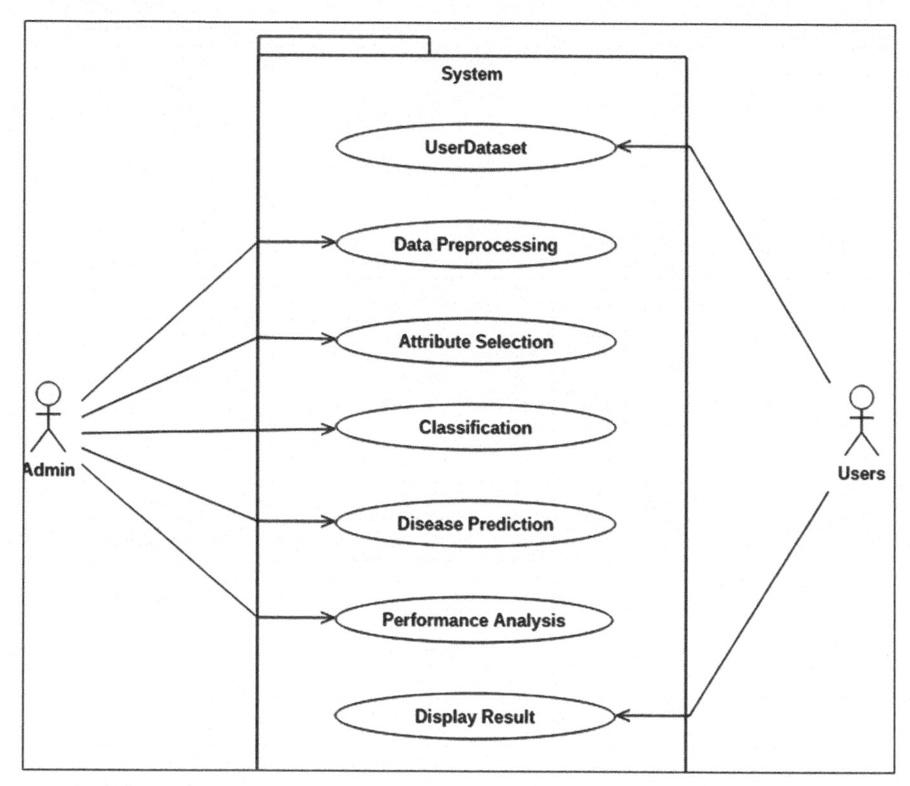

FIGURE 2.5 UML diagram.

Figure 2.5 shows the UML diagram and different stages of the process.

2.5 RESULTS AND CONCLUSION

2.5.1 DATA PREPARATION

Since the dataset consists of 4240 observations with 388 missing data and 644 observations to be risked for heart disease, two different experiments were performed for data preparation. First, we checked by dropping the missing data, leaving only 3751 data and only 572 observations risked for heart disease (Figs. 2.6 and 2.7).

The number of observations are reduced which further provides inappropriate training for the proposed model. So, we progressed with the imputation of data with the mean value of the observations and scaling those using Simple Imputer and Standard Scaler modules of Sklearn.

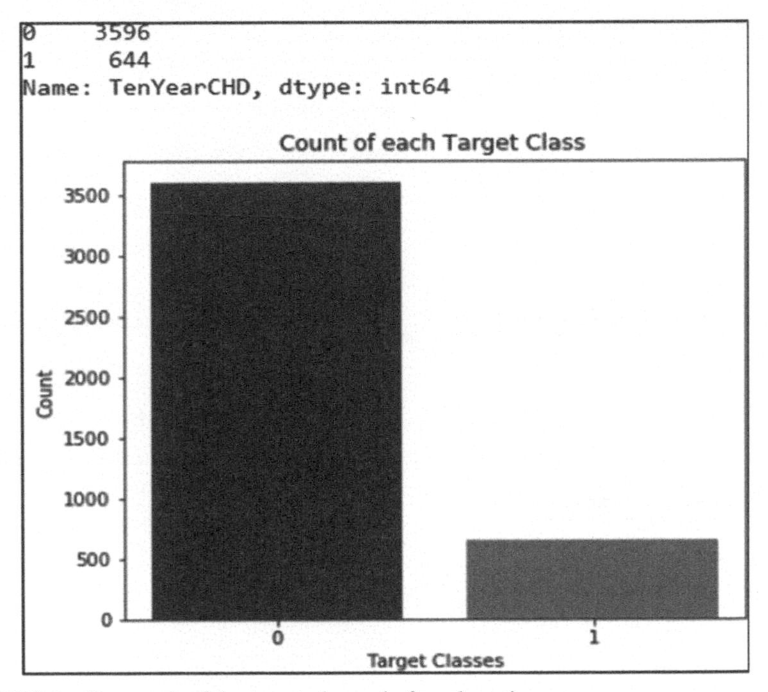

FIGURE 2.6 Bar graph of the target classes before dropping.

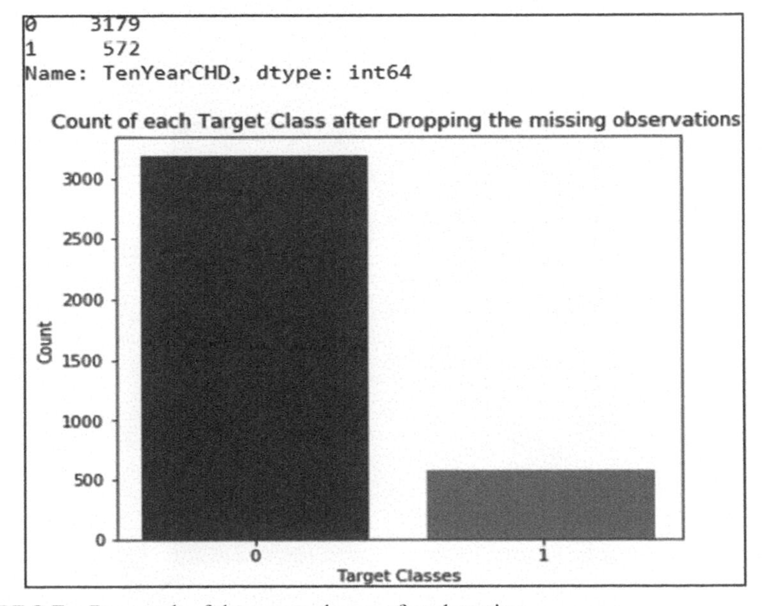

FIGURE 2.7 Bar graph of the target classes after dropping.

	male	age	currentSmoker	cigsPerDay	BPMeds	prevalentStroke	prevalentHyp	diabetes	totChol	sysBP	diaBP	BMI	heart
0	1.153113	-1.234283	-0.988276	-0.758062	-1.758000e-01	-0.077014	-0.671241	-0.162437	-0.940825	-1.196287	-1.083027	0.287258	0.34
1	-0.867217	-0.417664	-0.988276	-0.758062	-1.758000e-01	-0.077014	-0.671241	-0.162437	0.300085	-0.515399	-0.159355	0.719668	1.59
2	1.153113	-0.184345	1.011863	0.925410	-1.758000e-01	-0.077014	-0.671241	-0.162437	0.187275	-0.220356	-0.243325	-0.113213	-0.07
3	-0.867217	1.332233	1.011863	1.767146	-1.758000e-01	-0.077014	1.489778	-0.162437	-0.263965	0.800946	1.016227	0.682815	-0.90
4	-0.867217	-0.417664	1.011863	1.177931	-1.758000e-01	-0.077014	-0.671241	-0.162437	1.089756	-0.106878	0.092555	-0.663554	0.75
...	...	...	...	...	...	...	...	...	...	...	...	...	...
4235	-0.867217	-0.184345	1.011863	0.925410	2.059493e-17	-0.077014	-0.671241	-0.162437	0.254961	-0.061487	-0.915087	-0.933810	0.67
4236	-0.867217	-0.650984	1.011863	0.504542	-1.758000e-01	-0.077014	-0.671241	-0.162437	-0.602395	-0.265747	0.344486	-1.631584	0.84
4237	-0.867217	0.282295	-0.988276	-0.758062	-1.758000e-01	-0.077014	-0.671241	-0.162437	0.728764	0.051991	0.008585	-1.064025	0.34
4238	1.153113	-1.117623	-0.988276	-0.758062	-1.758000e-01	-0.077014	1.489778	-0.162437	-1.166445	0.392425	1.268138	-0.049334	-0.73
4239	-0.867217	-1.234283	1.011863	1.767146	-1.758000e-01	-0.077014	-0.671241	-0.162437	-0.918263	0.029296	0.260496	-1.201610	0.75

4240 rows × 14 columns

FIGURE 2.8 Dataset after scaling and imputing.

2.5.2 EXPLORATORY ANALYSIS

Correlation matrix visualization before feature selection shows:

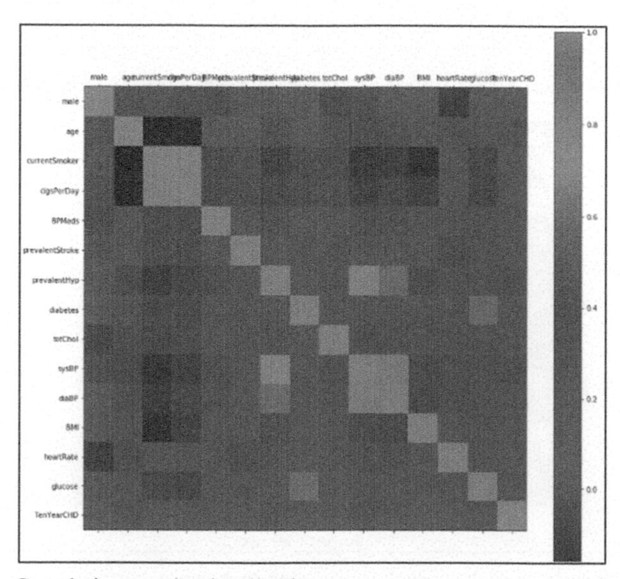

FIGURE 2.9 Correlation matrix visualization.

It shows that there is no single feature that has a very high correlation with our target value. Also, some of the features have a negative correlation with the target value (Figs. 2.7, 2.8, and 2.9).

2.5.3 PERFORMANCE EVALUATION

Accuracy

Accuracy is the proportion of various right forecasts given by the model to the all out number of occurrences.

$$Accuracy = \frac{(\mathrm{TP} + \mathrm{TN})}{(\mathrm{TP} + \mathrm{FP} + \mathrm{FN} + \mathrm{TN})}$$

Precision

Precision in this work gauges the extent of people anticipated to be in danger of creating CHD and had a danger of creating CHD.

$$Precision = \frac{\mathrm{TP}}{(TP + FP)}$$

Recall

Recall, in this work, gauges the extent of people who were at a danger of creating CHD and were anticipated by the calculation to be in danger of creating CHD.

$$Recall = \frac{\mathrm{TP}}{(TP + FN)}$$

Specificity

Specificity here measures the extent of people who were not in danger of creating CHD and were anticipated by the calculation to be not in danger of creating CHD.

$$Specificity = \frac{\mathrm{TN}}{(\mathrm{TN} + \mathrm{FP})}$$

F1 Score

Symphonious mean of accuracy and review is called F1 Score.

$$F1Score = \frac{2(Precision \times Recall)}{(Precision + Recall)}$$

ROC (Receiver Operator Characteristic)

It is a likelihood bend showing the capacity of a model to recognize classes. The ROC bend shows the exchange off between FPR and TPR. As per Ref. [9], ACU like 1 would have the option to impeccably separate the two classes on account of double arrangement. Accordingly, ACU closer to 1 is better prescient measure (Figs. 2.10, 2.11, and 2.12).

$$\text{TPR} = \frac{\text{TP}}{(TP + FN)}$$

$$\text{FPR} = \frac{\text{FP}}{(FP + TN)}$$

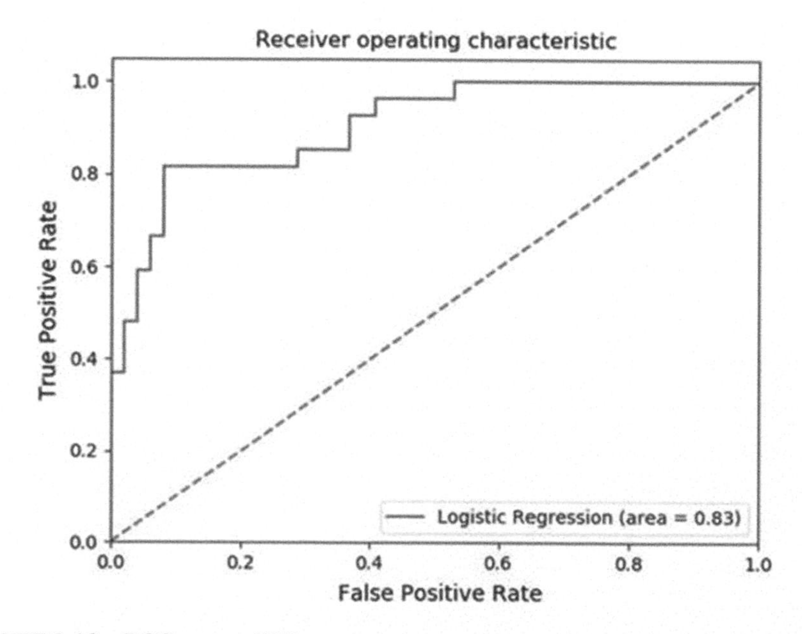

FIGURE 2.10 ROC curve of DT.

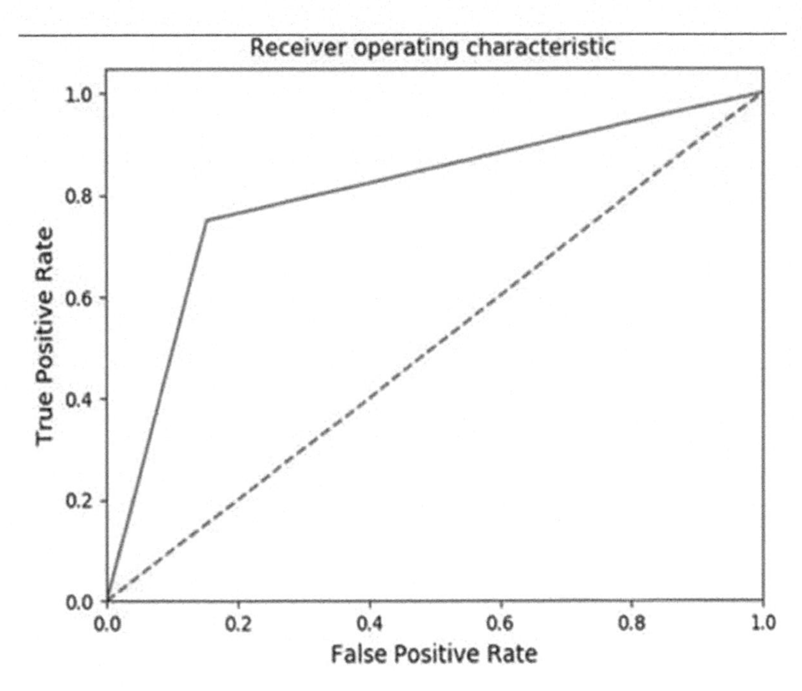

FIGURE 2.11 ROC curve of LR.

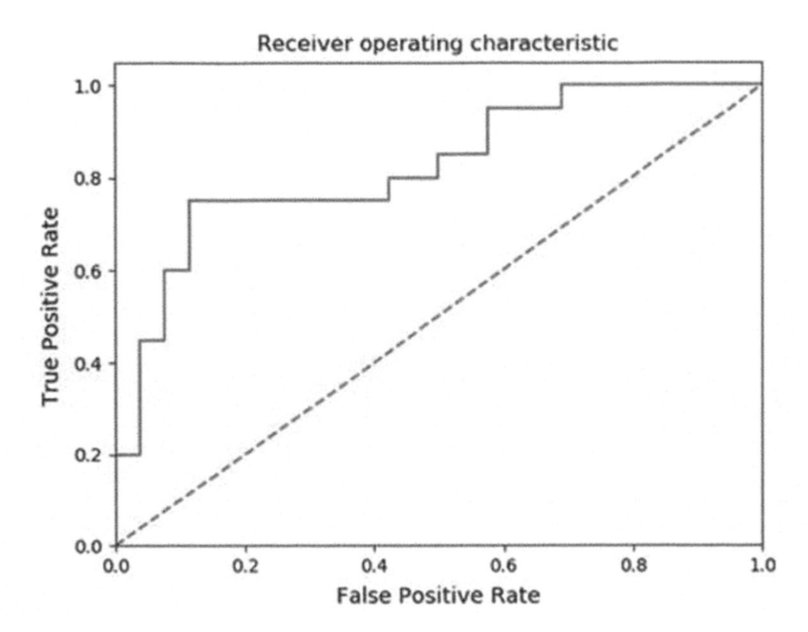

FIGURE 2.12 ROC curve of NB.

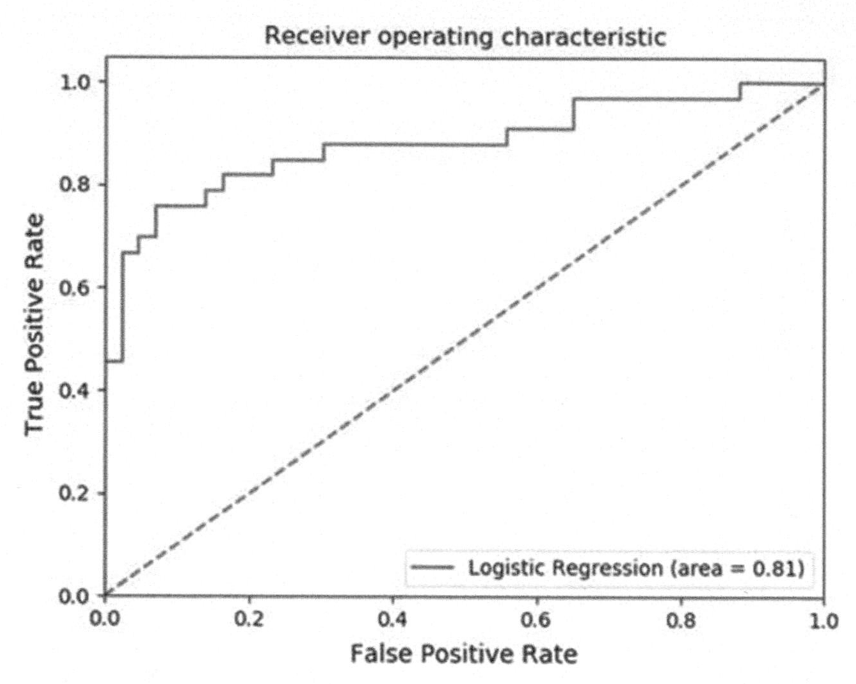

FIGURE 2.13 ROC curve of SVM.

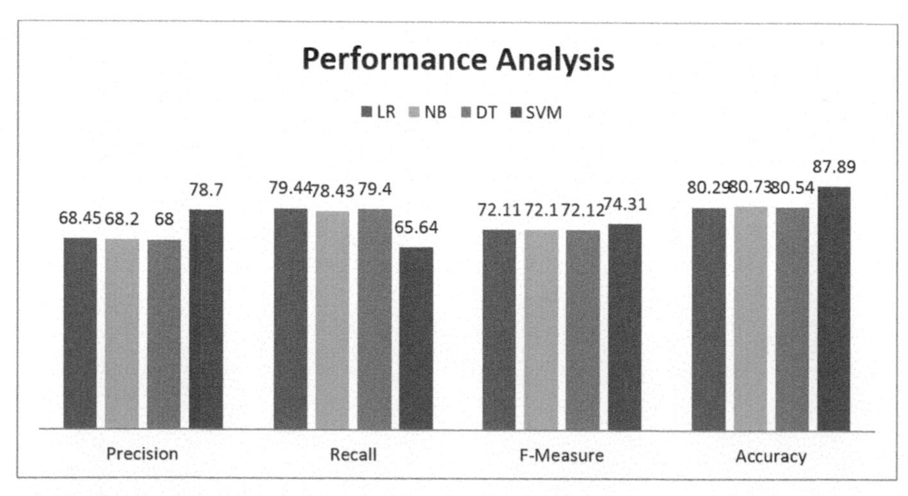

FIGURE 2.14 System analysis.

Figure 2.14 shows the comparison of all parameters and the better value in each parameter.

2.6 CONCLUSION

Expectation of cardiovascular illness is a significant test in medicinal services frameworks. The goal of the proposed work is to give an investigation of various information mining grouping procedures with their upsides and downsides. Informational index of 304 records and 14 properties is gathered from UCI. Results show the exactness for SVM and Naïve Bayes, calculated regression, and decision tree calculation with various number of preparing dataset and testing dataset. Precision chart shows that SVM calculation is better than Naïve Bayes since exactness of SVM is not underneath half in any preparation and testing dataset. SVM calculation performs better for huge dataset utilizing radial bias function (RBF). In light of writing survey, just two calculations specifically SVM, logistic regression and Naïve Bayes order, strategic relapse and DT have been actualized up until now. There is still extension for development in exactness, particularity, and affectability. So other order approaches can be executed and tried.

KEYWORDS

- **heart disease**
- **machine learning**
- **logistic regression**
- **backward elimination**

REFERENCES

1. Krishnan, S.; Geetha, S. Prediction of Heart Disease Using Machine Learning Algorithms. In *1st International Conference on Innovations in Information and Communication Technology (ICIICT)*, April 25–26, 2019.
2. Maratha A. P.; Shaji, S. P. Prediction and Diagnosis of Heart Disease Patients Using Data Mining Technique. In *International Conference on Communication and Signal Processing*, April 4–6, 2019, IEEE, India, 2019. 978-1-5386-7595-3/19/$31.00 ©2019.
3. Dinesh Kumar, G.; et al. Prediction of Cardiovascular Disease Using Machine Learning Algorithms. In *Proceeding of 2018 IEEE International Conference on Current Trends toward Converging Technologies*, Coimbatore, India, 2018.
4. Raju, C.; et al. A Survey on Predicting Heart Disease Using Data Mining Techniques. In *Proc. IEEE Conference on Emerging Devices and Smart Systems (ICEDSS 2018)*, March 2–3, 2018, Mahendra Engineering College, Tamil Nadu, India, 2018.

5. Gavhane, A.; et al. Prediction of Heart Disease Using Machine Learning. In *Proceedings of the 2nd International Conference on Electronics, Communication and Aerospace Technology (ICEC 2018), IEEE Conference Record # 42487*; IEEEXplore, 2018. ISBN: 978-1-5386-0965-1978-1-5386-0965-1/18/$31.00.
6. Manikandan, S. Heart Attack Prediction System. In *International Conference on Energy, Communication, Data Analytics and Soft Computing (ICECDS-2017)*, 2017. 978-1-5386-1887-5/17/$31.00.
7. Kalaiselvi, C.; Nasira, G. M. Classification and Prediction of Heart Disease from Diabetes Patients Using Hybrid Particle Swarm Optimization and Library Support Vector Machine Algorithm. *Int. J. Comput. Algorithm* **2015,** *4*, 1403–1407. ISSN: 2278-2397.
8. Palaniappan, S.; Awang, R. Intelligent Heart Disease Prediction System Using Data Mining Techniques. *Int. J. Comput. Sci. Network Secur.* **2008,** *8* (8), 343–350.
9. UCI Machine Learning Repository and archive.ics.uci.edu/ml/datasets.html.
10. Cleveland Database: http://archive.ics.uci.edu/ml/datasets/Heart+Disease.
11. Prasad, D. V. R.; Mohanji, Y. K. V. Agriculture Crop Recommendation Tool Using Machine Learning. *Int. J. Cntrl. Automat.* **2019,** *12* (5), 618–622.
12. Degadwala, S. D.; Solanki, A.; Vyas, D.; Mahajan, A. Offline Gujarati Word Categorization Using Machine Learning Approach. *Int. J. Cntrl. Automat.* **2019,** *12* (Special Issue), 38–45.
13. Shaik, M. A. A Survey on Text Classification Methods through Machine Learning Methods. *Int. J. Cntrl. Automat.* **2019,** *12* (6), 390–396. Retrieved from www.scopus.com.
14. Puri, I. K. AI and the Future to Successfully Exploit AI Engineer Must Discern between Uncertainty ABD Risk. *Des. Eng. (Canada)* **2018,** *64* (1), 20–21.
15. Bindu Madhavi, G.; Rakesh Reddy, J. Detection and Diagnosis of Breast Cancer Using Machine Learning Algorithm. *Int. J. Adv. Sci. Technol.* **2019,** *28* (14), 228–237. Retrieved from www.scopus.com.
16. Aljuaid, L.; Wei, K. T.; Sharif, K. Machine Learning: Tasks, Modern Day Applications and Challenges. *Int. J. Adv. Sci. Technol.* **2019,** *28* (2), 329–340. Retrieved from www.scopus.com.
17. Luminoso, L. Creative Engineering. *Des. Eng. (Canada)* **2017,** *63* (1), 30–31. Retrieved from www.scopus.com.
18. Sultana, M.; Haider, A.; Uddin, M. S. Analysis of Data Mining Techniques for Heart Disease Prediction. In *2016 3rd International Conference on Electrical Engineering and Information Communication Technology (ICEEICT)*, 2016; pp 1–5.
19. Ali, M.; Khan, M. D.; Imran, M. A.; Siddiki, M. *Heart Disease Prediction Using Machine Learning Algorithms*. Doctoral Dissertation, BRAC University, 2019.

CHAPTER 3

A Study on Fusion Methods and Their Applications in Medical Image Processing

P. NISCHITHA[1] and CHANDRA SINGH[2*]

[1]*Department of ECE, Mangalore Institute of Technology & Engineering, Moodabidri, India*

[2]*Department of ECE, Sahyadri College of Engineering & Management, Mangalore, India*

[*]*Corresponding author. E-mail: chandrasingh146@gmail.com*

ABSTRACT

Multimodal image analysis is obtaining high significance in the field of medicine because of the truth that a huge amount of images having the medical data must be examined to evaluate different types of outcomes. Image fusion is carried to merge each and every significant data from one or multiple multimodal images into a solo informative image. It offers better diagnosis applicability of therapeutic images. Magnetic resonance imaging (MRI) and computed tomography (CT) images help in obtaining high quality anatomical details, whereas positron emission tomography (PET) images present functional details about organs, such as blood flow, metabolic process, and heart beat rate, etc. But because of their small temporal resolution, they do not provide regional information. The ideal solution to overcome these limitations is to combine or merge these images to provide both time domain and transform domain details in one image. The primary

Intelligent Interactive Multimedia Systems for e-Healthcare Applications. Shaveta Malik, PhD Amit Kumar Tyagi, PhD (Eds.)

purpose of image fusion is joining together the complementary and additionally excess data from different pictures to generate a fused image with greater amount of exact portrayal. This combined picture is more appropriate for machine recognition or further image processing tasks. Another objective is that it declines the space required to store the image and reduces the expense by storing just those solitary combined image, as opposed to the distinctive modality pictures. Image fusion aides physicians to effectively extricate the features that might not be typically distinctive for single image. Calibre is a parameter that processes apparent image degradation, that is, in association with the best image. The principle of an image fusion procedure is that all useful and efficient data should be retained. Performance evaluation of the fusion methods is conducted using various parameters. Assessment forms a vital role in the improvement of image fusion methods. This chapter explains the comparative analysis of the fusion of multimodal images, such as CT, MRI, and PET, etc. by using various fusion algorithms, such as Laplacian Pyramid fusion rule, discrete wavelet transform (DWT) and multi-resolution singular value decomposition (MSVD) method. It also explains the analysis of quality metrics which includes the analysis of entropy, standard deviation, fusion factor, SSIM and correlation measure to analyze the fused image performance to identify the best fusion algorithm for the chosen disease.

3.1 INTRODUCTION

Therapeutic imaging is the significant equipment in the present medical field. Various imaging methods, for example, X-ray imaging, ultrasonography (USG) and computed tomography (CT) are broadly utilized in clinical analysis for different sort of sicknesses. Efficient medical image analysis plays a very important role for the treatment of diseases. These are also used in learning fields which will aid students in their studies. However, it would be time-consuming to observe the images physically. Furthermore, as there is constantly a skewed element connected with the pathological inspection of the image by physicians, a computerized method offers an important support for physicians.[1]

Medical images will help in providing various characteristic information of internal organs and tissues in a human body. Multimodal medical images include their own appropriate application ranges since they provide various information regarding human organs and tissues. For instance, imaging techniques, such as MRI, CT, USG, MRA, etc., provide high-resolution structural images with anatomical details. Imaging techniques, such as PET, SPECT, etc., will offer low-spatial resolution images which include the details of the functionality of the organs. It is not possible to obtain complete and accurate information from a single modality image. Hence, merging functional and anatomical medical images through image fusion is essential and it has become the main focus under image processing research area.[2,23]

3.2 LITERATURE REVIEW

Suthakar et al., in their work, "Study of Image Fusion-Techniques, Method and Applications,"[6] have described the survey of image fusion techniques. Fusion algorithms like intensity-hue saturation (IHS), Brovey transform, principal component analysis (PCA), Laplacian transform, wavelet transform (WT), and its applications are explained in detail.

Muttan et al., in their paper, "Discrete Wavelct Transform (DWT) based Principal Component Averaging fusion for Medical Images,"[7] proposed a new fusion means using the weightage of multiscale observations. The fusion algorithms, such as PCA and DWT are explained in detail.

Himabindu et al., in their paper, "DWT Based Medical Image Fusion with Maximum Local Extrema,"[8] projected a new algorithm to improvise the image superiority by means of DWT using local extrema fusion method.

Bahri et al., in their work, "Improved Performance of Image Fusion by MSVD,"[9] proposed the multiresolution singular value decomposition (MSVD) algorithm to enhance the quality of image fusion, and the evaluation of image fusion performance metrics is explained.

Bhavana et al., in their paper, "Multi-Modality Medical Image Fusion using DWT,"[10] projected a fusion way for the fusion of positron emission tomography (PET) as well as magnetic resonance imaging (MRI) brain samples using DWT through a smaller amount of color deformation.

Kanagasabapathy et al., in their paper, "Image Fusion Based on WT,"[11] proposed a new scheme for fusing images by means of WT by making use of maximum coefficient rule on the decomposed images to attain frequency coefficients in the fused image.

Rencan Nie et al., in their paper, "Multi-source Information Exchange Encoding with PCNN for Medical Image Fusion,"[24] have proposed a novel fusion framework for multimodal medical images based on multisource information exchange encoding (MIEE) by using pulse coupled neural network (PCNN). The information embedded in an image is exchanged and encoded into another image. Then, the quantitative analysis is performed to estimate the fusion contributions according to a logical comparison of exchanged information.

Even though all methods attained best results, but directional information, spatial, and spectral information which are used to infer the preferred orientation of structures and frequency-related information present in the input images in order to perform fusion are retained by DWT and other image fusion algorithms. Thus, various image fusion methods are identified to retain the directionality and to reduce the ringing effect and are projected in this chapter.

3.3 IMAGE FUSION

Fusing different images is considered as the practice of combining two images which represent the similar prospect to obtain a solo image retaining the useful details. The primary objective of image fusion is joining together complementary and additionally excess data from different pictures to generate a fused image with greater amount of exact portrayal. This combined picture is more appropriate for machine recognition or further image transforming tasks. Another importance is that it declines the space required to store the image and reduces the expense by saving or storing just those solitary combined image, as opposed to the distinctive modality pictures. Image fusion aides physician/radiologists to effectively extricate or perceive the abnormality features that might not be typically distinctive for a solitary image.[2,7]

Fusing images is a method of merging two different images to form a solitary image by preserving the significant characteristics from individual images. The function of image fusion is to accomplish improved eminence of fused image which will be appropriate for human ocular insight. Individual modality analysis fails to provide sufficient information for the diagnosis purpose. Thus, combining anatomical and functional details from the medical images to present better quality information is required. Thus, fusion of medical images has developed into the main interest of research investigations.[18,20]

Fusion of images in the spatial domain includes methods, such as PCA, high-pass filtering technique, etc. These methods are simple, but they create spatial deformation in the output image, which turns out to be an unconstructive factor for additional processing of an image. This difficulty can be overcome by means of higher level curvelet transform fusion technique. Image fusion enhances reliability, reduces ambiguity, computational time, and storage rate by restoring a single instructive image than storing multiple images.[2]

3.4 PROPOSED METHODOLOGY

While fusing the images, selection of fusion methods plays a very important role. Hence, it is obligatory to use the fusion methods that are appropriate for medical image fusion and analysis.

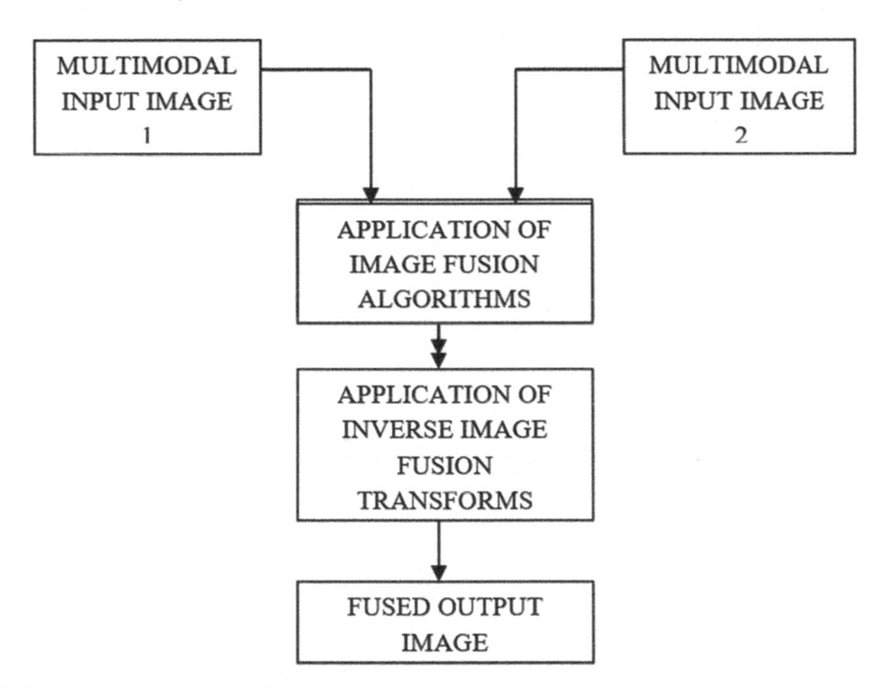

FIGURE 3.1 Overview of the system.

The overview of the algorithm is represented in Figure 3.1. Three stages are mainly involved. In the first stage, multimodal input images are considered. The second-stage involves combining the input images by applying image fusion methods, such as LPT, DWT, and MSVD to attain the fused final image. After the fusion volume coefficients are obtained, inverse image fusion transform is performed to get back the spatial domain fused image.

3.4.1 *LAPLACIAN PYRAMID TRANSFORM (LPT)*

Multiresolution analysis model is represented by image pyramids. Pattern selective approach is used by the Laplacian Pyramid method for fusing images, so that the fused image is constructed, not one pixel at a time. Each source image has to undergo a pyramid decomposition to obtain the decomposed coefficients, and these coefficients are further integrated so that a compound representation is attained, and reconstruction of the fused image is performed by using inverse LPT which is applied on the fused images. Figure 3.2 illustrates the LPT fusion method.

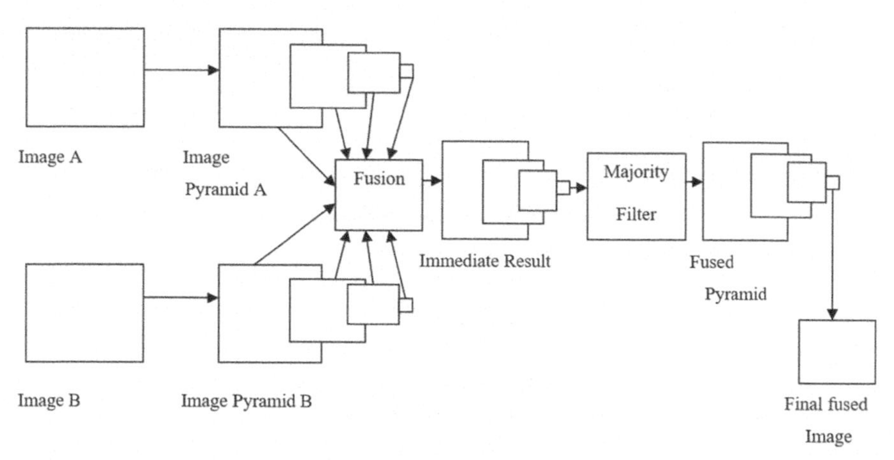

FIGURE 3.2 Overview of Laplacian pyramid fusion method.

3.4.2 *WAVELET TRANSFORM*

The Daubechies wavelets were introduced by Ingrid Daubechies. These wavelets are a family of orthogonal wavelets which define a DWT, and these wavelets are characterized by using a maximal number of vanishing moments for the given support width. The WT is related to the Fourier transform but with the difference in the function merit used. WT was evolved due to the requirement of the concurrent representation and localization of time and frequency for nonstationary signals, such as music, speech, and images. WT is one of the mathematical means to perform signal analysis, when the frequency of the signal varies with respect to time. Wavelet analysis helps in providing precise information about the signal data for certain types of signals as well as images when compared with other various techniques used for signal analysis. The WT comprises of two functions, first one is the scaling function $\varphi(t)$, which is known as the "father wavelet" and the second one is the wavelet function $\psi(t)$ which is also identified as the "mother wavelet." By dilating and translating the mother wavelet $\psi(x)$, a wavelet family can be generated, which is given by

$$\Psi_{(a,b)}(x) = \frac{1}{\sqrt{a}} \Psi\left(\frac{x-b}{a}\right) \tag{3.1}$$

where "a" represents the parameter used for scaling and "b" represents the parameter used for shifting

3.4.2.1 DISCRETE WAVELET TRANSFORM

Spatial domain images are hierarchically decomposed using a mathematical tool called DWT. DWT method will decompose an image into different scale frequency subbands. The final image can be reconstructed using these subbands. The low frequency portion of the image indicates the coarse data of signal, whereas high frequency portion of the image indicates the edge information.

Two-dimensional DWT analyzes an image by decomposing it into approximation and detailed coefficients at different frequency bands. Three levels of decomposition of 2-D image are represented in Figure 3.3. The DWT decomposition is performed using wavelet family "HAAR" which is real, orthogonal, symmetric, which makes DWT shift invariant.

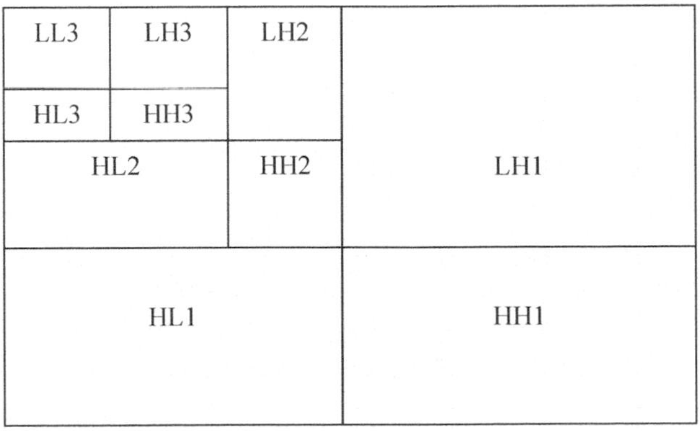

<table>
<tr><td>LL3</td><td>LH3</td><td rowspan="2">LH2</td><td rowspan="3">LH1</td></tr>
<tr><td>HL3</td><td>HH3</td></tr>
<tr><td colspan="2">HL2</td><td>HH2</td></tr>
<tr><td colspan="3">HL1</td><td>HH1</td></tr>
</table>

FIGURE 3.3 Representation of three-level decomposition of DWT.

3.4.3 MULTIRESOLUTION SINGULAR VALUE DECOMPOSITION

MSVD is a fusion algorithm which is very much analogous to WT, where the signals are separately filtered using finite impulse response filters, such as low-pass and high-pass filters and further to achieve the decomposition at the first level, the output obtained from each of the filters is decimated by a factor of 2. The decomposition at the second level is obtained by filtering the decimated low-pass filtered output separately using low-pass and high-pass filters followed by decimating it by considering 2 as the decimation factor.

MSVD involves the factorization of a rectangular matrix which is highly used for information retrieval by reducing the document vector space dimension. The decomposition of an image is explained as follows. Given a generic rectangular matrix *B of size* $n \times m$, the singular value decomposition of the matrix can be obtained as follows:

$$B = U \sum V^T \tag{3.2}$$

where U represents a matrix of size $n \times r$, V^T is a matrix of size $r \times m$, and Σ represents the diagonal matrix of size $r \times r$. U and V are unitary matrices, that is, $U^T U = I$ and $V^T V = I$. The singular values such as $\delta 1 > \delta 2 > \delta r > 0$ are the diagonal elements of the matrix Σ, where r indicates the rank of the matrix B.

3.4.4 WORKING PROCEDURE

The process of Image fusion is performed as per the block diagram in Figure 3.1 as the following steps:

i. Consider the input as two different image pairs from different modalities, such as CT/MRI, CT/PET.
ii. Apply image fusion transforms on the particular coefficient pair for all the coefficients of the image to obtain the coefficients of the output-fused image.
iii. Inverse image fusion transform is performed on fused image coefficients to get back fused image into the spatial domain.
iv. Apply performance metrics on fused image to examine the eminence of fusion method in comparison to original images.

3.5 PERFORMANCE METRICS FOR IMAGE FUSION

Quality is an attribute that describes perceived image deprivation, that is, in evaluation with the absolute image. Performance estimation is a significant step for the advancement of image fusion methods. It consists of full reference methods in which quality is considered in association with perfect image. Nonreference techniques do not involve reference image. Reference-based techniques include entropy and standard deviation. The reference parameter measures, such as PSNR, RMSE, fusion factor (FF), SSIM, and correlation (CR) measure are considered.

3.5.1 ENTROPY

Entropy quantifies the amount of data in a fused image. Higher the entropy value, superior will be the fusion performance. It is given by the following equation:

$$Ha = \sum_{i=0}^{L} h_{I_o}(i) \log_2 h_{I_o}(i) \tag{3.3}$$

where h_{Io} provides the count of histogram in a fused image

3.5.2 STANDARD DEVIATION

Standard deviation (SD) is used to calculate the contrast produced in the fused image. The standard deviation measures the amount of variation in a set of data values. It increases as the disparity among the images increases. It is given by

$$\sigma = \sqrt{\sum_{i=0}^{L} \left(i - \bar{i}\right)^2 h_{I_f}(i)} \tag{3.4}$$

where h_{I_f} is the number of histogram count in a fused image and "i" gives the summation index and "$\bar{i}$" is the mean of histogram.

3.5.3 FUSION FACTOR

It describes the level of reliance of the two images. A higher value implies increased fusion quality. It is given by the equation

$$\text{FMI} = I_{\text{AF}} + I_{\text{BF}} \tag{3.5}$$

where I_{AF} indicates joint information content between input image '*A*' and output image. Similarly, I_{BF} is the joint information content clarity between input image "*B*" and output image.

3.5.4 STRUCTURAL SIMILARITY INDEX MEASUREMENT (SSIM)

SSIM compares the confined patterns of pixel intensities among the original and fused images. The range alters between −1 and 1. Range 1 indicates that the original and the fused images have similarity. It is expressed as

$$\text{SSIM} = \frac{\left(2\mu_{I_a}\mu_{I_b} + S_1\right)\left(2\sigma_{I_aI_b} + S_2\right)}{\left(\mu^2_{I_a} + \mu^2_{I_b} + S_1\right)\left(\sigma^2_{I_a} + \sigma^2_{I_b} + S_2\right)} \quad (3.6)$$

where μ_{Ia}, μ_{Ib}, σ_{Ia}, σ_{Ia}, and σ_{IaIb}, indicates the local means, standard deviations, and cross-covariance for images I_a, I_b, $S_1 = (0.01 \times L)$. ^2, $S_2 = (0.03 \times L)$. ^2, where '*L*' is particular dynamic range of pixel.

3.5.5 CORRELATION MEASURE

The ability of fusion for an algorithm is computed by means of pixel gray level CR between the source and the output images. It computes the spectral feature similarities between the reference and output images. The value of cross CR should be near to +1, which implies that the reference and fused images are identical. Dissimilarity increases when the value of cross CR is −1. The CR among two images *s*(*a*, *b*) and *k*(*a*, *b*) is represented as follows:

$$\text{CR}(f,g) = \frac{\sum_{a,b}\left(s(a,b) - \overline{s}\right)\left(k(a,b) - \overline{k}\right)}{\sqrt{\sum_{a,b}\left(s(a,b) - \overline{s}\right)^2}\sqrt{\sum_{a,b}\left(k(a,b) - \overline{k}\right)^2}} \quad (3.7)$$

3.5.6 PEAK SIGNAL-TO-NOISE RATIO (PSNR)

PSNR is described as the ratio of the quantity of gray levels in the image to the number of pixels in the reference and the fused images. PSNR determines the eminence of reformation of lossy compressed images. Higher value indicates better fusion. It is given by

$$\text{PSNR} = 20\log_{10}\left(\max_i\right) - 10\log_{10}\left(M\right) \tag{3.8}$$

where max_i indicates the maximum pixel strength and "M" is the mean square error value.

3.5.7 ROOT MEAN SQUARE ERROR (RMSE)

It is determined as the square root of the mean square error of the corresponding pixels in the source image and the fused image. RMSE returns zero when the source and fused images are related. It is formulated as

$$\text{RMSE} = \sqrt{\frac{1}{\text{MN}}\sum_{x=1}^{M}\sum_{y=1}^{N}\left(I_r(x,y) - I_f(x,y)\right)^2} \tag{3.9}$$

where $I_r(x,y)$ is $M \times N$ reference image and $I_f(x,y)$ is $M \times N$ fused image.

3.6 SAMPLE RESULTS OF THE EXPERIMENTS CONDUCTED

Multimodal images of size 256 × 256 are considered to conduct the experiments. Multimodal images of a single person are collected from "The Cancer Imaging Archive" and the database is created. Simulation is performed using MATLAB R2015b software.

MRI and PET images representing the abdomen lesion are considered as the input image set 1 which is represented in Figure 3.4(a) and (b). Input image set 2 represents the CT and MRI images representing the abdomen lesion which is represented in Figure 3.4(c) and (d).

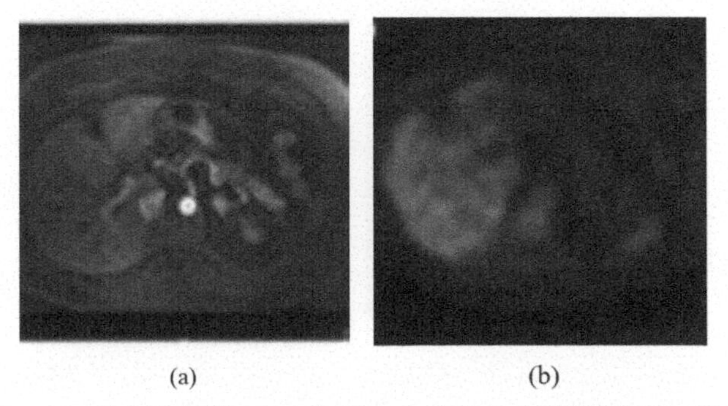

(a) (b)

FIGURE 3.4 Input set 1 (a) MRI image of abdomen lesion. (b) PET image of abdomen lesion.

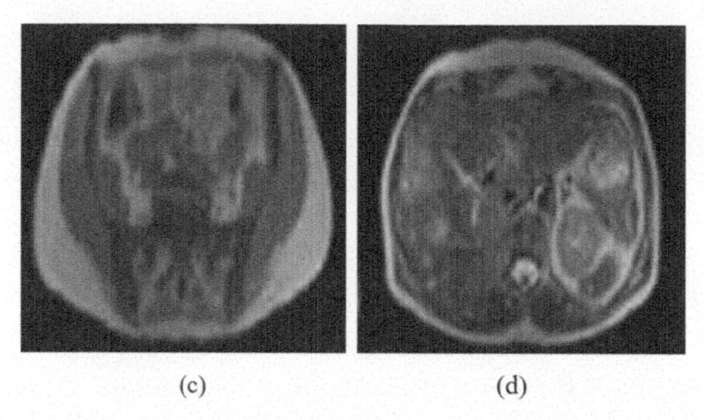

(c) (d)

FIGURE 3.4 (Continued) Input set 2(c) CT of abdomen lesion. (d) MRI of abdomen lesion.

At first, the input datasets are fused using Laplacian pyramid fusion algorithm and the fused images are obtained which are shown in Figure 3.5(a) and (b). Figure 3.6(a) and (b) shows the fused images obtained when DWT fusion algorithm is applied on the input datasets.

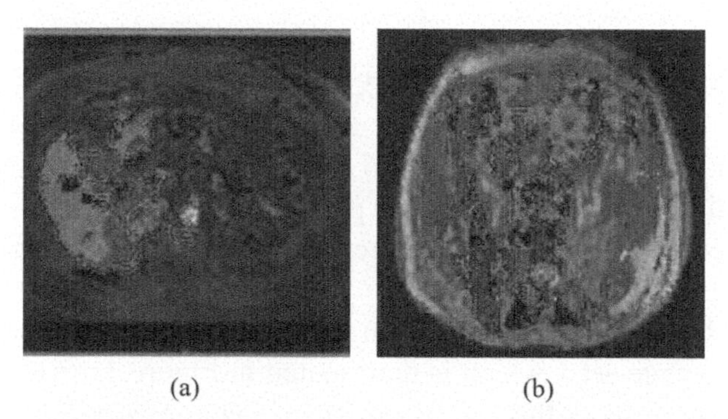

(a) (b)

FIGURE 3.5 Output images of Laplacian pyramid transform for (a) Image set 1. (b) Image set 2.

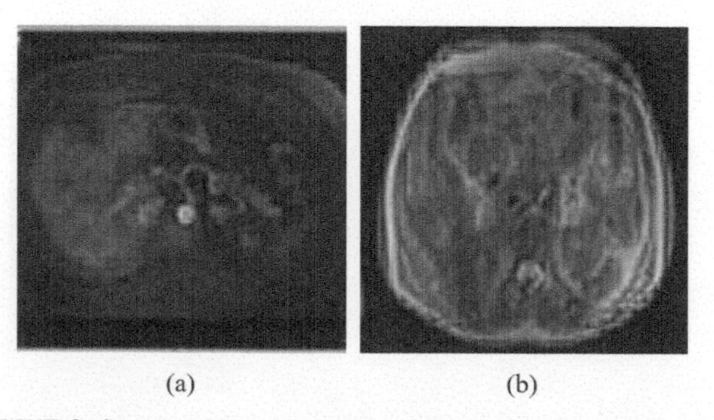

(a) (b)

FIGURE 3.6 DWT fusion algorithm output for (a) Image set 1. (b) Image set 2.

Figure 3.7 (a) and (b) shows the output-fused images attained by applying MSVD algorithm to the input datasets.

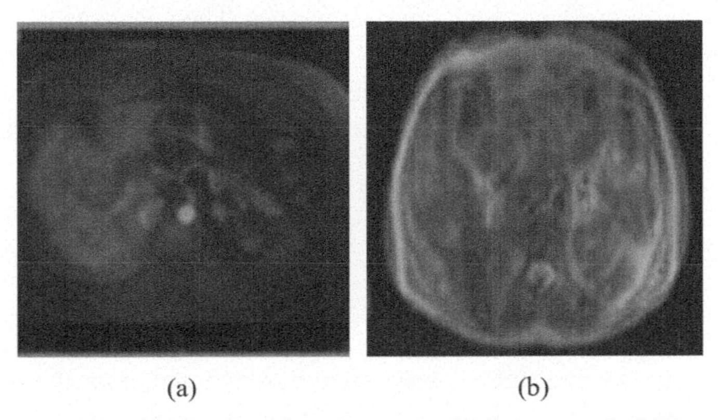

(a) (b)

FIGURE 3.7 MSVD fusion algorithm Output for (a) Image set 1. (b) Image set 2.

Various quality metrics, such as entropy, SD, FF, SSIM, and CR measure are used to analyze the performance of the fused images and the values attained for these quality metrics are tabulated which is represented in Table 3.1.

TABLE 3.1 Parameter Analysis for Different Image Fusion Techniques for Input Images.

Input sets	Parameters	Laplacian pyramid	DWT	MSVD
Image set 1	Entropy	6.5369	15.8950	6.2884
	SD	6.2929	0.1378	0.1047
	FF	0.0907	0.1353	0.0742
	CR	0.9332	0.9695	0.9910
	SSIM	0.9995	0.9998	0.9999
Image set 2	Entropy	7.0900	15.9302	6.6399
	SD	9.2693	0.1815	0.1659
	FF	0.2127	0.2582	0.2522
	CR	0.9539	0.9655	0.9706
	SSIM	0.9994	0.9988	0.9989

The value of entropy is high in DWT fusion technique for both input image datasets, which indicates that the information content is high in the fused image. The SD value is high in Laplacian pyramid fusion, which indicates that it produces high contrast images. The CR measure is high for MSVD fusion method which implies that the input image features are well mapped on to the output-fused image. If the SSIM value is approximately equal to "1" then, it indicates that there is high degree of similarity among the fused and original image.

3.7 CONCLUSION

The fusion of multimodal images is combining the useful and redundant data from the images into a solo image, which increases the quality of image perception, reduces the time, and also decreases the storage space. Two different images are fused by using different image fusion methods, such as Laplacian pyramid, DWT, and MSVD. The proposed algorithm achieves significant result. A relative study has been performed among these fusion methods by analyzing various fusion matrices and the results are tabulated. The entropy value obtained signifies the information clarity in the fused output image. The standard deviation indicates the contrast produced between input and the output-fused images. High value of CR measure indicates that the features in the input selected images are reflected correctly in the output-fused image also. SSIM value indicates the similarity among the fused and the original image.

KEYWORDS

- **image fusion**
- **discrete wavelet transform**
- **multi-resolution singular value decomposition**
- **entropy**
- **standard deviation**
- **fusion factor**
- **SSIM and correlation measure**

REFERENCES

1. Du, J.; Li, W.; Xiao, B. Anatomical-Functional Image Fusion by Information of Interest in Local Laplacian Filtering Domain. *IEEE Trans. Image Process.* **2017**, *26*, 5855–5866.
2. Pappachen, A.; Dasarathy, B. V. Medical Image Fusion: A Survey of the State of the Art. *Inform. Fusion* **2014**, *19*, 4–19.
3. Do, M. N.; Vetterli, M. The Contourlet Transform: An Efficient Directional Multiresolution Image Representation. *IEEE Trans. Image Process.* **2005**, *14*, 2091–2106.
4. Mehta, S.; Marakarkandy, B. CT and MRI Image Fusion Using Curvelet Transform. *J. Inform. Knowl. Res. Electr. Commun. Eng.* **2014**, *2* (2), 848–852.
5. Guruprasad, S.; Kurian, M. Z.; Suma, H. N. In *Fusion of CT and PET Medical Images Using Hybrid Algorithm DWT-DCT-PCA*, 2nd International Conference on Information Science and Security (ICISS), 2015; pp 1–4.
6. Johnson Suthakar, R.; Monica Esther, J.; Annapoorani, D.; Richard Singh Samuel, F. Study of Image Fusion-Techniques, Method and Applications. *Int. J. Computer Sci. Netw.* **2014**, *3* (11), 469–476.
7. Vijayarajan, R.; Muttan, S. Discrete Wavelet Transform Based Principal Component Averaging Fusion for Medical Images. *AEU – Int. J. Electr. Commun.* **2015**, *69* (6), 896–902.
8. Rajarshi, K.; Himabindu, C. In *DWT Based Medical Image Fusion with Maximum Local Extrema*, International Conference on Computer Communication and Informatics (ICCCI), 2016; pp 1–5.
9. Bahri, A.; Shiwani, S. Improved Performance of Image Fusion by MSVD. *Int. J. Innov. Sci. Res. Technol.* **2016**, *1* (6), 11–15.
10. Bhavana, V.; Krishnappa, H. K. Multi-Modality Medical Image Fusion Using Discrete Wavelet Transform. *Procedia Comput. Sci.* **2015**, *70*, 625–631.
11. Kanagasabapathy, A.; Vasuki, A. Image Fusion Based on Wavelet Transform. *Int. J. Biomed. Signal Process.* **2011**, *2* (1), 15–19.
12. Nischitha, P. K. Fusion and Segmentation of Abdominal Cancerous Images. *Int. J. Eng. Technol.* **2018**, *7*, 752–757.
13. Yin, M.; Liu, X.; Liu, Y.; Chen, X. Medical Image Fusion with Parameter-Adaptive Pulse Coupled-Neural Network in Nonsubsampled Shearlet Transform Domain. *IEEE Trans. Instrum. Meas.* **2019**, *68*, 1–16.
14. Yin, H. Tensor Sparse Representation for 3-D Medical Image Fusion Using Weighted Average Rule. *IEEE Trans. Biomed. Eng.* **2018**, *65*, 1–12.
15. Hu, Y.; Carter, T. J.; Uddin Ahmed, H.; Emberton, M.; Allen, C.; Hawkes, D. J.; Barratt, D. C. Modelling Prostate Motion for Data Fusion During Image-Guided Interventions. *IEEE Trans. Med. Imaging* **2011**, *30*, 1887–1900.
16. Jorge Cardoso, M.; Modat, M.; Wolz, R.; Melbourne, A.; Cash, D.; Rueckert, D.; Ourselin, S. Geodesic Information Flows: Spatially-Variant Graphs and Their Application to Segmentation and Fusion. *IEEE Trans. Med. Imaging* **2015**, *34*, 1–13.
17. Singh, S.; Anand, R. S. Multimodal Medical Image Fusion Using Hybrid Layer Decomposition with CNN-based Feature Mapping and Structural Clustering. *IEEE Trans. Instrum. Meas.* **2020**, *69*, 1–11.

18. Zhao, W.; Lu, H. Medical Image Fusion and Denoising with Alternating Sequential Filter and Adaptive Fractional Order Total Variation. IEEE *Trans. Instrum. Meas.* **2017**, *66*, 1–12.
19. Kuiken, T. A.; Miller, L. A.; Turner, K.; Hargrove, L. J. A Comparison of Pattern Recognition Control and Direct Control of a Multiple Degree-of-Freedom Transradial Prosthesis. *IEEE J. Transl. Eng. Health Med.* **2020**, *39*, 1–8.
20. Kumar, A.; Fulham, M.; Feng, D.; Kim, J. Co-Learning Feature Fusion Maps from PET-CT Images of Lung Cancer. *IEEE Trans. Med. Imaging* **2020**, *39*, 1–14.
21. Bhatnagar, G.; Jonathan Wu, Q. M.; Liu, Z. Directive Contrast Based Multimodal Medical Image Fusion in NSCT Domain. *IEEE Trans. Multimedia* **2013**, *15*, 1014–1024.
22. Qi, S.; Calhoun, V. D.; van Erp, T. G. M.; Bustillo, J.; Damaraju, E.; Turner, J. A.; Du, Y.; Yang, J.; Chen, J.; Yu, Q.; Mathalon, D. H.; Ford, J. M.; Voyvodic, J.; Mueller, B. A.; Belger, A.; McEwen, S.; Potkin, S. G.; Preda, A.; Jiang, T.; Sui, J. Multimodal Fusion with Reference: Searching for Joint Neuromarkers of Working Memory Deficits in Schizophrenia. *IEEE Trans. Med. Imag.* **2018**, *37*, 1–12.
23. Szpala, S.; Wierzbicki, M.; Guiraudon, G.; Peters, T. M. Real-Time Fusion of Endoscopic Views with Dynamic 3-D Cardiac Images: A Phantom Study. *IEEE Trans. Med. Imag.* **2005**, *24*, 1207–1215.
24. Nie, R.; Cao, J.; Zhou, D.; Qian, W. Multi-source Information Exchange Encoding with PCNN for Medical Image Fusion. *IEEE Trans. Circ. Syst. Video Technol.* **2020**, *30*, 1–14.

CHAPTER 4

A Study on an Elderly Caretaker Bot Using Computer Vision and Artificial Intelligence

CHANDRA SINGH[1,*], P. NISCHITHA[2], ASHWATH RAO[1], SHAILESH S. SHETTY[3], and PAVANALAXMI[1]

[1]*Department of Electronics & Communication Engineering, Sahyadri College of Engineering & Management, India*

[2]*Department of Electronics & Communication Engineering Mangalore Institute of Technology & Engineerning, Mangalore, India*

[3]*Department of Computer Science & Engineering, Sahyadri College of Engineering & Management, India*

[*]*Corresponding author. E-mail: Chandrasingh146@gmail.com*

ABSTRACT

Currently, 42% of the world's population is under the age of 25. A survey has revealed that the sector of people aged 65 and above will rise at a rapid rate and reach its maximum by 2050. Eventually, this will lead to shortage of caregivers. Providing proper care and caretaking becomes a matter of great concern for the families of such aged people. Hence, the need for companion robot becomes crucial. The companion robot is a particular type of robot designed in order to surveil and assist the elderly people, especially the ones who stay at home alone. The bot functions in two modes, the auto mode and the manual mode. In the manual mode, any relative, that is, the end user can instruct the bot and surveil the aged person through the video that will be live streamed. In the auto mode, the bot continuously monitors the person and

Intelligent Interactive Multimedia Systems for e-Healthcare Applications. Shaveta Malik, PhD
Amit Kumar Tyagi, PhD (Eds.)

assists him in various tasks. It includes voice-controlled activities, such as turning TV channels, calling a person, setting reminders for turning off gas stove and taking tablets. It also includes medical diagnosis and mood detection of the person. Thus, we are planning to come up with a cost-effective bot that will help the elderly people to increase their ease of living with regards to their day-to-day activities.

A set of surveys and studies have led us to a certain set of requirements that a companion bot must possess in order to assist the elderly person and facilitate easy communication with the caregiver. Based on the requirements, we have come to a conclusion that the robot must be capable of

- Being remotely operated by the manual mode end user or the care giver.
- Streaming a live video of the auto mode end user or the care receiver.
- Continuously monitoring the auto mode end user.
- Controlling electrical appliances based on simple voice commands.
- Providing easy dial-up service based on simple voice commands.
- Punctually reminding and providing the medicines to be taken.
- Providing a frequent basic medical check-up facility.
- Reporting any irregular observation in the medical diagnosis to the nearest doctor.
- Detecting the mood of the elderly person using image processing.
- Performing simple tasks, such as playing music, displaying family photos, or notifying the caregiver to make a call based on the mood detected. These requirements were set as the objectives to be achieved and then a deeper analysis of each one was conducted to understand the various methodologies we could follow and the problems we would face in each one of them, leading us to select the best solution.

4.1 INTRODUCTION

As per the World Health Organization (WHO) forecasts, the future during childbirth is as a rule modestly expanded on the planet. This development is considerably more recognizable in the gathering of created nations, where populace maturing is an across the board marvel whose sway on numerous fields (for the most part economy, family and human services) can be considered as a significant issue. In spite of the fact that the physical and mental states of every old individual could differ, older individuals experience the ill effects of specific maladies identified with age, for example, cardiovascular difficulties, torment, osteoarthritis, disintegration of intellectual capacities and general trouble, among others. Thus, numerous senior residents have exceptional necessities and prerequisites that make driving a free life troublesome. Much of the time, the family members need to assume on the liability of eldercare, and this circumstance frequently muddles adjusting work and natural life. Mechanical arrangements that consolidate administration robots and ambient-assisted living (AAL) frameworks could assist old with peopling and their parental figures to improve their everyday life. For instance, envision that an older individual needs to remain at home and is unsuitable for achieving certain assignments. The relative who goes about as parental figure needs to go out for a while, yet needs to effortlessly speak with the older individual if important. A savvy partner portable robot could be the ideal right hand that encourages such correspondence. Then again, the robot could likewise be utilized by old individuals themselves. For instance, in the event that they need to do certain straightforward household assignments, for example, shipping objects (when their physical abilities are diminished), or to collaborate with amusement applications that permit them to improve their physical or state of mind. The robot could likewise be customized to naturally make them move for some significant every day moves, for example, taking the medicine. Somebody could likely believe that there exist enough business brilliant cell phones intended to meet these objectives. In any case, these days, the utilization of this sort of electronic gadgets does not look recognizable for some possible senior clients. Therefore, the friend robot can be considered as a key component that carries innovation closer to old individuals, without the need of clients having to handle complex gadgets or programming programs that could lead them to a baffling and befuddling experience.

In current generation, computers and other many advanced technologies are widely being used. Among this, digital photogrammetry is a well-developed area in research. But there are some defects in digital photogrammetry, some of them are in analytical photogrammetry, they have very high photographic resolutions and low scanning resolutions. Because of this, there is a need of stereo coverage for stereo photogrammetry. To add to this defect, it has less accuracy, there is unavailability of algorithms, camera hardwares are very large in size and the quality of GCP's are also very low. LiDAR (light detection and ranging) is a revolutionary technology in the field of analytical photogrammetry, it is also faster than photogrammetric technology. LiDAR is a device which can efficiently monitor targets/objects. In the years 1960s and 1970s, the first remote sensing instrument using laser was developed. In

1980s and 1990s, the first altimetry system was developed. In 1995, the first airborne light detection and ranging system was developed. In 1994, this technology was used in SHOLS, In 1996, this technology was used by Mars Orbiter Laser Altimeter, and in 1997, by Shuttle Laser Altimeter. LiDAR is mostly used in survey applications, which requires to create high-resolution maps, they are also used in self-driving cars. The working of laser involves the emission of pulses of focused light and the light reflected is measured. The LiDar sensor transmits the laser beam to the object and later measures the time for the reflected light to return to the receiver. The reflectivity of the input beam is measured based on the intensity of the beam. LiDAR consist of four components, such as GPS, IMU, computer, and Laser. The characteristics of LiDAR include wavelength, duration emission, output, coherence, output power, power requirements. Laser converts electricity to light. The efficiency of laser source is the ratio of output to input. Max efficiency is 30% and a Min efficiency is 0.001%. the GPS Receiver tracks the z,y,x position, where x and y give the position and z gives the altitude. IMU unit: It measures the inertia which tracks the position of the vehicle in which the sensor is mounted on. Computer: It is a very important part of the LiDAR system, because there are no data without a computer. All the data from the sensor are stored in the computer system. Time of flight principle is used to calculate the distance. If pulse is triggered at t1 sec and then we receive signal at t2 sec, then for a given threshold, if the signal intensity, for example, is 50% more than PT, then the received signal can be considered, and all others can be ignored.

Therefore,

$$\text{TOF} = t2\ t1 = t \tag{4.1}$$

$$\text{Range} = c\ t / 2 \tag{4.2}$$

$$\text{Resolution} = (\text{accuracy of } t) / 2 \tag{4.3}$$

If the outputs are multiple (from equation 4.1, 4.2, and 4.3), that is, if there are two or more objects in the same path of light beam, we get information from many sources. This is possible because of the shape of the objects. Because of the properties of these objects, we can obtain 3D structures of the objects like tree.

4.2 BACKGROUND

In manual mode of the companion robot, live streaming of the video becomes the key element, enhancing the connectivity and the degree of communication between the two end users. With this benefit comes a need to concentrate on protecting the privacy of the elderly user. Ensuring that the video can only be accessed by the caregiver and carrying out the streaming process in a safe manner even when the internet connection is not very strong are the two challenges. Providing a balanced performance at a reasonable cost is the goal.[4] Several companies and research groups have developed models that can stream a live video. "Cybi: A smart companion robot for elderly people"[5] is one of the papers that we have referred and have drawn ideas from. Cybi also streams a live video for a similar purpose, wherein the bottom layer components are controlled with the help of Raspberry Pi. The intermediate component is a PC or a tablet, which provides interface for teleoperation using spoken commands. At the top layer is the caregiver, who can teleoperate the device using an application. This entire process is done using WebRTC and Google App Engine. All the layers included in transmitting the video ensure high-level privacy and makes it more reliable.

The path toward investigating in a home circumstance is not simply to move from direct A toward point B. It incorporates the front referenced practices that must be dependably fused, as searching for the CR, watching and following the CR, making a beeline for explicit spots inside the home, interfacing with the CR (for instance permit her to put and recuperate things from the robot's amassing box or advancement things to the arrangement), self-overseeing mooring to the empowering station, or being remotely controlled from outside. The middle parts for the course contain arranging and confinement computations and the pilot itself. To address the area natural variables of the robot, we use inhabitance maps made using the laser scanner, similarly as a visual procedure for monocular scene proliferation.

The made aides are used by a limitation approach to measure the robot zone in the overall orchestrate diagram by using a local guide of continuous discernments and the data on the overall guide. Like the overall computerized designing, we also use a deliberate programming plan for the guide with a couple of internal parts. The middle section of the pilot gets the current navigational endeavor from overlying layers like "Course Behaviors" or the "Application Layer" by methods for RPC (remote procedure calls) or using an inalienable scripting language. Both enable the control of the pilot from passed on applications.

After a task is gotten by the pilot, it is rotted into specific development and speed requests of the robot's wheels in order to process the perfect endeavor. For setting up the task, the development coordinator and the objectives expect a noteworthy activity here. We rot each navigational task and practices into targets. Each objective is an alternate programming module concentrated for explicit endeavors, for example, following a way or an individual, driving at a particular speed or heading, etc. The goals are recognized as detached programming modules. This grants us to incorporate new targets viably, when new tasks and practices become significant without changing existing bits of the pilot.

The improved ranges and stances can be used to deliver an inhabitance Map, which addresses nature as a probabilistic inhabitance framework at and the propelled act graph. Camshift technology is used as an auxiliary tracker, which ensures smooth and stable movement. When the basic tracker is stable, the color histogram of the target location is calculated and used when the basic tracker loses the track of the target. The histogram information is updated once in every 2 s. This winds up the basic planning of the robot in both automode and manual mode.[8]

While operating in the auto-mode, one subtask that the robot is supposed to do is detecting the mood of the elderly person. Once the user is detected, the robot must take an image of the elderly person's face from a fixed viewpoint by adjusting its location. This image is labeled at several regions that have higher deviation during the change in the expression. These sets of labels are referred to as learning set. And the labels from the images that are already fed into the processor are referred to as the testing set. Now the coordinates of the learning set is compared with the coordinates of the testing set and one image from the testing set which shows least deviation is tagged with an emotion which will be the result of the mood detector. Numerous algorithms have been proposed for facial expressions. This paper proposed by a group of people that can detect facial expressions even under large variations in lighting, that is, under any number of light sources and also for any intensity of the light sources.[9]

Another significant subtask that the robot does in auto-mode is reminding the client to take suitable measurements of medications in the time proposed by the specialist. "Plan and Implementation of an Automated Reminder Medicine Box for Old People and emergency clinic"[10] is a straightforward undertaking accomplished for a similar reason. The gadget contains 21 hermetically sealed compartments to keep the medication. The chaperon or attendant of the client needs to make a week after week arrangement for the gadget by keeping medications in these compartments, three compartments for every day. The time at which the meds must be taken must be set during the principal utilization of the gadget.

This information will be put away in the SD card and will consistently be contrasted and a real-time clock, as and when the time coordinates, the gadget plays a sound in the speaker, reminding the client to take medications and furthermore squints the LED of the compartment from which the medication must be taken.[11] This straightforward cost-effective structure has the ability to guarantee clinical well-being, proper medication measurements and forestall sedate abusement of older people.[12]

4.3 METHODOLOGY

The companion robot will be operating in two modes, one is the manual mode and the other is the automode. The process of switching between the two modes is controlled by the patient's caregiver. Manual mode, which is also called as the surveillance mode, allows the manual mode end user or the caregiver to keep track of the activities done by the automode end user or the care receiver by streaming a live video of the elderly person. This is done by using a web cam. Video recorded by the camera is given to Raspberry Pi, wherein it is temporarily stored. This video is then uploaded to Web RTC through cloud with the help of a Wi-Fi module, from where the caregiver can access the video from any location using a mobile application. Proposed block diagram is shown in Figure 4.1

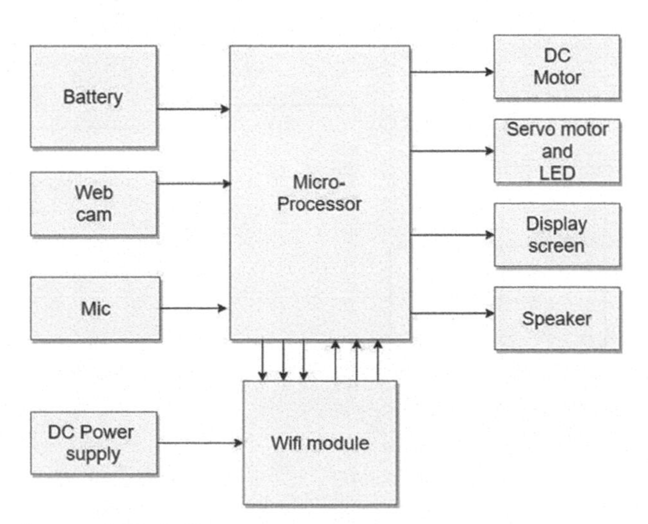

FIGURE 4.1 Proposed block diagram.

Automode allows the bot to perform certain tasks without human interpretation. To start with any of the tasks, the bot must first locate the user and must be capable of following the user. A sequence of images taken with the help of computer vision technologies helps the bot to track the elderly person. After getting the exact location of the person, the bot will continuously monitor the person and wait for any one of the four voice command inputs from the user, that is, TV, Call, Doctor, and Gas. Voice-controlled commands are then processed and the respective tasks of turning ON the TV, calling the required person, connecting with the doctor and setting a reminder to turn off the stove are performed by the bot.

In the same mode, the bot also performs the task of reminding the user to take the prescribed medicines at the mentioned time. This can be done using a simple design containing 21 compartments, with an LED attached to each one of them, which blinks when the respective compartment's medicines are to be taken. The bot must also detect the mood of the elderly person once in ever specified time interval. This can be done by reading the facial expressions through image processing. The bot must first align the camera in such a way that the image is taken from a preset fixed viewpoint.

The image obtained is labeled at several points, and the coordinates of these points are recorded and stored in a set. These values are then compared with a pre-stored image's set of coordinates. The image which shows least deviation between the two sets of co-ordinates is determined and the expression tagged to that image is said to be the mood of the elderly person. Based on the mood detected, the robot does some activities like playing a song, displaying family photos, or notifying the caregiver to make a call (Fig. 4.2).

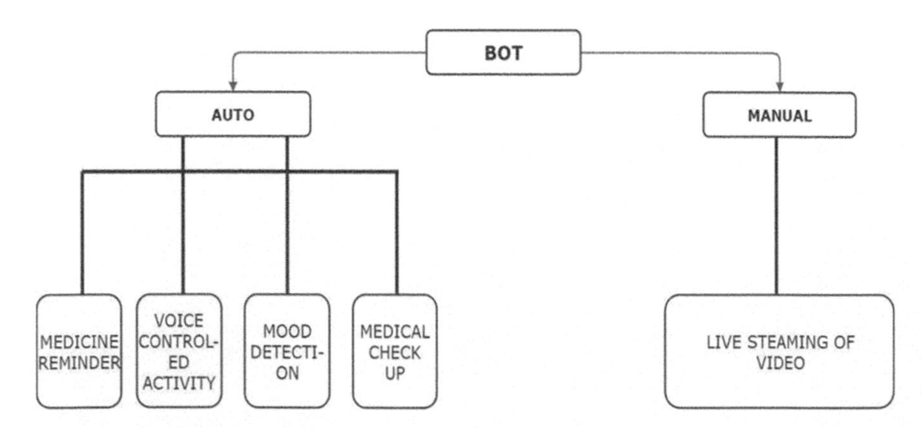

FIGURE 4.2 Implementation of bot in both modes.

In manual mode of the companion robot, live streaming of the video becomes the key element, enhancing the connectivity and the degree of communication between the two end users. With this benefit comes a need to concentrate on protecting the privacy of the elderly user. Ensuring that the video can only be accessed by the caregiver and carrying out the streaming process in a safe manner even when the internet connection is not very strong are the two challenges that need to be looked through. Providing a balanced performance at a reasonable cost is the goal of the design process. Several companies and research groups have developed models that can stream a live video. In automode, the companion robot must perform the task of human tracking and localization. This should be possible utilizing PC vision innovation to recognize, distinguish, and track the human with the assistance of an arrangement of images capable of following human movement without building the ecological guide or info of the objective area.

The essential advancement of the robot is to precisely find the situation of the objective and to move in time precisely. The skeletal following calculation is at first utilized as an essential following calculation for human discovery. The skeletal following calculation acquires the skeletal data of the considerable number of people in the view. At the point when the client gives an order to the robot through a signal as a banner of start, the robot bolts the area of the clients its last objective and starts playing out the undertaking as per the order received. Another significant subtask that the robot does in automode is reminding the client to take fitting measurements of medications in the time recommended by the doctor. While working in the automode, one subtask that the robot should do is identifying the disposition of the old individual.

Once the user is detected, the robot must take an image of the elderly person's face from a fixed viewpoint by adjusting its location. This image is labeled at several regions that have higher deviation during the change in the expression. First we have to do a excel file containing two columns medicine and time, and then we have to upload it to the microcontroller. After that there is a real-time clock module which will be connected to the microcontroller which will monitor the real time and remind for taking the medicines as per the time mentioned in the file.

For this, there are set of predefined commands. When the speech from the user matches the commands, then it will do the labeled work for that command. This includes calling person, setting up a reminder for stove on and off, changing the TV channel. All these will be connected with IoT Mood Detection: it involves facial expression detection. Whenever it detects a sad face, it will play some pre-stored videos or images of close ones. We can use a heartbeat sensor, when there is some irregularities in it.

The way toward exploring in a home situation is not just to move from direct A toward point B. It includes the front referenced practices that must be reliably incorporated, as looking for the CR, watching and following the CR, heading to specific spots inside the home, connecting with the CR (e.g., allow her to put and recover things from the robot's stockpiling box or promotion things to the plan), self-governing docking to the energizing station, or being distantly controlled from outside. The center parts for route comprise planning and limitation calculations and the pilot itself. To speak to the neighborhood environmental factors of the robot, we use inhabitance maps manufactured utilizing the laser scanner just as a visual methodology for monocular scene reproduction. The made guides are utilized by a restriction way to gauge the robot area in the worldwide arrange outline by utilizing a neighborhood guide of ongoing perceptions and the information on the worldwide guide. Like the general automated engineering, we additionally utilize a measured programming plan for the guide with a few inner parts. The center segment of the pilot gets the current navigational undertaking from the overlying layers like "Route Behaviors" or the "Application Layer" by means of RPC or by utilizing an inherent scripting language. Both empower the control of the pilot from the conveyed applications.

After an assignment is obtained by the pilot, it is decayed into particular movement and speed orders of the robot's wheels, so as to process the ideal undertaking. For task preparing, the movement organizer and the goals assume a significant job here. We decay every navigational assignment and practices into targets. Every goal is a different programming module concentrated for specific undertakings, such as following a way or an individual, driving at a specific speed or heading, and so forth. The destinations are acknowledged as isolated programming modules. This permits us to include new targets effectively when new errands and practices become important without changing the existing pieces of the pilot.

4.4 SLAM ALGORITHM

SLAM (Simultaneous Localization And Mapping) is utilized to estimate the posture of a robot what is more, the guide of the earth simultaneously. It is hard, in light of the fact that a guide is required for restriction and a decent posture gauge is required for planning. Limitation implies surmising a guide given area.. Planning implies construing a guide given areas.

To construct the guide of the earth, the SLAM calculation forms the LiDAR sweeps and assembles a diagram that connects these outputs. The robot perceives a formerly visited place through sweep coordinating and may build up at least one circle terminations along its moving way. The SLAM calculation uses the circle conclusion data to refresh the guide and modify the assessed robot direction.

(1) Loading the laser scanned data from the file
The disconnected SLAM data. MAT record contains the sweeps variable, which contains all the laser filters utilized in this model. load('offlineSlamData.mat').
A story plan and inexact way of the robot are accommodated for illustrative purposes. This picture shows the relative condition being planned and the estimated direction of the robot.
(2) Here the LiDAR SLAM item, set the guide goal, and the maximum LiDAR extend. Set the maximum LiDAR go marginally little than the maximum output extend (8m), or exact close to max go. Set the framework map goal to 20 cells for every meter, which gives a 5 cm exactness.

Utilizing a higher circle conclusion edge helps to reject bogus encouraging points in circle conclusion ID process. In any case, remember that a high-score match may at present be an awful match. For instance, checks gathered in a domain that has comparative or rehashed highlights are bound to deliver bogus positives. Utilizing a higher circle conclusion search span permits the calculation to look through a more extensive scope of the guide around momentum present gauge for circle terminations.

Gradually, add outputs to the hammer Alg object. Output numbers are printed whenever added to the guide. The item dismisses filters if the separation between checks is excessively little. Include the initial 10 outputs first to test your calculation.

Remake the scene by plotting the outputs and stances followed by the slamAlg.

Keep on including examines in a circle. Circle terminations ought to be naturally recognized as the robot moves. Posture diagram streamlining is performed at whatever point a circle conclusion is distinguished. The yield streamlining Info has a field, Is Performed, that demonstrates when posture diagram enhancement happens.

Plot the outputs and stances at whatever point a circle conclusion is recognized and check the outcomes outwardly. This plot shows overlaid filters and an advanced posture diagram for the principal circle conclusion. A circle conclusion edge is included as a red connection.

Plot the last constructed map after all sweeps are added to the slamAlg object.

The past for circle ought to have included all the sweeps notwithstanding just plotting the underlying circle conclusion and Outwardly Inspects it.

A picture of the outputs and posture diagram is overlaid on the first floor plan. We can see that the guide coordinates with the first floor plan well in the wake of including all the outputs and upgrading the posture chart.

Manufacture of Occupancy Grid Map

The enhanced sweeps and postures can be utilized to produce an inhabitance Map, which speaks to nature as a probabilistic inhabitance matrix. Picture the inhabitance matrix map populated with the laser. The diagram is depicted in Figure 4.3.

In this, authors have used LIDAR as the sensor, generally it can be a camera, RIDAR, etc.

Feature Match: This is basically the process of finding corners, centers of what the sensor is detecting. For example, if the LiDAR is looking at duster, then feature matching is finding the edges of the duster. There are many algorithms which can be used in feature match like, difference of Gaussian, shift, surf, etc. Here, the goal of this stage is to use this algorithm to identify the same object when it is viewed in the next frame, that is, with a different angle. Also, the outliers are removed so that the algorithm focusses only on the key feature points which are available from the features.

Pose Estimation: we take the features that we have found from the feature match stage and determine how they move from one frame to another, that is, it tells us how an object has moved from one frame to another frame.

Loop Closure: The goal of this step is to find out whether we have visited a point before or not. It is a very important process of discovering new environments. If we have visited the same area that we have visited before, then the model will inform us that we have already visited this area. By this process, we actually can get the position of viewing angle.

Bundle adjustment: It is a process of further refining the aggregation of errors done in the loop closure. Both the loop closure and the bundle adjustment give feedback to the pose estimation process. The goal of this feedback is constantly updating our state.

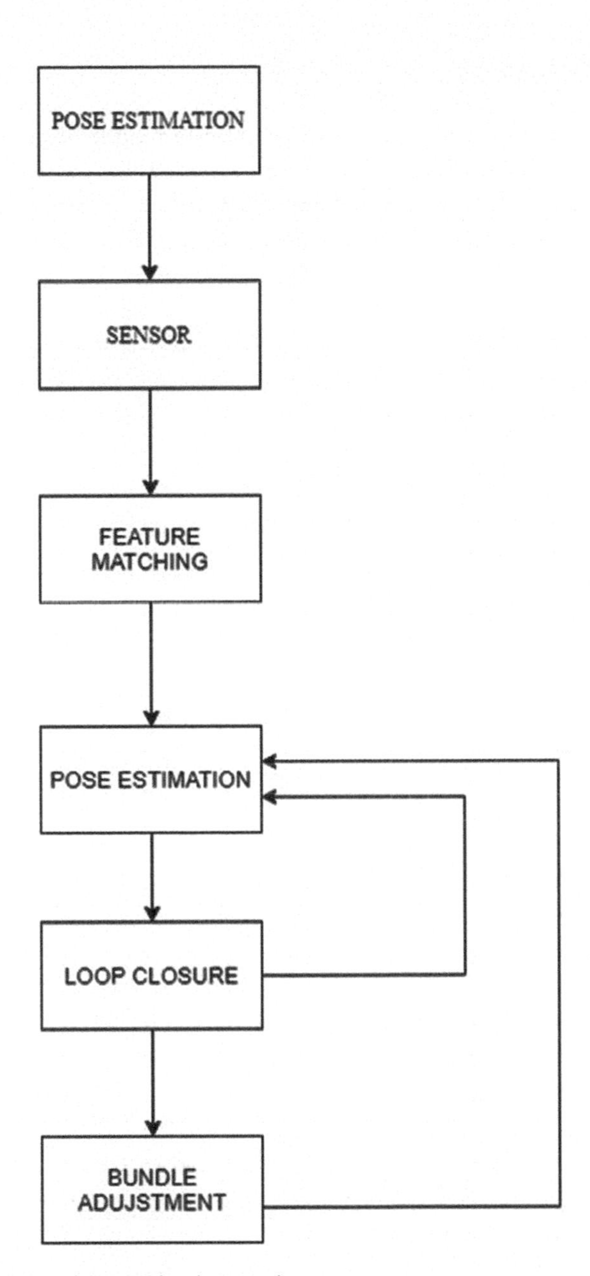

FIGURE 4.3 General SLAM implementation.

In the last, several interesting research work on Robotics and Automation (including raised issues) can be found in Ref. [13-18].

4.5 CONCLUSION

Providing proper care and caretaking becomes a matter of great concern for the families of such aged people. Hence, the need for companion robot has becomes crucial. The companion robot is a particular type of robot designed in order to surveil and assist the elderly people, especially the ones who stay at home alone. That is, the end user can instruct the bot and surveil the aged person through the video that will be live streamed. In the automode, the bot continuously monitors the person and assists them in various tasks. It includes voice-controlled activities like turning TV channels, calling a person, setting reminders for turning off the gas stove, and taking tablets. It also includes medical diagnosis and mood detection of the person. Thus, we are planning to come up with a cost-effective bot that will help the elderly people to increase their ease of living when concerned to their day-to-day activities.

KEYWORDS

- **manual mode**
- **auto mode**
- **surveil**
- **image processing**
- **light detection and ranging**

REFERENCES

1. Wu1, X.; Thomas, R. C.; Drobina, E. C.; Mitzner, T. L.; Beer, J. M. Telepresence Heuristic Evaluation for Adults Aging with Mobility Impairment. *Proc. Hum. Factors Ergon. Soc. Annu. Meet.* **2017**, *61* (1), 16–20. https://doi.org/10.1177/1541931213601499.
2. Wada, K.; Shibata, T.; Asada, T.; Musha, T. Robot Therapy for Prevention of Dementia at Home—Results of Preliminary Experiment. *J. Robot. Mechatron.* **2007**, *19* (6), 691–697.
3. Stern, C.; Konno, R. The Effects of Canine-Assisted Interventions (CAIs) on the Health and Social Care of Older People Residing in Long Term Care: A Systematic Review. *JBI Libr. Syst. Rev.* **2011**, *9* (6), 146–206.
4. Walters, M. L.; Koay, K. L.; Sverre Syrdal, D.; Campbell, A.; Dautenhahn, K. In *Companion Robots for Elderly People: Using Theatre to Investigate Potential Users' Views*, IEEE ROMAN: The 22nd IEEE International Symposium on Robot and Human Interactive Communication Gyeongju, Korea, 2013; pp 26–29.
5. Broekens, J.; Heerink, M.; Rosendal, H. Assistive Social Robots in Elderly Care: A Review. *Gerontechnology* **2016**, *8* (2), 94–103.

6. Helal, A.; Abdulrazak, B. TeCaRob: Tele-care Using Telepresence and Robotic Technology for Assisting People with Special Needs. *Int. J. Human-friendly Welfare Robot. Syst.* **2006**, *7* (3), 31–38.
7. Van Den Heuvel, H.; Huijnen, C.; Caleb-Solly, P.; Nap, H. H.; Nani, M.; Lucet, E. Mobiserv: A Service Robot and Intelligent Home Environment for the Provision of Health, Nutrition and Safety Services to Older Adults. *Gerontechnology* **2012**, *11* (2), 10–21.
8. Coradeschi, S.; Cesta, A.; Cortellessa, G.; Coraci, L.; Gonzalez, J.; Karlsson, L.; Furfari, F.; Loutfi, A.; Orlandini, A.; Palumbo, F.; Pecora, F.; von Rump, S.; Štimec, A.; Ullberg, J.; Östlund, B. In *GiraffPlus: Combining Social Interaction and Long Term Monitoring for Promoting Independent Living*, Proceedings of the 6th International Conference on Human System Interaction (HSI 2013), Gdansk, Poland, June 2013; pp 578–585.
9. Haidegger, T.; Benyó, Z. Surgical Robotic Support for Long Duration Space Missions. *Acta Astronautica* **2018**, *63* (7), 996–1005.
10. Cabibihan, J. J.; So, W. C.; Saj, S.; Zhang, Z. Telerobotic Pointing Gestures Shape Human Spatial Cognition. *Int. J. Social Robot.* **2012**, *4* (3), 263–272.
11. Pavón-Pulido, N.; López-Riquelme, J. A.; Ferruz-Melero, J.; Vega Rodríguez, M. A.; Barrios-León, A. J. A Service Robot for Monitoring Elderly People in the Context of Ambient Assisted Living. *J. Ambient Intell. Smart Environ.* December **2014**, *6* (6), 595–621.
12. Pavón, N.; Ferruz, J.; Ollero, A. In *Describing the Environment Using Semantic Labelled Polylines from 2D Laser Scanned Raw Data: Application to Autonomous Navigation*, IEEE/RSJ International Conference on Intelligent Robots and Systems (IROS 2010), Taipei (Taiwan), Sept 2010; pp 3257–3262.
13. Baturone, I.; Moreno-Velo, F.; Sanchez-Solano, S.; Ollero, A. Automatic Design of Fuzzy Controllers for Car-Like Autonomous Robots. *IEEE Trans. Fuzzy Syst.* Aug **2014**, *12* (4), 447–465.
14. Alexander, B.; Hsiao, K.; Jenkins, C.; Suay, B.; Toris, R. Robot Web Tools. *IEEE Robot. Autom. Mag.* December **2012**, *9* (4), 20–23.
15. Waibel, M.; Beetz, M.; Civera, J.; d'Andrea, R.; Elfring, J.; GalvezLopez, D.; Häussermann, K.; Janssen, R.; Montiel, J. M. M.; Perzylo, A.; Schiessle, B.; Tenorth, M.; Zweigle, O.; Van de Molengraft, M. J. G. RoboEarth—A World Wide Web for Robots. *Robot. Autom. Mag. IEEE* June **2011,** *18* (2), 69–82.
16. Kumar Tyagi, A.; Sreenath, N. Future Challenging Issues in Location Based Services. *Int. J. Comput. Appl.* Mar **2015**, *114* (5), 0975–8887.
17. M. M. Nair, A. K. Tyagi and N. Sreenath, "The Future with Industry 4.0 at the Core of Society 5.0: Open Issues, Future Opportunities and Challenges," 2021 International Conference on Computer Communication and Informatics (ICCCI), 2021, pp. 1-7, doi: 10.1109/ICCCI50826.2021.9402498.
18. Tyagi A. K., Fernandez T. F., Mishra S., Kumari S. (2021) Intelligent Automation Systems at the Core of Industry 4.0. In: Abraham A., Piuri V., Gandhi N., Siarry P., Kaklauskas A., Madureira A. (eds) Intelligent Systems Design and Applications. ISDA 2020. Advances in Intelligent Systems and Computing, vol 1351. Springer, Cham. https://doi.org/10.1007/978-3-030-71187-0_1

CHAPTER 5

Knowledge Shift for Candidate Categorization in Lung Nodule Detection Using 3D Convolutional Neural Network

P. MANJU BALA*, S. USHARANI, R. RAJMOHAN, M. PAVITHRA, and T. ANANTH KUMAR

Department of Computer Science and Engineering, IFET College of Engineering, Villupuram, Tamilnadu, India

**Corresponding author. E-mail: pkmanju26@gmail.com*

ABSTRACT

The instinctive identification mechanism for lung nodules allows for early airing and effective medical complications. However, still several candidates for lung nodules developed during this method by initial rough detection. Lung disease indicates the disorder that impacts the lungs. Lung diseases are probably the most widely recognized disease in the world. Millions of individuals suffer from lung disease due to smoking, pollution, heredity, and infections caused via blood vessels as indicated by the World Health Organization (WHO). In two-dimensional (2D) convolutional neural network (CNN), the precision and proficiency is exceptionally low. To overcome the shortcomings three-dimensional (3D) CNN is developed for lung nodule detection. By utilizing this technique, the precision and effectiveness is improved and the pictures are given in 3D with the goal that the influenced region can be plainly recognized. Furthermore, candidate categorization is performed to classify the images to extract the deep features. The experimental findings of the LIDC information collection indicate that 0.9645 and

Intelligent Interactive Multimedia Systems for e-Healthcare Applications. Shaveta Malik, PhD Amit Kumar Tyagi, PhD (Eds.)

0.9871 are precise in the suggested protocol. The suggested identification of the agent of the lung nodule demonstrates research observations that there is a higher degree of success over predefined structures. In comparison, the detection rate of all the three trials is over 95% while measuring for three measurements in sizes 28_28, 40 40, and 56_56, showing that the suggested framework is not responsive to angular measurements.

5.1 INTRODUCTION

Pulmonary tumor is one of the most dangerous forms of cancer. As per the original study, in the last two decades, the mortality rate triggered by emphysema has risen significantly relative to several ailments. Year after year, the casualty count continues to increase. Temporary surgical attention and elimination of cardiovascular cancer are efficiently managed by early cystic lesions. This will result in the development of a significant quantity of medicinal slices. Taking a look at those multiple photographs and analyzing them ultimately raises the productivity of doctors. As is well established, CAD programmes would use more than one professional physician's expertise and available tools to help clinicians accomplish cancer procedure. CADs also provide hard cystic examination and cyst assessment in the method of diagnosing pulmonary tumours.

Furthermore, the rugged cyst identification also ends in many patients for the primary cyst. Patients were ground-breakingly observed and classified with the help of the tomographic images. The preliminary power of the system helped to discriminate these cysts from the input images. Certain pulmonary tracts are somewhat close to the true cysts that can result in a rise in adverse events. In addition to these issues, a multistage convolutional neural network (CNN) for a pulmonary pathological cyst by means of information exchange is transmitted and enhanced. This approach makes it possible to distinguish both the tiny tumours in coarse contrast and larger tumours in apparently greater fidelity. In the input space, the proposed system will remove the functions from the neural network of deeper points at multiple dimensions and thresholds in order to maintain the primary goal defined in steep metrics and to improve the compromised information with elevated functionality.

The proposed method in the specific domain will also solve the complexities of image classification created by the varying location and composition of lungs and by real tumors and artificial tumors that appear as accurate substitutes for the radiographical heterogeneity cysts. Therefore, it is important to choose the correct lesions. Moreover, the task in choosing the actual lumps is to classify the tumor area as the recipient that wants to be studied. To alleviate tumor segmentation subject-class imbalance problem, the following two kinds of difficulties have to be considered: the radioactivity diversity, which could give rise to the concealment of certain cysts, and anti cysts. Usually a variety of clinicians use CAD to discourage medical tomography identification, polyp detection in the intestine, and respiratory diseases.

Lung nodulet is the fundamental reason of disease-related deaths around the globe, with reference to total of 1.6 million deaths.[1] The occurrence of harmful development lungs is unequivocally related to the period of end, with 5-year all in all continuance reducing from mastermind IA contamination (85%) to compose IV disease (6%).[2] Regardless, in routine clinical practice, some starting time lung threatening development were delayed in assurance, as a result of that most patients are every now and again asymptomatic. Lung danger screening is expected to decrease lung sickness-related mortality through diagnosing the disease at its starting period in high-chance partners with commonly limited naughtiness, to enable those patients to get mending assignments.[3,4] Early screening primers, all things considered, assessed the reasonability of chest radiography (CXR) got together with sputum cytology, and demonstrated no effect on lung threatening development mortality.[5] In any case, in the latest decade, technologic advances in handled tomography (CT) re-brought imaging-based screening into the center intrigue. The national lung screening trial (NLST) displayed a 20% abatement in lung harmful development mortality for low-parcel figured tomography (LDCT) differentiated and CXR screening, similar to a 6.7% all-cause mortality decline. Regardless of the way that the positive results are being given, it isn't yet evident whether the positive outcomes of NLST can be copied by other randomized controlled screening primers. Liquid biopsy is a noninvasive method to manage screening and early finish of lung harm. It may get one of the essential or elective approaches to manage thoracic imagery for the early end and screening for lung harmful development. Similarly, how to modify the preferences and harms of the screening is uptil now a questionable issue. Uncommon concerns have risen up out of lung handles recognized from screening, for instance, counterfeit positive results, over diagnosis, false negative results, and physical and mental issues. This chapter rapidly studies the latest progression in lung harmful development screening and the rule unanswered requests on lung handle area, to look at perfect techniques for use of lung disease recognizable proof. Computer-aided diagnosis PC-supported location (CADe), additionally called PC-helped finding (CADx), are frameworks that help specialists in the translation of clinical pictures. X-beam, MRI,[8] and ultrasound imaging techniques offer a vast volume of data to be examined and fully analyzed in a short time frame by the radiologist or other clinical specialist. Computer-aided design frameworks process advanced pictures for run-of-the-mill appearances and to feature obvious segments, for example, potential infections, so as to offer contribution to help a choice taken by the expert. Computer-aided design likewise has potential future applications in computerized pathology with the approach of entire slide imaging and AI calculations. So far, its application has been restricted to measure immune staining but, in contrast, is being researched for the standard Hue recolor. PC-aided configuration is an interdisciplinary invention that consolidates components of man-made thought and PC vision with the preparation of the radiological and pathological picture. For example, a few clinics use CAD[10] to assist in preventive clinical registration in mammography[11] (conclusion of bosom malignancy), identification of polyps in the colon, and lung disease. A common application is the tumor region. CNN[12] is a specific sort of phony neural framework that uses perceptrons, an AI unitcount, for coordinated learning, and to analyze data.

CNNs apply to a picture dealing with ordinary language getting ready and various sorts of mental tasks. A convolutional neural framework is in any case called a ConvNet.

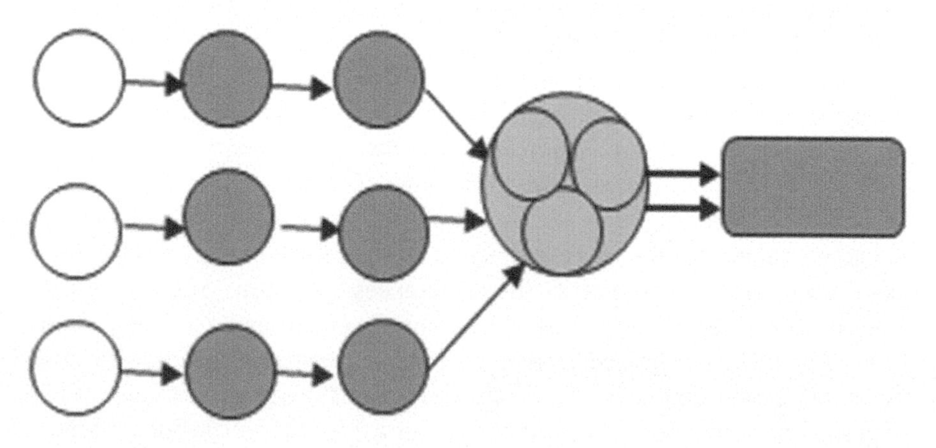

FIGURE 5.1A M-scale CNN model.[14]

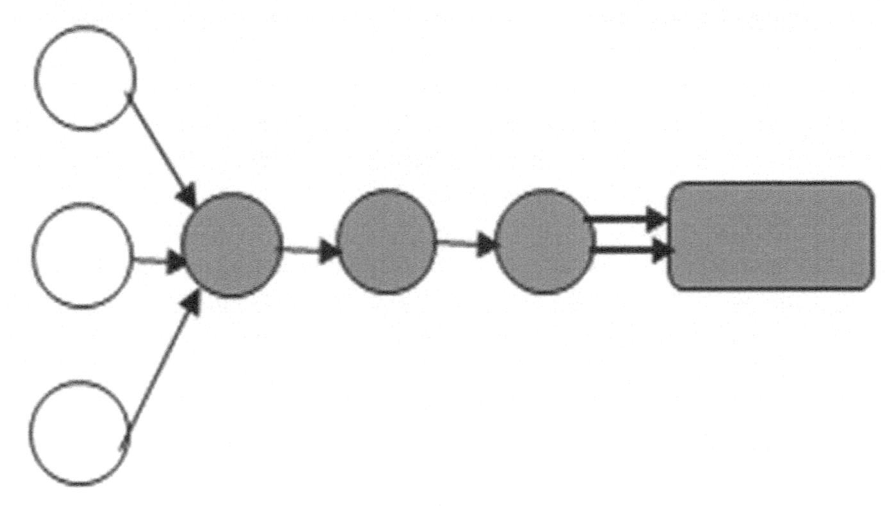

FIGURE 5.1B M-view CNN model.[15]

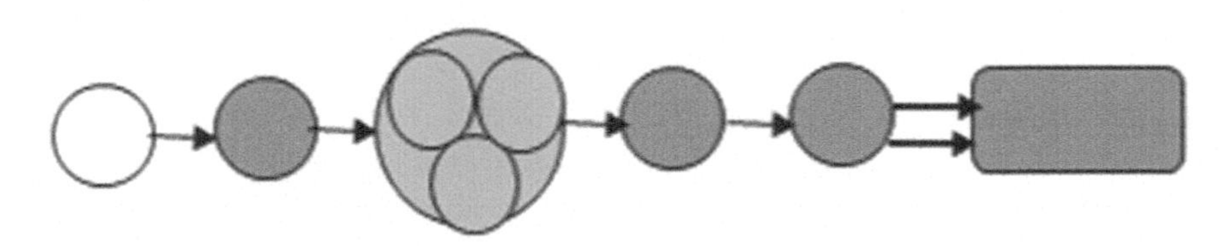

FIGURE 5.1C M-crop CNN model.[16]

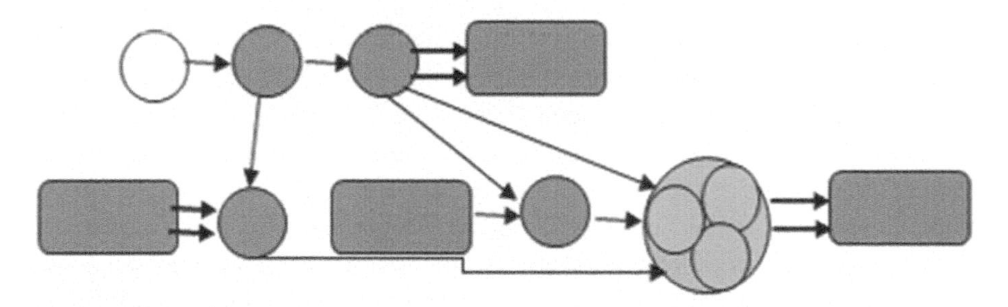

FIGURE 5.1D Proposed method.

Figure 5.1 Present CNN progress trends for the classification of lung nodules with several resolutions. (a) The M-scale CNN.[14] (b) The M-view CNN.[15] (c) The M-crop CNN.[16] (d) Our method.

5.2 RELATED WORK

In traditional methods of automatic cyst identification, handmade elements are typically utilized. Such characteristics provide the essential attributes, such as bronchus nodule structure, form, and scale. It can also contain increased elements, dissociated from key levels like minor frequency waves, discrete localized variations (DLV), and directed vector histograms (DVH), etc. The characteristics are indeed self-designed, without self-learning capabilities like software, and the conventional approaches are not smart enough to function in the result. Identifying minimal functionality is a 2D neural network process. Any tumors are not completely or heavily represented, and may be misidentified by cysts in the database.

There are fewer simple pictures in the current method. Toward the 2D CNN process, minimal attributes are captured. Certain gland forms in the database, which can contribute to the misidentification of cysts, really aren't adequately described and not completely illustrated.

Initially, a total lung knob is frequently disseminated on different cuts. Notwithstanding, in capturing the logical data between slices, the 2D CNN technique is limited. Therefore, in future works, a 3D CNN model could be considered. Second, the LIDC (Lung Image Database Consortium) information collection system is prepared for our model. However, a few types of knobs are not fully discussed in the information index or are not completely displayed, which could contribute to a fake nodule recognition. We accept that a bigger informational index for test planning and compelling example preprocessing before preparing will help recognize these applicants.[6,7] In capturing the logical data between slices, the 2D CNN technique is limited; therefore, a 3D CNN model need to be considered.[3] Again, the model is prepared for knowledge collection on LIDC. In any event, in the information index, a few types of knobs are not fully spoken about or not fully featured, which may prompt the fake identification of nodules. This chapter presents the diganosis instrument's highlights and demonstrates the way to create a deep learning model using the apparatus. The chapter identified the data availability technique using dermal cell images and its test application to identify disease cells in the DLS model. In distinguishing harmful development cells from the images of disease cells, the DLS models achieved an AUC of 98.71%.[9] LIDC Dataset: Data Extraction, Autoencoder: Feature Extraction are the methodologies used in this chapter The LIDC[9,22] has made publicly available a database containing thoracic CT images of 1010 lung tumor patients closely studied by up to four radiologists. The mess up between the data and yield l-2 norm is used for rebuffing. In autoencoder, typically sigmoid or hyperbolic concern limits are used as the incitation job. In autoencoder, the cost capacity can be precisely figured out, which results in a chance of utilizing progressively advance strategies for improvement, for example, L-BFGs for preparing the systems. One of the upsides of this methodology of utilizing the comment of the considerable number of radiologists when contrasted with numerous other comparative methodologies[13] that utilize the best comment for acutor the explanation with biggest region is that while testing in reality, on the off chance, one of the radiologists gives an incomplete comment of a knob (which can happen often); our framework would in any case work.

5.3 SYSTEM ARCHITECTURE

Right now the strategy utilizing CNN is used. As a progressively robotized approach, the CNN technique utilizes crude picture information as data and can be legitimately classified as relent. The counterfeit positive of the classification mission is thus significantly reduced and the ratings on classification estimates are increased. When utilizing this strategy, it will improve the productivity and accuracy. The 3D CNN depicted in Figure 5.2 is utilized for more description.

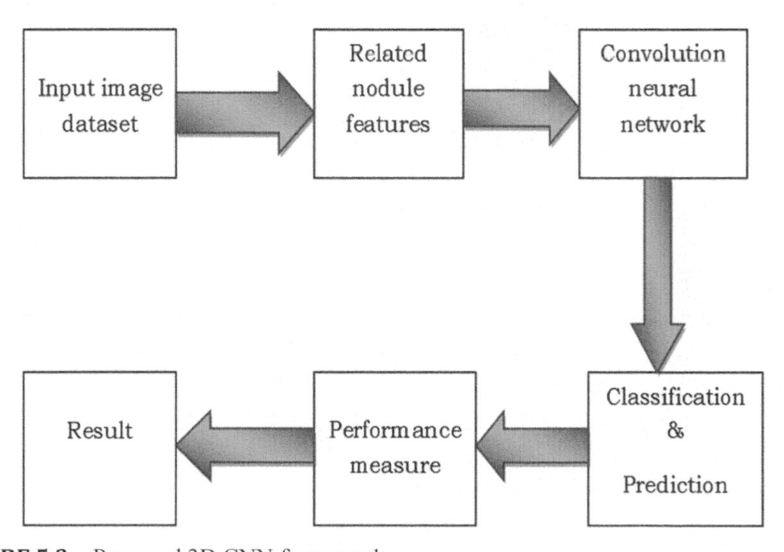

FIGURE 5.2 Proposed 3D CNN framework.

The CT picture is taken as an input; then utilizing the CNN calculation, it distinguishes the knob highlights and afterward it arranges and foresees the lung malignancy in the primary stage. A short time later all the related issues lungs can be anticipated and classified. Finally, by utilizing these forecasts, the outcomes is delivered in a 3D image though it gives more explanation of knobs and makes the framework increasingly effective.

5.3.1 DATA ACQUISITION

Figure 5.3 exemplifies the proposed model for lung knob detetction. Data help look at signals that calculate actual physical conditions and transform into automated numerical features over the subsequent models that can be limited by a PC. Hardware is used for signal molding, to convert sensor signals into a structure that can be modified to computerized values. Computerized imaging or advanced image acquisition is the development of a carefully encoded representation of an article's visual features, for example, a physical scene or an item's inside structure. Table 5.1 decsribes the nodule parameters realized for knob detection.

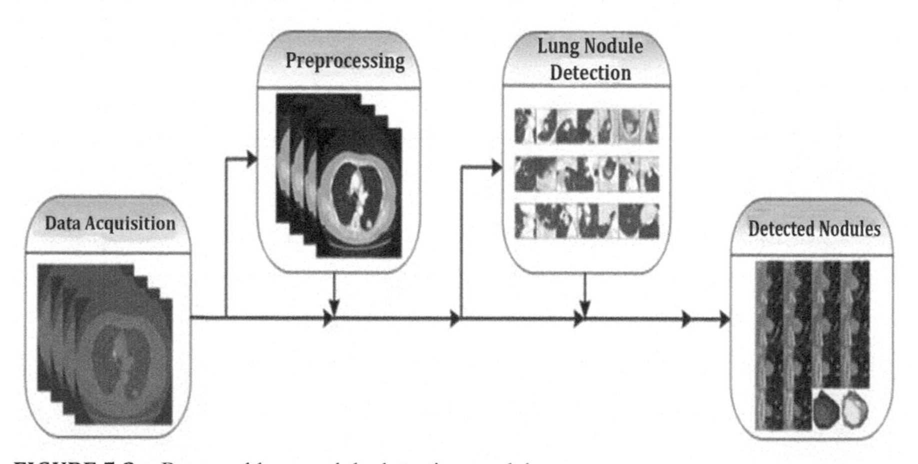

FIGURE 5.3 Proposed lung nodule detection model.

TABLE 5.1 Diameter distribution of the lung nodules.

Diameter(mm)	3=<d<6	6=<d<9	9=<d<12	12=<d<15	d>=15
Proportion (%)	44.32	27.35	11.55	6.07	10.71

5.3.2 PREPROCESSING

Preprocessing alludes to all the changes on the crude information before it is taken care of ay the AI or profound learning calculation. Picture prepreparation is the name of the process for images at the lowest degree of deliberation, which is an enhancement of image details that smotheres unwanted bends or upgrades certain image highlights that are necessary for additional handling. The duplication of images is used for image prepreparation. The planning of a convolutionary neural system on crude images, for instance, would most certainly trigger terrible grouping exhibitions. Data preprocessing is a technique that is used to change over the unrefined data into an ideal enlightening record. So to speak, at whatever point the data is amassed from different sources, it is accumulated in a rough game plan, which isn't commonsense for the assessment. In AI, CNNs or ConvNets are perplexing feed-forward neural systems. CNNs are utilized for picture arrangement and acknowledgement in view of its high exactness. It has 55,000 pictures—the test set has 10,000 pictures and the approval set has 5000 pictures. Picture prepreparation is the term for the process on images at the lowest degree of deliberation, which is an enhancement of image details that smotheres unnecessary bends or improves certain image highlights that are necessary for additional handling. Picture pre-preparation utilizes image repetition. Algorithm 5.1 elucidates the proposed enhancement for choosing the lobs for feature extraction.

Algorithm 5.1: Lung Nodule Identification

Input: Infected Lung Image

Output: Lung Nodules encircled

Step 1. Transform the feature vector into a boolean picture.

Step 2. Delete the image-boundary-connected clumps.

Step 3. Name the picture.

Step 4. Maintain the two prominent places marks.

Step 5. Disintegration of the circumference 3. The pulmonary nodes connected to the arteries are utilized.

Step.6 Circumference 12 disc removal process. The technique is to bind glands to the lungs.

Step.7 *In the 2D matrix of the pulmonary gland, cover the little gaps.*

Step.8 *Superimpose the binary mask on the input image.*

5.3.3 LUNG NODULE DETECTION

To organize each up-and-come group, different layers were checked in the bunch thickly and the preparations of the 3D CNN were carried out and the usual probabilistic expectations are addressed since the predicted probability of the bunch is a lung knob. Thick assessment is required to decrease the change of grouping and lift the last discovery exactness of our CAD framework. As in the preparation stage, every up-and-comer layer is lined up with the vital bearing. Principal to the analysis of cellular breakdown in the lungs in CT examination is the discovery and translation of lung knobs. As the capacities of CT scanners have progressed, more elevated levels of spatial goal uncover smaller lung irregularities.

5.4 EXPERIMENTAL RESULTS

Using Keras, we implemented our neural network and trained and tested our device with google colab. We have conducted a 14-kfold cross validation of the approach proposed. We obtained about 100 K examples from the training scans for each fold. Training the network for each fold required around 1 h. A high detection sensitivity of 90% was achieved by the proposed process. This finding was comparative to the numbers recorded in the substantial convolutionary network study, relative to the research findings of recent published experiments on LIDC datasets, and was much better than the newly stated 3D CNN tumor segmentation identification research.

The learning periods of the dual trials were calculated at 50 iterations and the failure estimates and precision rates were represented as the arcs of Figures 5.4 and 5.5. As shown in the graphs, while both of certain tests may coalesce gradually, the comprehension loss degraded, with its percentage error dropping after the empirical epoch. The learning precision of comprehension loss is also demonstrated to be a quicker increase in the testing precision graph. This is because the differential evaluation is done using the neural network learning process, which contributes to upgrading the metrics of its preceding layer.

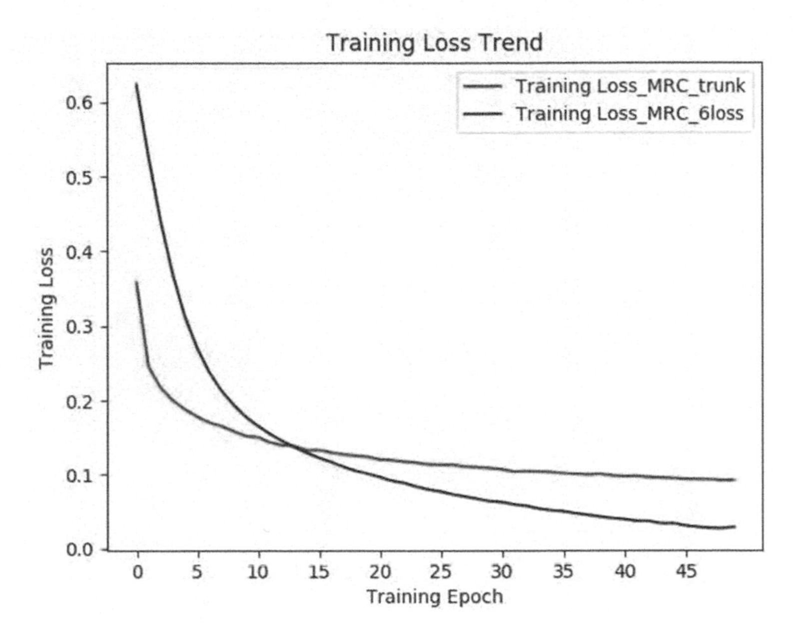

FIGURE 5.4 Training loss.

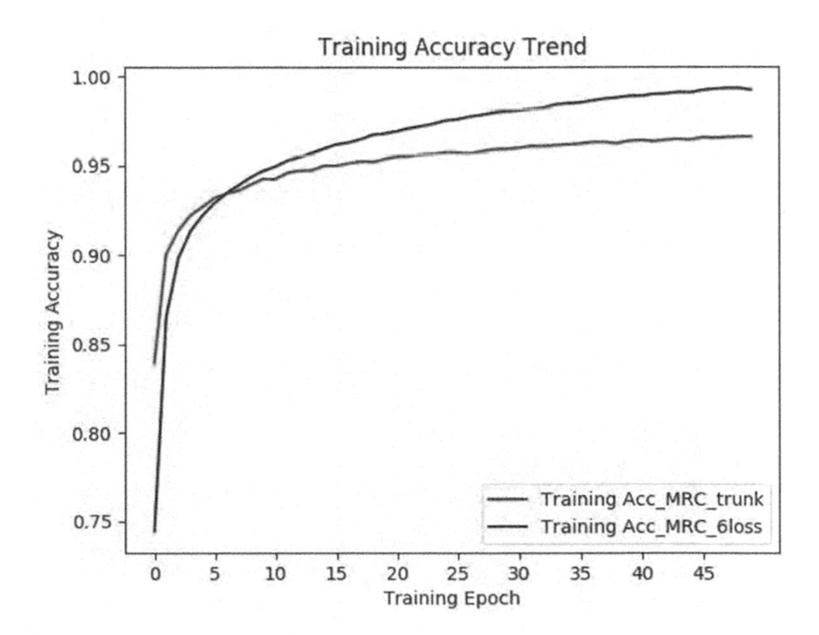

FIGURE 5.5 Training accurcay.

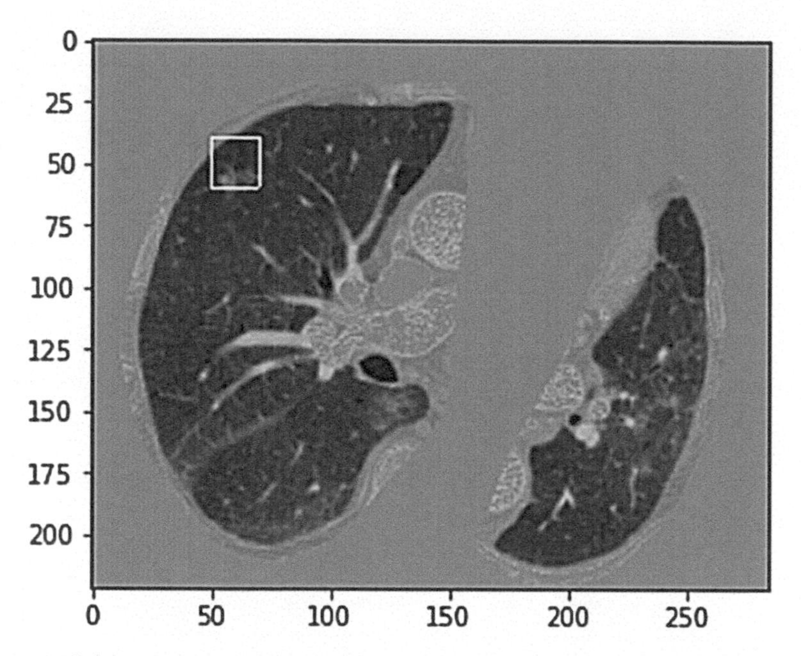

FIGURE 5.6(A) Lung knob true detection-1.

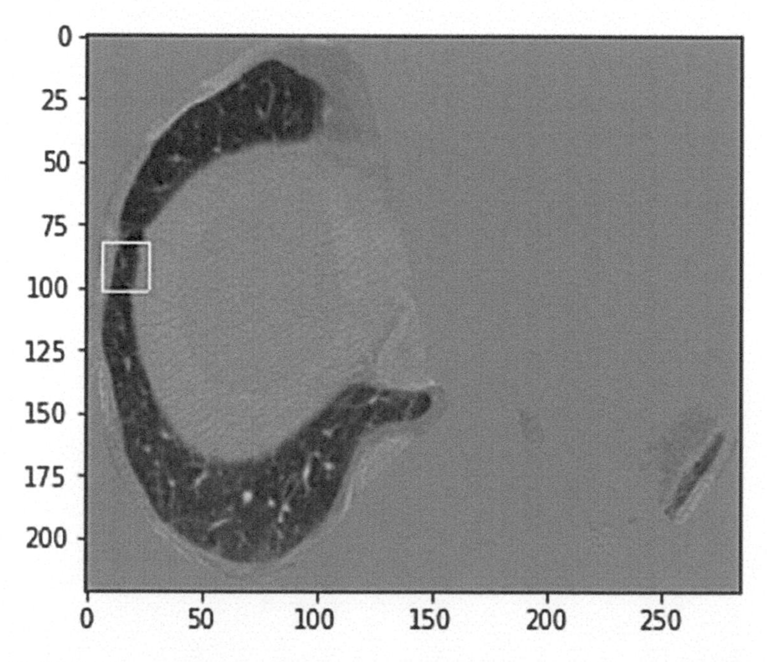

FIGURE 5.6(B) Lung knob false detection.

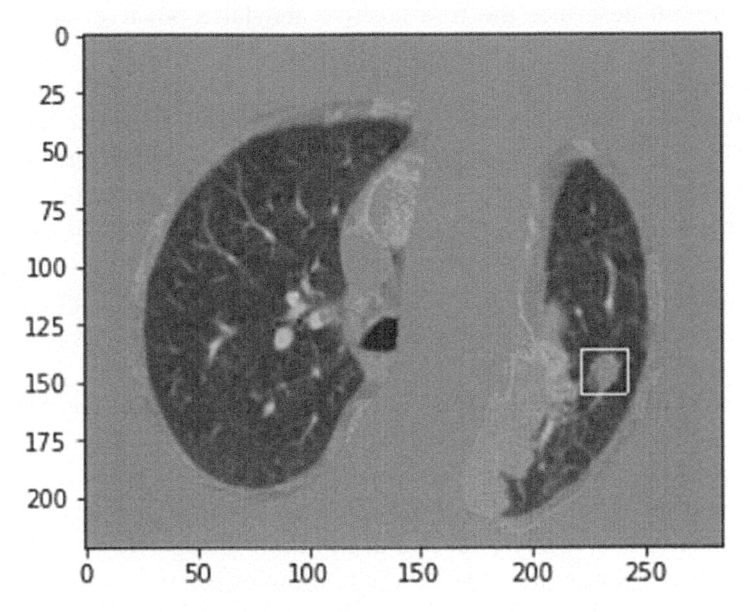

FIGURE 5.6(C) Lung knob true detection-2.

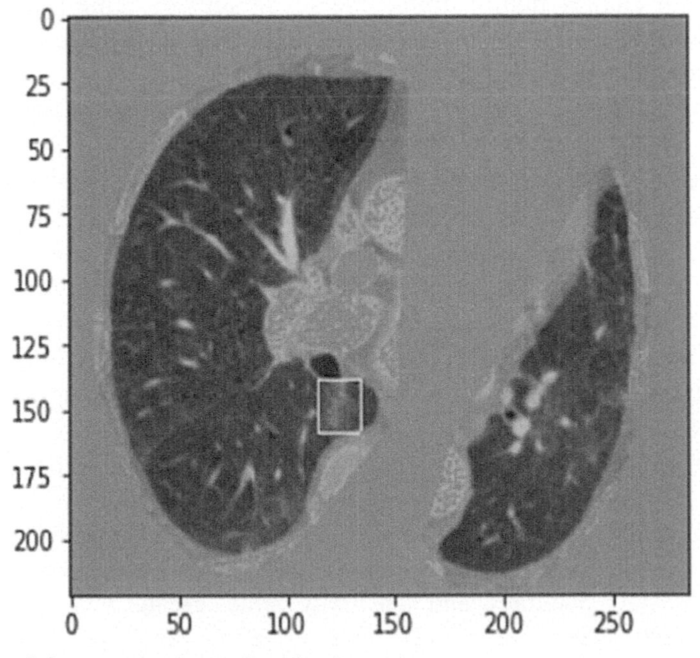

FIGURE 5.6(D) Lung knob true detection-3.

Figure 5.6 describes the true positive and false positive cases in lung knob detection in LIDC dataset using the proposed algorithm.

A lung knob is a little development on the lung and can be favorable or dangerous. The development for the most part must be littler than 3 cm to qualify as a knob. Generous knobs are noncancerous, ordinarily not forceful, and don't spread to different parts of the body. Dangerous knobs are carcinogenic and can develop rapidly. If we have any more established chest X-beams, we should let our primary care physician see them to decide the development pace of the knob. When all is said it's done, dangerous knobs two-fold in size each one to a half year. Knobs with a slower or quicker development rate are less inclined to be destructive. Contrast and the past electronic wellbeing records apply the CNN calculation and distinguish the exact issue.

In reviewing the outcomes with the solely geometrical design-based approach, it noticed that the modern machine learning algorithm suggested was incredibly effective in eliminating negative outcomes from the standard framework. The CAD framework of the cyst might have a analogous composite architecture in the integration of feature selection with a knowledge-based error reduction. In the baseline method, the suggested solution significantly reduced the inaccurate positive rate. This major change demonstrated the advantages of using a deep neural network system to learn features for the particular purpose of detecting lung nodules.

These designs significantly reduced the complexity of the problem, allowed us to train an efficient deep CNN with a relatively small amount of data, and boosted the performance of the proposed CAD system. Our device achieved state-of-the-art efficiency in the experiment and outperformed significantly by using the standard shallow learning, the baseline system. The findings also demonstrated the benefits of CNN and alignment of candidate principal path and dense assessment. Our findings showed that it is promising to train big, deep CNN with limited data to significantly improve CAD system performance with the aid of a priori understanding, augmentation of data, and regularization.

Next, we compared our approach variable multi-image technologies.[14–17] Table 5.2 displays the performance. However, a realistic approximation is challenging due to variations in data set, our proposed mix-resolution remains highly realistic throughout all measurements.[19] Our approach obtains its precision, AUC, accuracy, and sensitivity values greater than 95% of these previous techniques.

TABLE 5.2 Classifiction Perfromance Comparison.

Methods	Accuracy	AUC	Sensitivity	Specificity
Afshar[15]	0.8452	0.8715	0.7792	0.8014
Dong[16]	0.8936	0.8923	0.8526	0.8923
Shulong[17]	0.9489	0.9563	0.9069	0.9548
Juan[18]	0.9581	0.9746	0.9287	0.9702
Our method	0.9645	0.9871	0.9518	0.9725

5.5 CONCLUSION

In this chapter, a lot of dataset for improving the exactness and effectiveness level in lung-related issues and the principal phase of lung malignant growth recognition the 3D picture is to improve the lucidity. Improving the precision and adequacy level in lung-related issues and the chief period of lung harmful development acknowledgment the 3D picture is to improve the clarity. As a lung knob competitor characterization approach, our methodology will sufficiently enhance the precision of lung knob selection pooled with the convincing starting harsh discovery approach. The findings of the LIDC sample group indicate that the approach proposed accomplishes a successful score in the identification performance assessment index. Our technique can easily improve the precision of tumor segmentation testing, coupled to successful first-hand detection procedure, as knowledge pertaining for a target cyst.

KEYWORDS

- **lung nodule**
- **3D CNN**
- **candidate categorization**
- **LIDC**

REFERENCES

1. Liu, F.; Lee, D.-H.; Lagoa, R.; Kumar, S. Computer-aided Detection of Lung Nodules Using Outer Surface Features. *Bio-Med. Mater. Eng.* **2015**, *26* (s1), 1213–1222.

2. Dou, Q. et al. Automatic Detection of Cerebral Microbleeds from MR Images via 3D Convolutional Neural Networks. *IEEE Trans. Med. Imag.* May **2016**, *35* (5), 1182–1195.
3. Shen, W. et al. Multi-crop Convolutional Neural Networks for Lung Nodule Malignancy Suspiciousness Classification. *Pattern Recogn.* Jan. **2017**, *61*, 663–673.
4. Liu, W. et al. In *SSD: Single Shot Multibox Detector, Proceedings of the ECCV*; Amsterdam: The Netherlands, 2016; pp 21–37.
5. Setio, A. A. et al. Validation, Comparison, and Combination of Algorithms for Automatic Detection of Pulmonary Nodules in Computed Tomography Images: The LIDC Challenge. *Med. Image. Anal.* Dec. **2017**, *42*, 1–13.
6. Glorot, X.; Bengio, Y. Understanding the Difficulty of Training Deep Feedforward Neural Networks. *Aistats* **2010**, *9*, 249–256.
7. Valente, I. R. S.; Cortez, P. C.; Neto, E. C.; Soares, J. M.; de Albuquerque, V. H. C.; Tavares, J. M. R. Automatic 3D Pulmonary Nodule Detection in CT Images: A Survey. *Comput. Methods Programs Biomed.* Feb. **2016**, *124*, 91–107.
8. Dou, Q. et al. Automatic Detection of Cerebral Microbleeds from MR Images via 3D Convolutional Neural Networks. *IEEE Trans. Med. Imag.* May **2016**, *35* (5), 1182–1195.
9. Reeves, A. P. et al. The Lung Image Database Consortium (LIDC): A Comparison of Different Size Metrics for Pulmonary Nodule Measurements. *Acad. Radiol.* Dec. **2007**, *14* (12), 1475–1485.
10. Camarlinghi, N.; Gori, I.; Retico, A.; Bellotti, R.; Bosco, P.; Cerello, P.; Gargano, G.; Torres, E. L.; Megna, R.; Peccarisi, M. et al. Combination of Computer-Aided Detection Algorithms for Automatic Lung Nodule Identification. *Int. J. Comput. Assist. Radiol. Surg.* **2012**, *7* (3), 455–464.
11. van Ginneken, B.; Armato, S. G.; de Hoop, B.; van Amelsvoort-van de Vorst, S.; Duindam, T.; Niemeijer, M.; Murphy, K.; Schilham, A.; Retico, A.; Fantacci, M. E. et al. Comparing and Combining Algorithms for Computer-Aided Detection of Pulmonary Nodules in Computed Tomography Scans: The Anode09 Study. *Med. Image Anal.* **2010**, *14* (6), 707–722.
12. Ngiam, J.; Coates, A.; Lahiri, A.; Prochnow, B.; Le, Q. V.; Ng, A. Y. In *On Optimization Methods for Deep Learning*, Proceedings of the 28th International Conference on Machine Learning (ICML-11), 2011; pp 265–272.
13. Lampert, T. A.; Stumpf, A.; Ganc¸arski, P. An Empirical Study into Annotator Agreement, Ground Truth Estimation, and Algorithm Evaluation. arXiv preprint arXiv:1307.0426, 2013.
14. Afshar, P.; Oikonomou, A.; Naderkhani, F. et al. 3D-MCN: A 3D Multi-scale Capsule Network for Lung Nodule Malignancy Prediction. *Sci. Rep.* **2020**, *10*, 7948. https://doi.org/10.1038/s41598-020-64824-5
15. Dong, X.; Xu, S.; Liu, Y. et al. Multi-view Secondary Input Collaborative Deep Learning for Lung Nodule 3D Segmentation. *Cancer Imag.* **2020**, *20*, 53. https://doi.org/10.1186/s40644-020-00331-0.
16. Li, S.; Xu, P.; Li, B.; Chen, L. et al. Predicting Lung Nodule Malignancies by Combining Deep Convolutional Neural Network and Handcrafted Features. *Phys. Med. Biol.* **2019**, *64* (17).
17. Lyu, J.; Bi, X.; Ho Ling, S. Multi-Level Cross Residual Network for Lung Nodule Classification. *Sensors* **2020**, *20*, 2837. doi:10.3390/s20102837.

18. Roth, H. R.; Lu, L. et al. In *Deeporgan: Multi-level Deep Convolutional Networks for Automated Pancreas Segmentation*, International Conference on Medical Image Computing and Computer-Assisted Intervention; Springer, 2015; pp 556–564.
19. Narmadha, S.; Gokulan, S.; Pavithra, M.; Rajmohan, R.; Ananthkumar, T. In *Determination of Various Deep Learning Parameters to Predict Heart Disease for Diabetes Patients*, 2020 International Conference on System, Computation, Automation and Networking (ICSCAN); Pondicherry, India, 2020; pp 1–6. doi: 10.1109/ICSCAN49426.2020.9262317.
20. Adimoolam, M.; John, A.; Balamurugan, N. M.; Ananth Kumar, T. Green ICT Communication, Networking and Data Processing. In *Green Computing in Smart Cities: Simulation and Techniques*; Balusamy, B., Chilamkurti, N., Kadry, S., Eds.; Green Energy and Technology; Springer: Cham, 2021. https://doi.org/10.1007/978-3-030-48141-4_6K.
21. Bala, P. M.; Hemamalini, S. In *Efficient Query Processing with Logical Indexing for Spatial and Temporal Data in Geospatial Environment*, 2019 IEEE International Conference on System, Computation, Automation and Networking (ICSCAN); Pondicherry, India, 2019; pp 1–6. doi: 10.1109/ICSCAN.2019.8878743.
22. Armato, S. G.; McLennan, G. et al. The Lung Image Database Consortium (LIDC) and Image Database Resource Initiative (IDRI): A Completed Reference Database of Lung Nodules on CT Scans. *Med. Phys.* **2011**, *38* (2), 915–931.

PART 2

Methods and Techniques for Supporting Multimedia Systems for e-Healthcare

CHAPTER 6

Data Mining Techniques: Risk-Wise Classification of Countries toward Prediction and Analysis of Worldwide COVID-19 Dataset

SACHIN KAMLEY*

Department of Computer Applications, S.A.T.I., Vidisha (M.P.), India

**E-mail: skamley@gmail.com*

ABSTRACT

For the last couple of months, the novel corona virus infections disease (COVID-19) has captured almost all the countries globally like the USA, UK, Italy, China, India, France, Russia, Brazil, and many more. Due to this fatal virus, more than one crore people have been infected and over 500,000 people have died till date. The main reason of death is that doctors and medical staff were unaware from this pandemic and no medications have been successful at this stage to fight against the virus. Moreover, there is less chance of developing medicines and antidotes for the next 1–2 years. In this direction, there is a need to design an effective automated system, which would be helpful for medical stockholders like doctors, nurses, pathologists, etc. for accurate decision making. Therefore, data mining system has the mastery to analyze huge dataset and the capability to track and control COVID-19 infections. In this study, worldwide data of more than 150 countries from the Base lab and Worldometer data sources for the period of January 10, 2020 to September 6, 2020 employed for the study purpose.

Intelligent Interactive Multimedia Systems for e-Healthcare Applications. Shaveta Malik, PhD Amit Kumar Tyagi, PhD (Eds.)

Dataset employed in this study consisted of 20 various important attributes like number of infected, new cases, new deaths, recovery rate, death rate, serious cases, old age people, cases per million population, and tests per million population, etc. For experimental purpose, the Matlab R2015a data mining tool is utilized and the most prominent data mining techniques like Naïve Bays, Random Forest, Logistic Regression, and Scaled Conjugate Gradient (SCG) are selected for prediction and analysis of COVID-19 dataset. Finally, SCG method has obtained the highest performance accuracy, that is, 96.21%. This study mainly focuses on analyzing and predicting the COVID-19 dataset for the next 2–3 weeks ahead as well as listed the countries based on risk classification level, that is, low to high and vice versa. As a result, countries are classified in these respective levels and some new countries are also recognized where COVID-19 case rates are high. This study would be beneficial for police staff, government and medical professionals, etc. to manage hospital beds, surgical masks, automatic sanitizer machine, and inferred cameras, etc. on right time in order to prevent further occurrence of this serious outbreak worldwide.

6.1 INTRODUCTION

In today's world, the novel corona virus infection (COVID-19) has captured almost all the countries in the world. The virus was firstly recognized by Chinese physician in Wuhan, China, in December 2019.[1] As it spread to more than 18 countries, the WHO declared this outbreak a Public Health Emergency of International Concern (PHEIC) on January 30, 2020 and called it as COVID-19 on February 11, 2020.[2]

On March 11, WHO declared it as a "global pandemic" due to more than 118,000 cases in 120 countries and over 4200 deaths. As on August 23, 2020, 23,380,569 infected cases were reported out of which 15,906,479 recovered and 808,697 died.[3]

The common level symptoms of virus include fever, dry cough, tiredness, cold, breath shortage, etc. and the virus can enter in human body by physical contacts, respiratory droplets in the air, or by touching the contaminated surfaces.[4]

The major issue with this virus is that without showing symptoms, a person can possess a virus for several days so millions of people are getting affected day-by-day due to this fatal virus.

To prevent this outbreak, most of the countries in the world have implemented a partial or strict lockdown in affected regions.[4,5]

In this direction, medical stockholders' are continuously trying to develop an appropriate medicine or antidote but no country has succeeded to develop such medicines till date.

For the last couple of decades, data mining methods and algorithms have solved numerous complicated problems including areas like natural language processing (NLP), medical and healthcare, agriculture, education, finance, intelligent robots, and many more. Therefore, these methods have the ability to extract hidden or meaningful information from the vast dataset size.[6]

To design better policy and fruitful decision making, the knowledge or patterns generated from these datasets could be helpful for stakeholders like government, medical, or business people for preventing COVID-19 cases or deaths.

One of the most prominent areas of data mining technique is prediction where several researchers have contributed in this area including weather forecasting, earthquake prediction, disease prediction, and share forecasting, etc.[7] In this direction, various prediction and classification models like Naïve Bays (NB), Neural Network (NN), Random Forest (RF), Regression, Association Rule Mining (ARM), etc. have been developed.

This study mainly focuses on analysis and prediction of COVID-19 dataset based on the output parameters like recovery rate (RR), case fertility rate (CFR), death rate (DR), and cases per million population (CPMP) and listed the countries based on risk level, that is, low, medium, and high, respectively. Moreover, performances of proposed data mining techniques are assessed based on the accuracy measure.

After a background discussion, Section 6.2 describes the significant research in brief, Section 6.3 describes the data preprocessing process, Section 6.4 discusses about proposed methodology in detail, followed by Section 6.5, which discusses about experimental result analysis, and the last Section 6.6 discusses about the conclusion and future scopes of the study.

6.2 LITERATURE REVIEW

Ferreira et al.[8] have analyzed to improve neonatal jaundice diagnosis among newborns. For this purpose, they have considered the 227 healthy newborns data consisting of 70 variables for experimental purpose. Therefore, results were obtained by using four most important data mining techniques like classification and regression trees (CART), Naïve Bays (NB) classifier, logistic Regression (LR) and Multi Layer Perceptron (MLP). At last, the NB classifier has recorded the highest performance accuracy than others.

Alazab et al.[9] have presented a deep CNN model for the analysis of the incidence of COVID-19 distribution across the world. Therefore, chest X-ray images are used as an input to identify COVID-19 patients, and the Prophet algorithm (PA), autoregressive integrated moving average (ARIMA), long-short-term memory NN (LSTMNN) model were adopted to predict the number of confirmed cases, recoveries, and death over the next 7 days. Finally, experimental results had shown that Australia and Jordan countries had average accuracy 94.80% and 88.43%, respectively, and the f-measure range 95–99% was achieved. Moreover, they have also concluded that the number of cases is significantly higher than in non-coastal areas.

Iwendi et al.[10] have used fine-tuned boosted RF model for the prediction of COVID-19 patient health. The proposed model was designed based on the parameters like patient geographical, health, travel and demographic data to predict the death rate (DR), recovery rate (RR) or severity of the case. At last, their experimental results had shown that proposed model recorded the 94% accuracy and 0.86 F1 score on the dataset used. They also added that there is a positive correlation between deaths and patients' genders as well as majority of patients are aged between 20 and 70 years.

Bayes and Valdivieso[11] have adopted Bayesian approach for predicting the number of deaths in Peru for the next 70 days. Therefore, they have used empirical data from the country China for the study purpose.

Beck et al.[12] have implemented the artificial intelligence (AI) model to identify the commercially available drugs in order to treat COVID-19 patients. However, the Bidirectional Encoder Representations for the Transformers (BERT) Framework is used as a core of a model.

Sujatha et al.[13] have presented various data mining techniques like linear regression, MLP and vector autoregressive (VAR) model for predicting the stretch of COVID-19. For this study, they have collected dataset from Kaggle data source and experimental results stated that proposed techniques could be helpful to know the increasing rate of COVID-19 cases in India as well as to anticipate the epidemiological peak of the disease.

Volpert et al.[14] have used data analytics tool as well as suggested the efficiency of quarantine model in order to prevent the spread of corona virus infections. Therefore, they designed mathematical model to assess the placed quarantine people.

Kutia et al.[15] have designed e-health framework in the Ukraine and e-health application Ezdorovya in China in order to provide useful knowledge and implemented e-health application mainly concern for health care benefits.

At last, this study is augmented by using NB, RF, LR, and scaled conjugate gradient (SCG) methods, respectively.

6.3 DATA PREPROCESSING

In this study, the real dataset of more than 150 countries from the Worldometer and Base lab data sources for the period of January 10, 2020 to September 6, 2020 are considered for study purpose.[16] Therefore, dataset contain various attributes like country, number of infected, number of deaths, number of cured, DR, RR, CFR, case per million population, new case, new deaths, tests per million population, serious cases, population density, median age people, hand wash facility, and old-age people, etc. The main purpose of data preprocessing process is to clean data, that is, removing noise, missing values, and redundant information from dataset.[17] The sample of data description is shown by Table 6.1.

TABLE 6.1 Sample of Data Description.

S. no.	Attribute	Value {min-max}	Description
1	Country	String	–
2	NOI	3–5798983	No. of infected
3	NOD	0–179240	No. of deaths
4	NOC	0–3127418	No. of cured
5	DR	0–28.44	Death rate
6	RR	0–100	Recovery rate
7	CFR	0–19.2	Case fertility rate
8	CPMP	3–41485	Cases per million population
9	NC	0–6194	New cases
10	ND	0–504	New deaths
11	TPMP	4–1778619	Tests per million population
12	SC	1–16552	Serious cases
13	PD	3.2–7915.73	Population density
14	MAP	7.15–48.2	Median age population
15	OAP	0.526–245.46	Old age population

TABLE 6.1 *(Continued)*

S. no.	Attribute	Value {min-max}	Description
16	HWF	2.6–99	Hand washing facility
17	HBF (per 100K)	0.3–13.05	Hospital beds facility
18	TR	0–81800	Total recovered
19	FS	0.3–44.4	Female smokers
20	MS	14.5–70.2	Male smokers

Table 6.2 is too long, so here we are showing only sample of pre-processed data.

TABLE 6.2 Sample of Pre-processed Data.[16,18]

Country	NOI	NOD	NOC	DR	RR	CPMP	NC	ND	TPMP
USA	5,798,983	179,240	3,127,418	3.09	53.93	17,505	2256	40	225,547
Brazil	3,536,488	113,454	2,670,755	3.21	75.52	16,621	0	0	66,476
India	2,979,562	55,950	2,223,202	1.88	74.62	2156	6194	22	24,959
Russia	951,897	16,310	767,477	1.71	80.63	6522	4921	121	231,597
South Africa	603,338	12,843	500,102	2.13	82.89	10,155	0	0	58,986
Peru	576,067	27,245	384,908	4.73	66.82	17,438	0	0	88,251
Mexico	549,734	59,610	376,409	10.84	68.47	4257	5928	504	9597
Colombia	522,138	16,568	348,940	3.17	66.83	10,246	0	0	47,288
Spain	407,879	28,838	0	7.07	0	8723	0	0	170,147
Chile	393,769	10,723	367,897	2.72	93.43	20,574	0	0	111,932

6.4 PROPOSED METHODOLOGY

In this section, several learning techniques (including our proposed work) are discussed in detail.

6.4.1 NAÏVE BAYS

One of the well-known and popular classifications algorithms that is based on the principle of Bays theorem, the Naïve Bays (NB) algorithm states that occurrences of particular features of a class are independent of the presence or absence of other features.[19] However, it is a powerful classification algorithm that is being extensively used for healthcare prediction. In NB, the posterior probability of each class is calculated as, like conditional probability, it is being used to classify the datasets. Equation 6.1 shows NB formula, which helps to predict the class level.[20]

$$P(C|V) = \frac{P(V|C) \times P(C)}{P(V)} \tag{6.1}$$

where C = class label for instance,
V= instances for prediction

Figure 6.1 shows the flowchart of NB classification model.

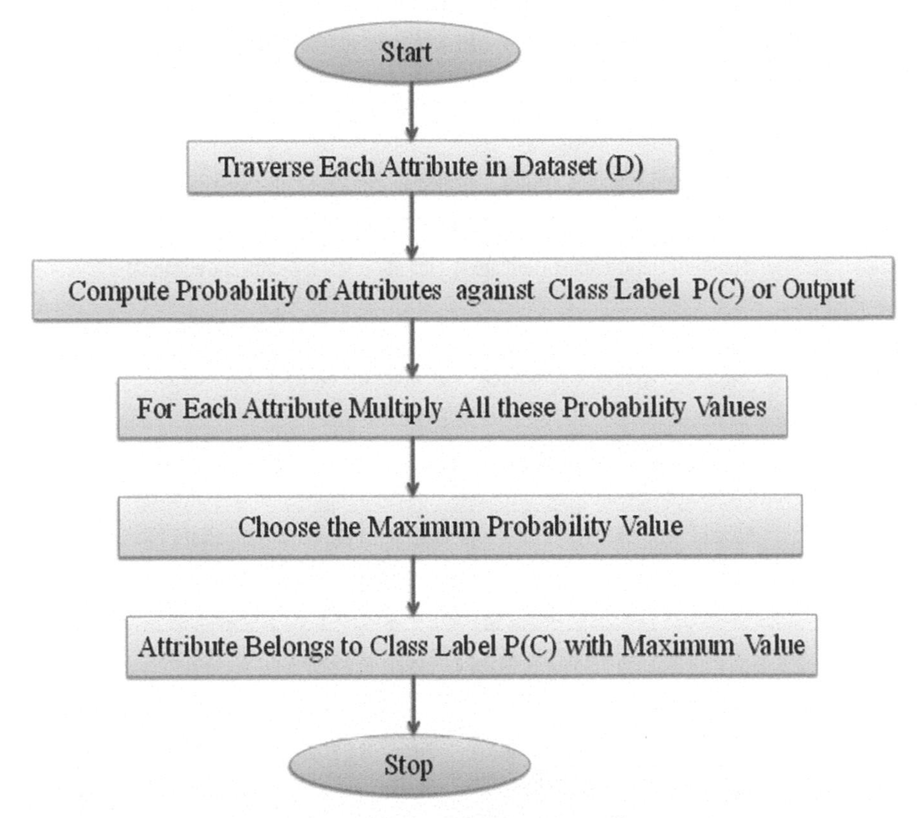

FIGURE 6.1 Flowchart of Naïve Bays classification model.[21]

Figure 6.1 shows that the procedure of NB classification model where each attribute is used to calculate the probability value against the class level and, after calculation of these values, maximum probability value is selected, that is, attribute belongs to particular class level.

6.4.2 *RANDOM FOREST*

One of the most usable data mining ensemble classification algorithm, which is based on Decision Tree methodology is known as RF. Due to its simplicity and flexibility, it can be used for both classification and regression purpose. To get more accurate and stable prediction, RF randomly creates multiple decision trees using portion of a dataset and combined them together the generated trees (forest). Therefore, bagging method is used to train the data.[22] In the testing phase, class label for each test dataset is predicted for each tree as well as majority votes of the class label is assigned to test data. The biggest advantage of this algorithm is that it gives reasonable performance than traditional decision tree algorithm. Figure 6.2 shows the flowchart of the RF classification model.

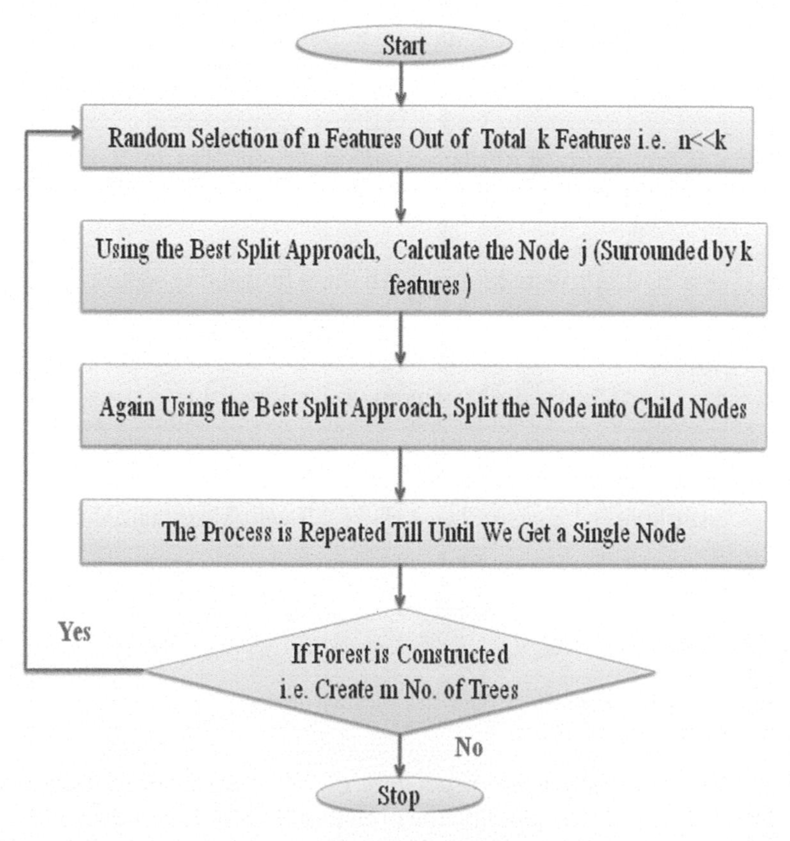

FIGURE 6.2 Flowchart of the random forest classification model.[23]

Figure 6.2 shows the procedure of RF classification model where best-split approach is used to construct the node based on some random features. The process is repeated till whether we get a single node and, at last, big tree or forest is generated.

6.4.3 *LOGISTIC REGRESSION*

One of the well-known and popular methods in regression family that is used to create the relationship between independent variables as against the categorical dependent variables named as LR.[24] Generally, it is also called as binary LR because dependent variable is binary in nature, that is, true and false, 1 and 0, yes and no. When the dependent variable has more than two values then multinomial LR (MLR) is adopted. Equation 6.2 is used to denote the LR mathematically where multiple independent variables are used to predict the dependent variable value.[25]

$$\text{Logestic Regression}(\sigma) = \ln\left(\frac{\sigma}{1-\sigma}\right) \tag{6.2}$$

In eq 6.2, σ is the proportion of observations with an outcome of 1, $1-\sigma$ is the probability of an outcome of 0,

Ratio of $\frac{\sigma}{(1-\sigma)}$ is called the odds, ln is the logarithm of the odds.

6.4.4 *SCALED CONJUGATE GRADIENT*

Simple back propagation algorithm always performs the negative of the gradient, that is, adjusts the weights in the steepest descent direction, which is the major drawback of the algorithm. However, performance function is decreasing most rapidly in this direction and unable to produce the fastest convergence. To provide fastest convergence, SCG algorithm is adopted where a search is performed along with conjugate directions than steepest descent directions.[26]

The major advantage of SCG algorithm is that it minimizes the error through conjugate gradient direction as well as it is used to determine the step size. The flowchart of the SCG method is shown by Figure 6.3.

Figure 6.3 clearly shows the process of SCG method. However, step size is calculated iteratively and weight vector is reduced until optimal conjugate direction is not found. At last, the process is terminated when mean square errors are reduced and there are no changes in weight vectors.

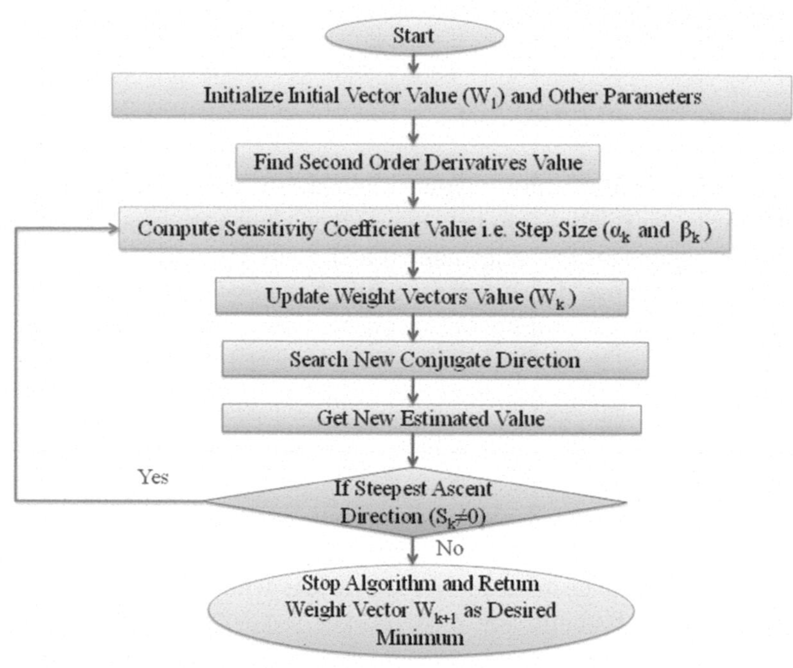

FIGURE 6.3 Flowchart of the scaled conjugate gradient (SCG) method.[27]

6.5 EXPERIMENTAL RESULTS

This study contains real dataset of more than 150 countries for the period of January 10, 2020 to June 30, 2020 (142 samples) and testing dataset of July 1, 2020 to September 6, 2020 (68 samples) for study purpose. To find optimum classification and prediction performance, five-fold cross validation technique is considered where each experiment is performed at five times. At last, the optimum classification results are considered. This study mainly focuses on analyzing and prediction of the global COVID-19 dataset based on proposed classification methods. Thus, four different classification methods like NB, RF, LR, and SCG are utilized and for experimental purpose, Matlab R2015a, machine learning software is used. The following accuracy measures are considered to assess the performance of the proposed methods, which is shown by Table 6.3.

TABLE 6.3 Performance Assessment Measure for Proposed Techniques.[18]

S. No.	Accuracy Measure	Formula	Description
1	Mean absolute error (MAE)	$\text{MAE}=\frac{1}{N}\sum_{i=1}^{N}\left\|p_i-p\right\|$	Average of all absolute errors
2	Root mean square error (RMSE)	$\text{RMSE}=\sqrt{\sum_{i=1}^{N}\frac{(P_i-Q_i)^2}{N}}$	Squared root of mean squared error
3	Accuracy (%)	$\text{Accuracy(A)}=\frac{\text{TP}+\text{TN}}{\text{TP}+\text{TN}+\text{FP}+\text{FN}}$	Measures the performance of correctly classified instances over entire dataset

TP, true positive; TN, true negative; FP, false positive; FN, false negative.

Table 6.4 shows the sample of performance comparison based on actual and predicted infected cases for NB model.

TABLE 6.4 Sample of Country-wise Performance Comparison Based on Hospital Beds Facility (HBDF) Available for Naïve Bays (NB) Model.

S. No.	Country	Actual class	Predicted class	Accuracy
1	Afghanistan	Poor	Poor	Correct
2	Argentina	Good	Good	Correct
3	Algeria	Poor	Poor	Correct
4	Belgium	Poor	Good	Incorrect
5	Brazil	Poor	Good	Incorrect
6	Canada	Good	Good	Correct
7	Chile	Good	Good	Correct
8	Germany	Good	Good	Correct
9	India	Poor	Poor	Correct
10	Iran	Poor	Poor	Correct
11	Israel	Good	Medium	Incorrect
12	Italy	Medium	Medium	Correct
13	Japan	Good	Good	Correct
14	Mexico	Poor	Poor	Correct
15	Netherland	Medium	Medium	Correct
16	Philippines	Poor	Poor	Correct

TABLE 6.4 *(Continued)*

S. No.	Country	Actual class	Predicted class	Accuracy
17	Poland	Poor	Good	Incorrect
18	Russia	Poor	Poor	Correct
19	South Africa	Poor	Poor	Correct
20	Spain	Poor	Medium	Incorrect
21	South Korea	Good	Good	Correct
22	Sweden	Poor	Good	Incorrect
23	Switzerland	Good	Good	Correct
24	Turkey	Medium	Medium	Correct
25	UK	Poor	Poor	Correct
26	Uganda	Good	Poor	Incorrect
27	Thailand	Poor	Poor	Correct
28	Taiwan	Poor	Poor	Correct
29	Vietnam	Medium	Medium	Correct

Table 6.4 clearly shows that correctly vs. incorrectly classified instances based on NB model. However, some countries like Afghanistan, Algeria, India, Iran, Mexico, Philippines, Russia, South Africa, UK, Thailand, Taiwan, Myanmar, Kenya, Greece, and Egypt are identified where hospital beds facility (HBDF) is found to be poor. Table 6.5 shows confusion matrix for the NB model.

TABLE 6.5 Confusion Matrix for NB Model.

Instances=68	Poor	Good	Medium
Poor	26	7	2
Good	5	6	7
Medium	6	7	2
Total	37	20	11

There are so much random tree (forest) is created but due to the size limit we are showing only samples of the tree. Figures 64–6.7 show sample of random tree model, respectively.

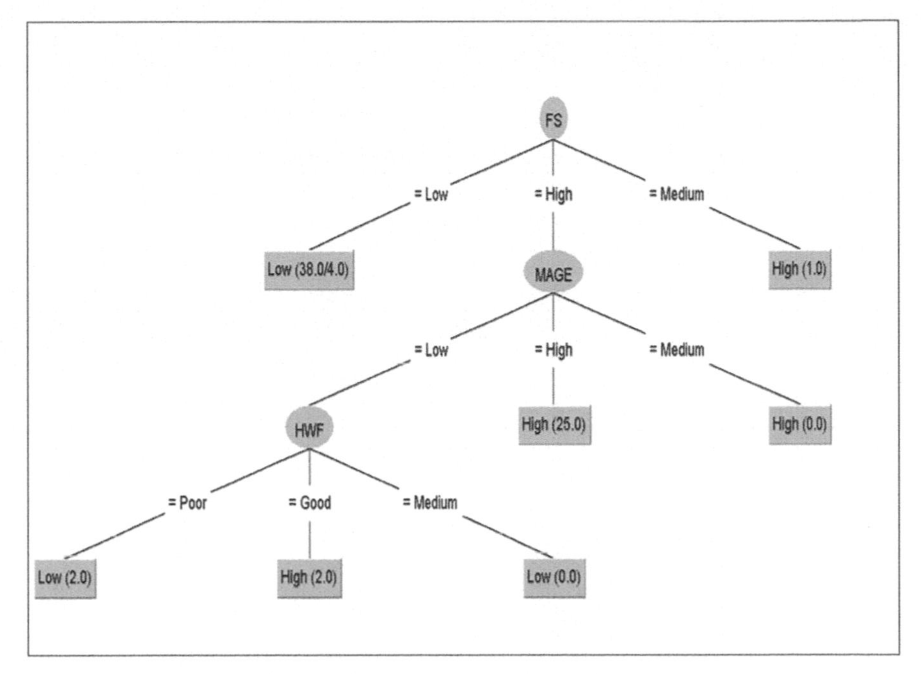

FIGURE 6.4 Random tree-generated based on OAGE class.

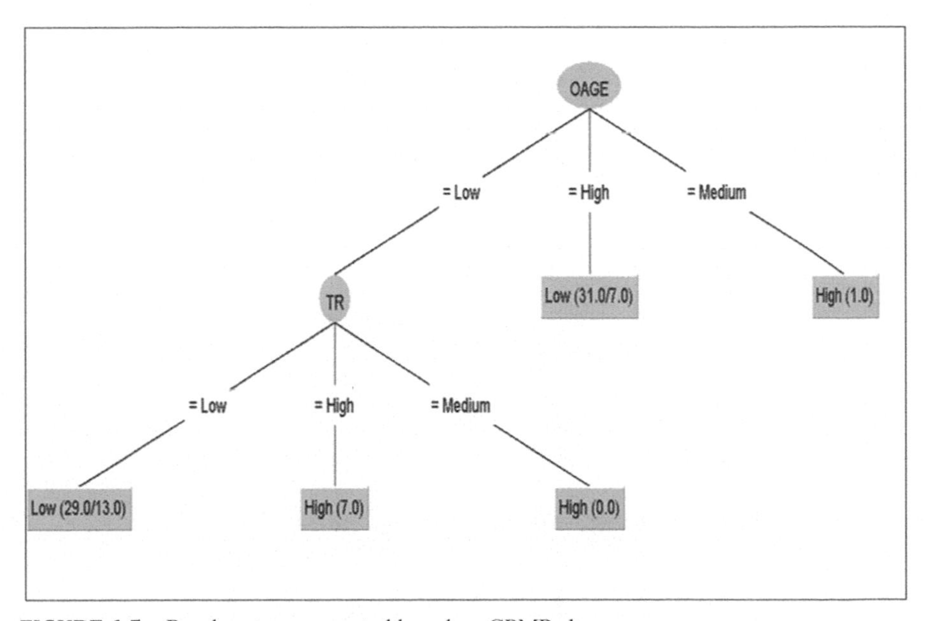

FIGURE 6.5 Random tree-generated based on CPMP class.

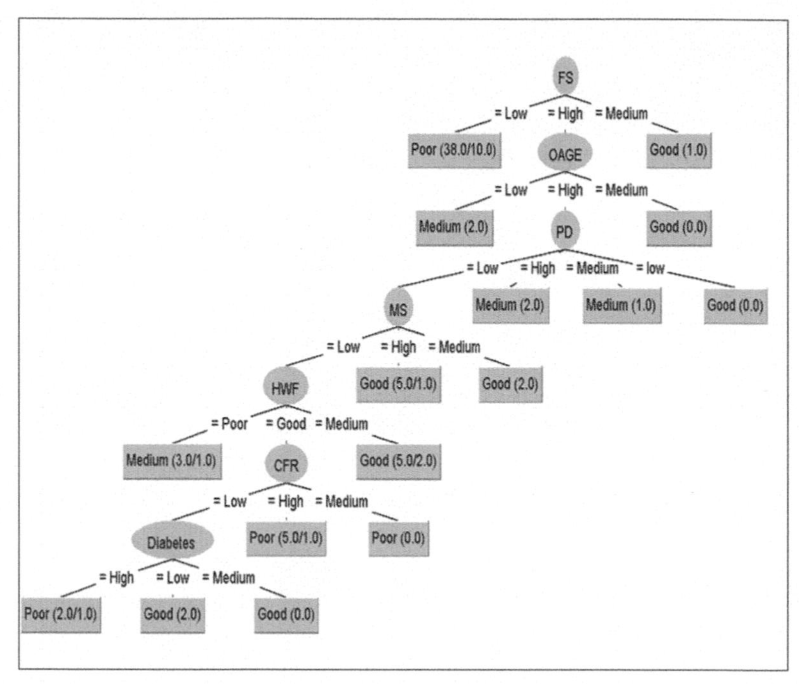

FIGURE 6.6 Random tree-generated based on HBDS class.

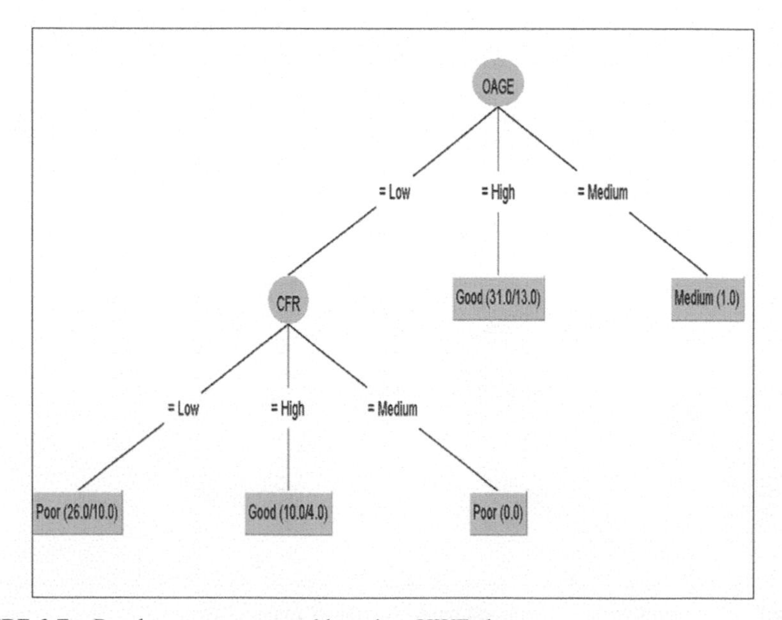

FIGURE 6.7 Random tree-generated based on HWF class.

Figures 6.4–6.7 show multiple decision trees based on output class OAGE, CPMP, HBDS, and HWF, respectively. However, the attributes are categorized into low, medium, and high category and classified based on the respective class level. In Table 6.6, all the decision trees are converted into decision rule format.

TABLE 6.6 Conversion of Random Tree (RT) into Decision Rule Format.

ID	Rules
1	If FS = high, MAGE = low, and HWF = poor, then OAGE = low.
2	If OAGE = low and TR = low, then CPMP = high.
3	If PD = medium, OAGE = high, and MS = high, then HBDS = medium.
4	If MS = low, HWF = good, CFR = low, diabetes = high, and HBDS = poor, then NOI = high.
5	If OAGE = high and PD = high, then DR = high.
6	If CFR = low and OAGE = low, then HWF = poor.
7	If diabetes = high and HWF = poor, then CFR = medium.
8	If NOI = high and RR = poor, then DR = high.
9	If TTMP = low and SC = high, then CPMP = high.
10	If ND = low and NC = medium, then PD = low.

Based on Table 6.6 results, if female smokers are high, median age is low, and hand-washing facility (HWF) is poor, then old age is low (ID-1). Similarly, based on ID-3, if population density is medium, old age is high, and male smokers are high then HBDF is poor. According to ID-8, if the number of infected cases are high and RR is poor, the DR is high. Moreover, if new deaths are low and new cases are medium, then population density is low.

Table 6.7 shows confusion matrix for RF model.

TABLE 6.7 Confusion Matrix for RF Model.

Instances=68	Low	High	Medium
Low	32	4	0
High	5	26	0
Medium	0	1	0
Total	37	31	0

Table 6.8 shows sample of odds ratio calculation for LR model.

TABLE 6.8 Sample of Odds Ratio Calculation for Logistic Regression Model.

		Class		
Attributes		**Low**	**High**	**Medium**
RR	Low	0.0918	0.5946	0.2911
	High	29.187	66.16	0.4316
	Medium	0	0	3.005
DR	Low	0.004	12.432	1.4774
	High	18.12	0	0
	Medium	0	0	7.891
NOI	Low	3.856	56.75	0.74
	High	0	0.001	23.543
	Medium	0	0	6.604
CFR	Low	12.35	3.51	4.186
	High	0.7984	0.564	0.0393
	Medium	0	0	6.874
MAGE	Low	0.001	0.0005	34.521
	High	73.13	84.23	0
	Medium	60.43	0	52.32
PD	Low	0.4081	13.13	0.2378
	High	9.1786	0.4984	1.5189
	Medium	0	0	12.34

Table 6.9 states that risk-wise classification of countries based on output parameter CPMP using LR model.

TABLE 6.9 Risk-wise Classification of Countries Based on Output Parameter CPMP Using LR Model.

S. No.	Low	High	Medium
1	Norway	Qatar	Singapore
2	Bangladesh	French Guiana	Albania
3	Estonia	Bahrain	Spain
4	Philippines	San Marino	Sweden
5	Poland	Chile	Belarus
6	Finland	Panama	Kazakhstan
7	Ghana	Kuwait	Monaco

TABLE 6.9 *(Continued)*

S. No.	Low	High	Medium
8	Venezuela	USA	Netherland
9	Morocco	Peru	Palestine
10	Pakistan	Brazil	Malta
11	Uzbekistan	Oman	Serbia
12	Afghanistan	Armenia	UAE
13	Australia	Aruba	Canada
14	Egypt	Andorra	Ecuador
15	Algeria	Luxembourg	Iceland
16	Greece	Maldives	Ireland
17	Kenya	Mayotte	Singapore
18	Indonesia	Israel	Turkey
19	Barbados	Colombia	Denmark
20	Hungary	South Africa	Germany
21	Guyana	Spain	Austria
22	Nepal	Portugal	Bermuda
23	Tajikistan	North Macedonia	Ukraine
24	Haiti	Russia	Bulgaria
25	Congo	Gibraltar	India
26	Hong Kong	Moldova	Namibia
27	Trinidad and Tobago	Saudi Arabia	France
28	Japan	Argentina	Romania
29	Zimbabwe	Belgium	Italy
30	South Korea	Iran	UK

*Green color=low risk; *red color=high risk; *sky blue color=medium risk.

In Table 6.9, countries like Belgium, Russia, Saudi Arabia, Moldova, Columbia, Israel, Luxemburg, Peru, USA, South Africa, Argentina, Spain, Portugal, Chile, Brazil, Denmark, India, France, Germany, Turkey, Kazakhstan , Singapore, and Austria, etc. have to be recognized as highest- and medium-risk level, respectively, based on CPMP. Confusion matrix for the LR model is shown in Table 6.10.

TABLE 6.10 Confusion Matrix for LR Model.

Instances=68	Low	High	Medium
Low	32	4	0
High	5	26	0
Medium	0	1	0
Total	37	31	0

Figure 6.8 states the training status of SCG method.

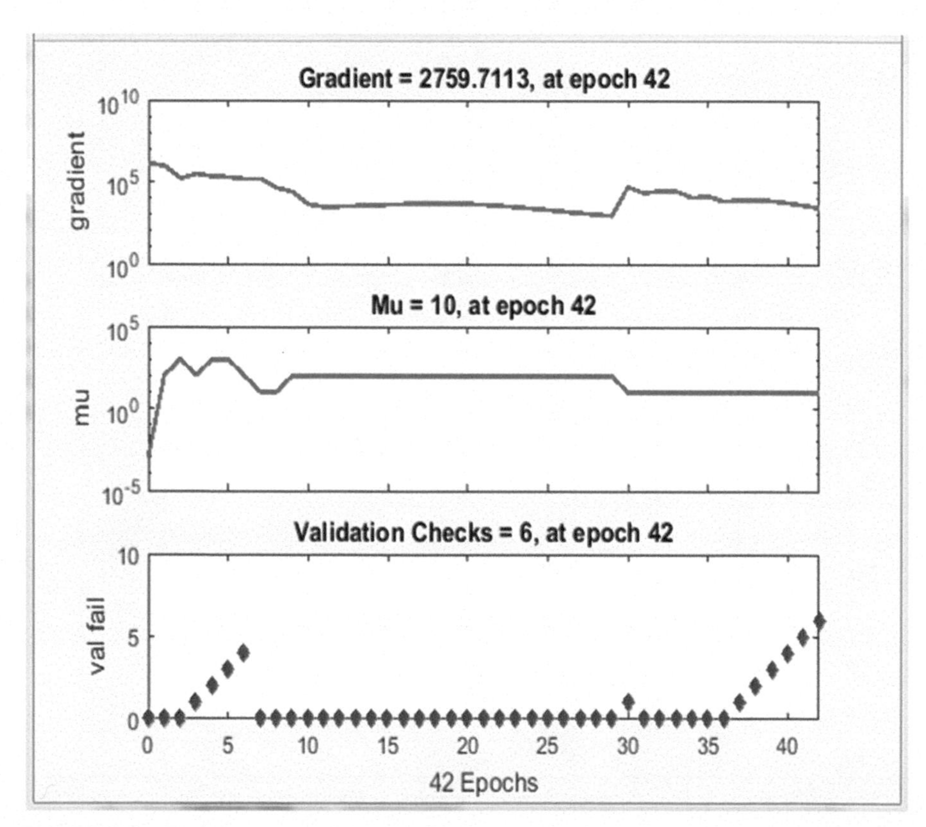

FIGURE 6.8 Training status of SCG method.

Figure 6.8 clearly states about the network training parameters like mutation, epochs, and validation checks. However, network started to achieve the best validation and stable performance after epoch no. 35. Histogram for training, validation, and testing error is shown in Figure 6.9.

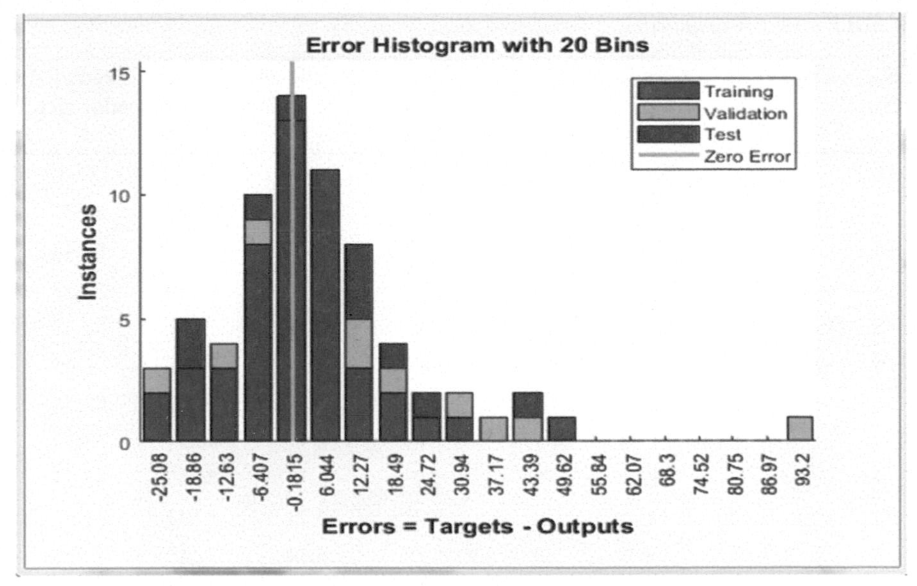

FIGURE 6.9 Histogram for training, validation, and testing error.

Figure 6.9 clearly states the error calculation (actual output-target output). Therefore, we have recorded the zero error after 20 bins. Table 6.11 shows country-wise performance comparison of actual and predicted output based on CFR, DR, and RR.

TABLE 6.11 Country-wise Performance Comparison of Actual and Predicted Output based on Case Fertility Rate (CFR), Death Rate (DR), and Recovery Rate (RR).

S. No.	Country	Actual value (CFR)	Predicted value (CFR)	Actual value (DR)	Predicted value (DR)	Actual value (RR)	Predicted value (RR)
1	Yemen	28.02	31.2	28.85	27.53	60.36	60.47
2	Italy	14.21	13.54	12.86	11.32	75.85	77.67
3	UK	14.38	14.96	12.07	11.12	9.32	12.43
4	Belgium	16.34	15.38	11.28	13.21	21.13	19.76
5	Mexico	10.76	12.34	10.7	10.35	69.71	72.21
6	France	20	17.65	9.67	10.24	27.52	26.76
7	Netherlands	13	14.21	8.45	7.98	8.32	9.21
8	Chad	8.43	8.21	7.45	7.16	88.39	84.56
9	Hungary	13.8	12.65	7.44	9.56	47.19	50.21
10	Canada	8.0	9.87	6.95	8.42	88.32	80.76

TABLE 6.11 *(Continued)*

S. No.	Country	Actual value (CFR)	Predicted value (CFR)	Actual value (DR)	Predicted value (DR)	Actual value (RR)	Predicted value (RR)
11	Sweden	12.15	13.76	6.87	6.47	6.54	8.98
12	Fiji	4.47	5.41	6.45	5.65	77.42	80.23
13	Liberia	6.41	7.23	6.28	8.43	89.05	88.43
14	Sudan	2.12	2.54	6.21	6.76	50.16	50.32
15	Ireland	6.9	7.32	6.02	6.12	79.11	80.12
16	San Marino	5.9	4.78	5.87	6	92.18	90.45
17	Niger	6.25	6.13	5.86	6.21	92.69	88.54
18	Iran	5.7	4.97	5.76	4.91	86.34	87.34
19	Ecuador	7.17	8.86	5.7	5.34	86.67	85.23
20	Spain	12	15	5.69	6.23	10.43	11.67
21	Egypt	4.8	5.65	5.53	5.34	77.43	77.56
22	China	4.5	5.12	5.44	6.52	94.34	90.23
23	Bermuda	5.9	5.54	5.14	5.46	90.29	87.23
24	Switzerland	1.17	2.32	4.58	5.3	84.4	90.21
25	Bolivia	3.7	4	4.49	4.85	57.86	61.67
26	Mali	4.9	6.21	4.45	3.97	78.82	74.43
27	Andorra	6	5.58	4.36	4.78	76.38	76.65
28	Peru	3.8	3.65	4.34	5.12	74.07	75.23
29	Slovenia	5.8	5.46	4.27	4.18	78.45	79.21
30	Brazil	3.8	4.32	3.06	4.76	79.96	83.21
31	Vietnam	0.12	0.87	3.34	3.86	77.69	82.87
32	Columbia	3.41	4.42	3.21	4.43	77.12	73.64
33	Germany	4.7	4.64	3.74	3.74	90.1	94.45
34	Denmark	5	6.53	3.54	3.57	88.36	88.76
35	India	3.0	3.54	1.72	3.56	77.31	78.66
36	USA	6	6.32	3	4.32	57.64	62.21
37	Poland	4.9	5.31	2.99	2.89	76.61	73.37
38	Russia	1.2	1.43	1.74	2.13	82	79.21
39	Greece	5.9	6.75	2.46	2.42	33.41	33.76
40	Jordan	1.01	2.78	0.68	1.23	72.25	73.78

Table 6.12 shows risk assessment based on output parameters.

TABLE 6.12 Risk Assessment Based on Output Parameters.

Output parameters	Low	Medium	High
CFR	0–4.99	5–7.99	≥8
DR	0–3.99	4–6.99	≥7
RR	0–40	41–69	≥70

Sample of output parameters based on risk-wise classification of country (Table 6.12) is shown in Table 6.13.

TABLE 6.13 Sample of Output Parameters Based on Risk Wise Classification of Country.

S. No.	CFR	DR	RR
1	Yemen	Yemen	Greece
2	France	Belgium	France
3	Belgium	Italy	Belgium
4	Spain	UK	UK
5	UK	Mexico	Spain
6	Netherlands	France	Netherlands
7	Sweden	Hungary	Sweden
8	Italy	Liberia	USA
9	Hungary	Canada	Bolivia
10	Mexico	Netherlands	Yemen
11	Canada	Chad	Sudan
12	Ecuador	Sudan	Hungary
13	Chad	China	Germany
14	Ireland	Sweden	San Marino
15	Liberia	Spain	China
16	Greece	Niger	Switzerland
17	USA	San Marino	Niger
18	Mali	Fiji	Liberia
19	Niger	Bermuda	Iran
20	Egypt	Ecuador	Bermuda
21	Andorra	Egypt	Ecuador
22	Bermuda	Switzerland	Chad
23	Slovenia	Peru	Brazil
24	Fiji	Iran	Vietnam
25	Poland	Bolivia	Canada
26	China	Andorra	Fiji
27	San Marino	Columbia	Slovenia
28	Germany	USA	Russia

TABLE 6.13 *(Continued)*

S. No.	CFR	DR	RR
29	Columbia	Slovenia	India
30	Brazil	Mali	Italy
31	Peru	Germany	Andorra
32	India	Denmark	Peru
33	Jordan	India	Mali
34	Sudan	Poland	Jordan
35	Switzerland	Greece	Columbia
36	Russia	Russia	Poland
37	Vietnam	Jordan	Mexico

*Red color= high risk; *sky blue=medium risk; *green color=low risk.

Table 6.14 shows list of top 20 countries based on output parameters.

TABLE 6.14 List of Top 20 Countries based on Output Parameters.

S. No.	CFR	DR	RR
1	Yemen	Yemen	Germany
2	France	Belgium	San Marino
3	Belgium	Italy	China
4	Spain	UK	Switzerland
5	UK	Mexico	Denmark
6	Netherlands	France	Niger
7	Sweden	Hungary	Liberia
8	Italy	Liberia	Iran
9	Hungary	Canada	Bermuda
10	Mexico	Netherlands	Ecuador
11	Canada	Chad	Chad
12	Ecuador	Sudan	Brazil
13	Chad	China	Vietnam
14	Ireland	Sweden	Canada
15	Liberia	Spain	Fiji
16	Greece	Niger	Ireland
17	Denmark	Ireland	Slovenia
18	USA	San Marino	Russia
19	Mali	Fiji	India
20	Niger	Bermuda	Italy

Table 6.14 clearly identified the top 20 countries based on output parameters. However, some new countries are also recognized based on Table 6.14 results like Yemen, Hungry, Ecuador, Chad, Liberia, Mali, Niger, Sudan, Ireland, San Marino, Fiji, Bermuda, Vietnam, and Slovenia. Based on Table 6.14 results, we can also state that the world's largest economy countries like France, UK, China, USA, India, and Russia have widely infected by COVID-19. To fight against COVID-19, the government of these countries would implement strong policy and rules like maintaining social distance, wearing surgical mask in offices and public places, keeping hand wash, or sanitizer etc. Table 6.15 shows accuracy comparison of proposed techniques.

TABLE 6.15 Accuracy Comparison of Proposed Techniques.

S. No.	Proposed techniques	MAE	RMSE	Accuracy (%)
1	NB	0.126	0.245	88.24%
2	RF	0.0851	0.193	92.13%
3	LR	0.16	0.381	85.29%
4	SCG	0.0432	0.116	96.21%

Table 6.15 clearly states that based on accuracy comparison SCG method had recorded the highest performance accuracy, that is, 96.21% than others.

6.6 CONCLUSION AND FUTURE SCOPE

Now days, COVID-19 has become a challenging problem to the society. At present, over 200 countries in the world are fighting with this fatal virus. Due to the increasing number of COVID-19 patients day-by-day, data mining researchers and experts have contributed toward the analysis and prediction of COVID-19 dataset for the last four to five months.

In this study, four effective data mining techniques like NB, RF, LR, and SCG are applied on COVID-19 dataset. Therefore, correct and incorrect classified instances are used to evaluate the performance of proposed techniques. Finally, SCG method has recorded the highest performance accuracy, that is, 96.21%.

Based on the NB results (Table 6.4), we have listed some countries where HBDF is found to be poor. These countries are Algeria, India, Mexico, Philippines, Russia, South Africa, Kenya, Greece, and Egypt.

In RF results (Table 6.6), we have converted RF into decision rule format and found some interesting decision rules. Based on the ID-2, if Old Age

Population (OAP) and total recovery are low, then CPMP is high. According to ID-8, if the number of infection (NOI) is high and RR is poor, then DR is high. Similarly, ID-4 states that if male smokers are low and HWF is good, and CFR is low and diabetes is high and HBDF is poor, then NOI is high.

In LR results (Table 6.9), we have recognized the countries based on the output parameter CPMP classified as high risk and medium risk level, respectively. These countries are Qatar, French Guiana, Bahrain, San Marino, Panama, Gibraltar, North Macedonia, Mayotte, Aruba, Peru, Oman, Kuwait, Columbia, Belarus, Monaco, Palestine, Malta, Serbia, Ecuador, and Bermuda.

Last, but not the least, SCG results (Tables 6.11 and 6.13) have identified some countries classified as high-risk level based on the output parameters like CFR, DR, and RR. These countries are Yemen, France, Belgium, Spain, Hungry, Chad, Netherlands, Italy, UK, Mexico, Liberia, Greece, Sweden, Sudan, Bolivia, Mali, Slovenia, Andorra, Switzerland, Fiji, Niger, USA, and Poland.

Moreover, we have also listed top 20 countries (Table 6.14 results) based on the output parameters like CFR, DR, and RR, respectively. The countries like Yemen, Germany, France, Belgium, San Marino, Italy, China, Spain, UK, Switzerland, Mexico, and Denmark comes under the top five positions in the CFR, DR, and RR category.

Due to increasing world-wide COVID-19 cases, in future, countries might change their levels, that is, low to high and vice versa. To fight against this serious pandemic, government should adopt and implement proper policies with planning.

To stop COVID-19 infections, government of countries worldwide must take some preventive actions like implementing curfew or lock-down rule in a two or four days in a week, use inferred camera to maintain social distance, manage automatic sanitizer machine in government offices, schools, colleges and public places, and compulsorily allow wearing surgical masks to save the life of the people.

The biggest advantage of this study is that it could be fruitful for government, medical professionals, police staff, and businesspersons in order to know the worldwide status of COVID-19 countries in advance. Moreover, it would be helpful for government to arrange the hospital beds, surgical masks, automatic sanitizers' machines, enhancing COVID test center, and last but not least, medical staff, so that proper actions would be taken on right time to fight against this serious pandemic.

For better analysis and results, some other popular machine learning techniques like Bayesian optimization, Neuro-fuzzy, and Fuzzy-genetic algorithm will be adopted in near future.

KEYWORDS

- **data mining**
- **COVID-19**
- **classification**
- **prediction**
- **base lab**
- **Worldometer**
- **Matlab R2015a**

REFERENCES

1. WHO Situation Report-94 Corona Virus Disease 2019 (COVID-19). https://www.who.int/docs/default-source/coronaviruse/ situation-reports/20200423-sitrep-94-covid-19.pdf?sfvrsn=b8304bf0_4 (accessed Aug 10, 2020).
2. Topcuoglu, N. Public Health Emergency of International Concern: Corona Virus Disease 2019 (COVID-19). *Open Dent. J.* **2020**, *14* (1), 71–72. https://doi.org/10.2174/1874210602014010071.
3. Read, J. M.; Bridgen, J. R.; Cummings, D. A.; Ho, A.; Jewell, C. P. Novel Corona Virus (COVID-19): Early Estimation of Epidemiological Parameters and Epidemic Predictions, 2020. https://doi.org/10.1101/2020.01.23.20018549.
4. BBC. Corona Virus Sharp Increase in Deaths and Cases in Hubei. https://www.bbc.co.uk/news/worldasiachina-51482994 (accessed August 12, 2020).
5. Wang, C.; Horby, P. W.; Hayden, F. G.; Gao, G. F. A Novel Corona Virus Outbreak of Global Health Concern. *Lancet* **2020**, *395* (10223), 470–473.
6. Shamsollahi, M.; Badiee, A.; Ghazanfari, M. Using Combined Descriptive and Predictive Methods of Data Mining for Coronary Artery Disease Prediction: A Case Study Approach. *J. Artif. Intell. Data Mining* **2019**, *7* (1), 47–58.
7. Shirsath, S. S.; Patil, S. Disease Prediction Using Machine Learning over Big Data. *Int. J. Innov. Res. Sci. Eng. Technol.* **2018**, *7* (6), 6752–6757.
8. Ferreira, D.; Oliveira, A.; Freitas, A. Applying Data Mining Techniques to Improve Diagnosis in Neonatal Aundice. *BMC Med. Inform. Decis. Mak.* **2012**, *12* (1), 143–149.
9. Alazab, M.; Awajan, A.; Mesleh, A.; Abraham, A.; Jatana, V.; Alhyari, S. COVID-19 Prediction and Detection Using Deep Learning. *Int. J. Comput. Inform. Syst. Ind. Manage. Appl.* **2020**, *12*, 168–181.

10. Iwendi, C.; Bashir, A. K.; Peshkar, A.; Sujatha, R.; Chatterjee, J. M.; Pasupuleti, S.; Mishra, R.; Pillai, S.; Jo, O. COVID-19 Patient Health Prediction Using Boosted Random Forest Algorithm. *Front. Public Health* **2020**, *8* (357), 1–9.
11. Bayes, C.; Valdivieso, L. Modeling Death Rates Due to COVID-19: A Bayesian Approach. arXiv. (2020) 2004.02386. https://arxiv.org/abs/ 2004.02386 (accessed Aug 5, 2020).
12. Beck, B. R.; Shin, B.; Choi, Y.; Park, S.; Kang, K. Predicting Commercially Available Antiviral Drugs That May Act on the Novel Corona Virus (2019-NCoV), Wuhan, China through a Drug-Target Interaction Deep Learning Model. *BioRxiv.* https://www.biorxiv.org/content/10.1101/2020.01. 31.929547v1.abstract (accessed August 5, 2020).
13. Sujatha, R.; Chatterjee, J. M.; Hassanien, A. E. A Machine Learning Forecasting Model for COVID-19 Pandemic in India. *Stoch. Environ. Res. Risk Assess.* **2020**, *34*, 959–972. doi: 10.1007/s00477-020-01827-8.
14. Volpert, V.; Banerjee, M.; Petrovskii, S. On a Quarantine Model of Corona Virus Infection and Data Analysis. *Math. Model. Nat. Phenom.* **2020**, *15* (24), 1–6.
15. Kutia, S.; Chauhdary, S. H.; Iwendi, C.; Liu, L.; Yong, W.; Bashir, A. K. Socio Technological Factors Affecting User's Adoption of E-health Functionalities: A Case Study of China and Ukraine E-health Systems. *IEEE Access* **2019**, *7*, 90777–90788. doi: 10.1109/ACCESS.2019.2924584.
16. Data Source. http://www.worldometer.com and http://www.baselab.com (accessed June 30, 2020).
17. Hong, S. J.; Weiss, S. M. Advances in Predictive Models for Data Mining. *Pattern Recognit. Lett.* **2001**, *22*, 55–61.
18. Han, J.; Pei, J.; Kamber, M. *Data Mining: Concepts and Techniques*, 3rd ed.; Elsevier, 2011.
19. Tomar, D.; Agarwal, S. A Survey on Data Mining Approaches for Healthcare. *Int. J. Bio-Science Bio-Technology* **2013**, *5* (5), 241– 266.
20. Romero, C.; Ventura, S.; Garcia, E. Data Mining in Course Management Systems: Moodle Case Study and Tutorial. *Comput. Educ.* **2018**, *51* (1), 368–384.
21. Chen, M. S.; Han, J.; Yu, P. S. Data Mining: An Overview from a Database Perspective. *IEEE Trans. Knowl. Data Eng.* **1996**, *8* (6), 866–883.
22. Haque, M. R.; Islam, M. M.; Iqbal, H.; Reza, M. S.; Hasan, M. K. Performance Evaluation of Random Forests and Artificial Neural Networks for the Classification of Liver Disorder. In *2018 International Conference on Computer, Communication, Chemical, Material and Electronic Engineering (IC4ME2)*; IEEE, 2018; pp 1–5.
23. Berk Richard, A. *Regression Analysis: A Constructive Critique*, 1st ed.; Sage Publications, 2003.
24. Friedman, H. T. *The Elements of Statistical Learning*, 2nd ed.; Springer Series in Statistics, 2009.
25. Alpaydin, E. *Introduction to Machine Learning*, 2nd ed. The MIT Press, 2010.
26. John, M.; Shaiba, H. Main Factors Influencing Recovery in MERS Co-V Patients Using Machine Learning. *J. Infect. Public Health* **2019**, *12* (5), 700–704.
27. Bayram, C.; Atalay, B. Speeding up the Scaled Conjugate Gradient (SCG) Algorithm and Its Application in Neuro-Fuzzy Classifier Training. *Soft Comput.* **2010**, *4* (4), 365–378.

CHAPTER 7

Mathematical Model of COVID-19 Diagnosis Prediction Using Machine Learning Techniques

V. KAKULAPATI* and M. NAGARAJU

Sreenidhi Institute of Science and Technology, Yamnampet, Ghatkesar, Hyderabad, Telangana 501301, India

Corresponding author. E-mail: vldms@yahoo.com

ABSTRACT

The most dangerous and infectious disease, COVID-19, is affected by a coronavirus, and infection is unknown before the epidemic commenced in Wuhan, China, in December 2019. Many researchers are using mathematical modeling. In this chapter, the anticipated corona-patient type of the healthy aged, including an enhanced region of operative features of the obtaining region curve (AUC) using the SVM (Support Vector Machines), has been significantly differentiated the important mathematical model. Empirical and analytical assessments have shown that the clinical diagnosis of COVID-19 may help ML algorithms' use to research the language biomarkers from the verbal utterances of older people. This chapter illustrates that the disease community's best ML model blends essential syntactic and leading characteristics. But predictive methods have to be trained on broader data sets, enhancing AUC and eventual COVID-19 medical assessment. A significant use resulting from mathematical modeling is that it claims transparency and accurateness about our model. These techniques can help in decision-making

Intelligent Interactive Multimedia Systems for e-Healthcare Applications. Shaveta Malik, PhD
Amit Kumar Tyagi, PhD (Eds.)

by useful predictions about substantial issues such as interference that bring changes in the spreading of COVID-19.

7.1 INTRODUCTION

Statistical or mathematical techniques are useful when joining a group to comprehend illness's actions and examine the circumstances in which it is removed or sustained. The high rate of disseminating the infection and the substantial number of deaths present great concern for science, governments, and all citizens. Coronavirus is a recent infectious infection caused by a coronavirus is recognized in Wuhan in the month of December 2019. The pandemic is primarily transmitted by droplets when an infected individual plays, sneezes, or exhales. In 216 countries, nearly 4 million coronaviral confirmations were recorded, with about 1,034,932 livelihoods loss caused by this virus.

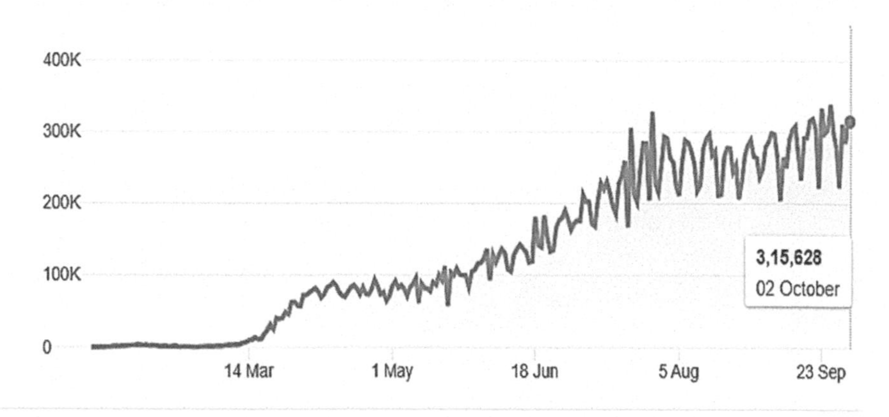

FIGURE 7.1 Total coronavirus cases worldwide.

The coronavirus pandemic (COVID-19, known as 2019–nCoV) evolved in December 2019, in China Wuhan. No vaccine has been approved by WHO for combating this virus.[1] The spread of this virus is significantly faster when people are in close contact with an infected person. The only way to control the broadly spread of this disease is by imposing traveling and movement restrictions. Also, frequent handwashing with warm water prevents humans from not touching their eyes, mouth, and nose with the virus. The typical symptoms of this disease are cough, chest discomfort, sputum development, and sore throat.[2]

Numerous people affected by the virus are over 20 million worldwide now, with the number of reported deaths amounting to over 8000. Countries with many affected COVID-19 patients are the USA, Russia, Brazil, Spain, Italy, and India.[3] But the number of recovering is auspicious at the moment.

Since the disease was reported, it has spread to over 211 countries worldwide and is becoming an international concern. Most of the research has been conducted in various laboratories with others using conspiracy theories to find a solution to this disease. An investigation conducted revealed that the death rate of this pandemic[4] is 4.6% across the globe. The death rate is high in the age range 69–80 years, 10%, while the patients over 80 years are 15%. Patients above the age of 50 years with chronic disease are not safe.[5] The rapid increase if this spread of disease is a significant threat, with an estimated 7–15 people can get infected when they are in contact with a person infected with COVID-19. This indicates that COVID-19 can affect a vast number of people within the shortest possible time.

With the image analysis, ML (machine learning) offers a lot of assistance in detecting the infection. ML can be used to classify novel coronavirus infection. It can also estimate the existence of the virus worldwide. For analysis purposes, ML needs a massive volume of information for pathogen detection or prediction. Categorizing the image into various groups, supervised ML techniques require annotated data. Over the past decade, a tremendous amount of progress has been made in this area in attempting specific investigations. To overcome this, a new pandemic has drawn many researchers around the world. In the form of X-ray images, data generated by several investigators developed ML models, which categorize X-ray images into COVID-19 or not.

Many researchers intend to use imaging modalities to accurately forecast patient outcomes. These include the introduction of a prevalence assessment,[6] the prediction of the requirement for continuing hospitalization depending on X-ray image data,[7] and the disparities of individual risk analysis on X-ray images.[8] The above methods may better distinguish patients who can require specialized and long-term diagnosis, make it easier for clinicians to prepare and handle their services more efficiently, and monitor patient status and understand whenever patient health deteriorates. Although these trials are restricted including both reach and in results, they represent the valuable path of study that can be augmented by new cases in the area and supplemented by clinical evidence, thus potentially enhancing the diagnosis and treatment of all clinicians and reducing the death rate of those who are severely infected.

Currently, the standard diagnostic method to detect the virus is RT-PCR (reverse transcription-polymerase chain reaction) method,[9] which is a laboratory test involving deoxyribonucleic acid (DNA) and ribonucleic (RNA) to predict specific ribonucleic acid volume using fluorescence. A lot of countries are using the RT-PCR method to diagnose COVID-19 disease. Other laboratories also further the test result with Computed Tomography (CT) scans.[10] Since several studies have argued that using a CT scan and X-ray is better than RT-PCR owing to its limited availability in developing countries.

Identifying COVID-19, symptoms in the lower parts of the lungs have superior accuracy when using a computed tomography scan rather than PCR. In the majority of situations, CT scan and X-ray test can be exchanged with RT-PCR test. However, these methods cannot exclusively address the problem owing to the relatively insufficient radiologists and the high volume of re-examination of infected persons who wish to know their illness condition. This can be addressed by designing and an advanced diagnostic system that use ML algorithms. This aims to reduce the time and effort required to perform image data of pandemic positive patients and measure disease growth rate.

This chapter advocate using mathematical models and ML techniques to predict coronavirus cases and diagnose patients via X-ray images using automated intelligent systems. The proposed approach can differentiate between COVID-19 patients from non-patient through training to find hidden patterns on X-ray images. Finally, this analysis significantly enhances the predictive X-ray images by identifying image features and developing the correct mathematical model.

7.2 RELATED WORK

For the past few months, detection and analysis of COVID-19 have been inquired. Due to its widespread around the globe, many diagnosing methods have been imposed to detect COVID-19 patients. Examples of these methods are RT-PCR, next-generation sequencing (NGS), LAMP, rapid serological, and CT scan. Several studies have implemented using statistical for diagnoses in healthcare.[11] One merit of using this method is that it is straightforward for implementation. Another research area, such as artificial intelligent, is also merging into design models to predict COVID-19 patents.[12] Ref. [13] emphasizes two main concepts. Their initial study was focused on COVID-19 studies and related to predicting the number of infected persons in the future.

They suggested that all data or information on COVID-19 should be public for other researchers to analyze their work. All the best models could only be developed based on real data, not simulation.

It consists of analyzing a sequence of parameters that simulates reality that is then overcome for detailed measured data in the equivalences.[14] Numerical methods help to quantitate conceptual models. Algorithms and statistics integrate knowledge and recognize concerns across knowledge subsets. For identifying the progression of an outbreak, that's concept effects are utilized. And some of the most common statistical models for a person-to-person propagation are the SIR (Susceptible–Infectious–Recovered) models.[15]

Conversely, while creating or validating a methodology,[16] models must acquire one dataset of specific data sources. Statistical models may be troublesome for large countries as they introduce interconnected thread-epidemics.[17] Multiple parameters, such as anthropological features and sample sizes, contribute substantially to the model's prediction.

ML algorithms were utilized to investigate[18] the result of coronavirus on the economy after the pandemic. Their main target was the outcome after curbing this pandemic. The economy with low capital per income will suffer from their studies since the lockdowns and job collapsing. They did not focus on designing a model to effectively diagnose coronavirus patients. Various advantages of the coronavirus diagnosis methods are described. They recommended the RT-PCR test for testing patients. Biosensors and other alternative devices can become the best way of diagnosing COVID-19.[19] CNN (Convolutional Neural Network) was used to develop a model to detect coronavirus patients by means of X-ray images.[20] Their study is similar to ours using ML algorithms.

In the Chinese epidemic of coronavirus, a mathematical model was developed.[21] The severity and period of the epidemic peak and final epidemic scale are estimated according to different preventative measures depending on a transformed parametric Sensitive Exposed Infectious Recovered (SEIR). This is a standard instance of using mathematical computational methods in the analysis of COVID-19 propagation and diffusion.

Statistical models have provided detailed knowledge of disease control and policy formulation and have presented valuable guidance. SEIR model,[22] based on reports from Dec 31, 2019, to Jan 28, 2020, explains the propagation characteristics of the COVID-19 and forecasts the worldwide distribution of the infection. A value for the fundamental ability to learn and adapt to the earlier epidemic was started by observing that the regular time distribution through Poisson improves its data fitting. The clinical progression[23] involved human epidemiologic, and intervention interventions in a paradigm that ultimately decreases influence propagation and the risk of complications through intervention techniques such as intense tracing of touch followed through quarantine and isolation. Mathematical models of probable outbreak equations[24] utilized in Wuhan, China, for predicting the outbreak scale, showed that prevention strategies need to prevent even more than 60% spread to manage the outbreak effectively. The use of the SEIR and Bayes' inference meta-populations[25] in China has shown that about 86% of any infection unknown by Jan 23, 2020, method, and the virtual epidemiologies world have been identified. It measured the transmissibility and seriousness and simulated the possible effects of the easing prohibitions in anticipation of China's second epidemic outbreak in China at mainland Chinese sites outside the Hubei provinces.[26]

In[27] Support Vector Machine was used to extract features from X-ray images for classification analysis on coronavirus patients. They used only 25 patients' X-ray images as their dataset, with 16 X-rays as infected and 9 X-ray images as non-infected. Their model achieved 89.6% accuracy with ResNet 50. The proposed efficient diagnosis method uses a mathematical model of the ML algorithm and supports Vector Machine and Area under the curve analysis.

7.3 BACKGROUND

Some researchers investigate[28] in the field of electronic medical records have been developed statistical models, it is perseverance to make them clinically operational. COVID-19 activities will have a roadmap to potential ML clinical transformation. To optimize the performance of a case in an appropriate ethical context with the highest number of patients.

Learning algorithms allow multi-layer treatment strategy to learn data presentation across more abstract layers.[29] Training models to perform classification is well-performed using text, pictures, and sounds. Deep learning models provide higher accuracies and improve human output in a specific instance.

7.3.1 X-RAY DIAGNOSIS USING DEEP LEARNING

Radiowaves are used as therapy by X-ray devices for the study of infected areas of the body with lung disease, dislocated bone, and cancer. Computed

Tomography used X-ray machines to examine the soft structure of active body parts to better view the primary soft tissue and organ.[30] Using an X-ray is quicker and faster than CT. X-ray is less harmful to humans than CT.[31]

Further, a classification application developed[32] uses multi-layer thresholding and regression to perceive coronavirus in X-ray images. The system utilized a few images for the validating, as well as images blur. Their system achieved a specificity of 97.9%, a sensitivity of 96.7%, and 97.4% accuracy.

7.3.2 COMPUTED TOMOGRAPHY SCAN DIAGNOSIS

Developed a CT scan method in the year 1972. It utilizes advanced X-ray technology to carefully diagnose fragile internal organs. It does not cause much pain, is easy and very fast, and produces three-dimensional images.[33] But it is costly as compared with X-rays.

Many studies have criticized the detection of COVID-19, detection using RT-PCR sensitivity, and specificity. Meanwhile, PCR is the standard method for diagnosing patients with COVID-19 around the globe now. Its side of generating a large amount of false-negative owe several reasons, such as disease stage, methodological drawbacks, delay in diagnosis. Hence, the RT-PCR test is not sufficient for assessing disease status, such as COVID-19. Their achievement of accuracy is not more than 50%.[34]

Also, infectious lung diseases such as bacterial pneumonia, tuberculosis were classified using a system with ResNest-50 designed.[35] They used a selected scan from 870 patients from a laboratory with 150 COVID-19 patients and the rest non-COVID-19 viral pneumonia patients.

7.3.3 COVID-19 INFECTION PREDICTION USING MACHINE LEARNING TECHNIQUES

The science of training machines using complex mathematical models is term as ML.[36] Our proposed system used an ML algorithm. With the design, data will be analyzed and find interesting patterns. The validation of information is then categorized according to the patterns learned during the learning process. Decision-makers rely on imposing lockdowns on infected cities and nations to monitor disease transmission. Statistically, using ML can forecast the behavior of novel cases to stop the disease from spreading.

COVID-19 perdition on most 15 affected countries was done using ARIMA (Autoregressive integrated moving average) model. The prediction shows that circumstances would worsen in Europe and South America, while China, Japan, and Taiwan remain stable.

Two statistical algorithms, regression model, and SEIR were used to evaluate and forecast the distribution of coronavirus in India. They used the dataset gathered by John Hopkins University repository. Their prediction result shows that report cases would be 5300 and 6148 from regression and SEIR, respectively.

7.4 METHODOLOGY

Proposes ML COVID-19 detection involves many stages, as displayed in Figure 7.2. The stages are briefly explained here.

1. Retrieve COVID-19 X-ray images of patients and healthy patients.
2. Generate 1500 chest X-ray images Accumulate X-ray images of coronavirus persistent and well persistent.
3. Generate 1500 chest X-ray images
4. Provide feature selection on images applying an ML algorithm
5. Divide the dataset into two groups, training group, and validation group
6. Evaluate the performance of the algorithm on the validation dataset.

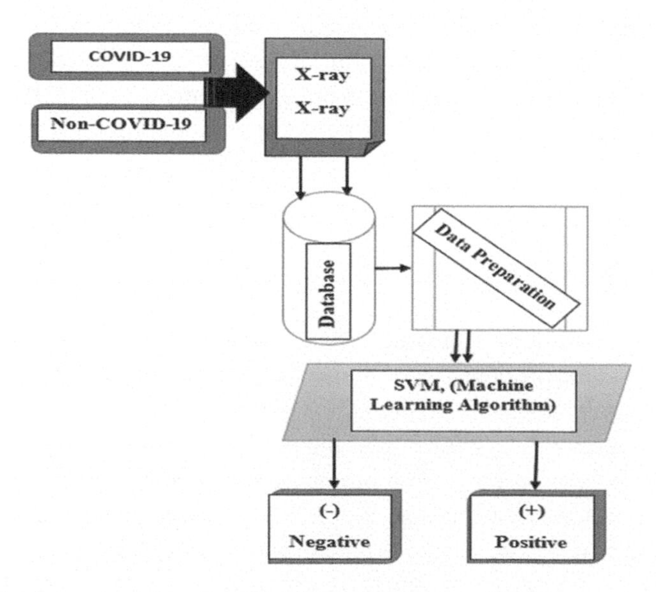

FIGURE 7.2 Propose machine learning diagnosis system architecture.

7.4.1 *DATASET*

The dataset used for this project was obtained from a Kaggle database, which contains an X-ray test on COVID-19 patients. These data set were used for evaluation involving a patient with and without COVID-19. Table 7.1 below shows a summary of the data. Only the X-ray images of both persons were employed—the total images for the study 140 images. The imbalance was controlled using the data augmentation method and improving convergence to contribute a better outcome.

TABLE 7.1 Dataset.

X-ray images	Frequency
COVID-19	40
Healthy	100
Total	**140**

Pre-processing X-ray images: The images are pre-processing of data was resize the image as an input for the algorithm are varied. Pre-processed some image Techniques for optimizing machine efficiency and training time acceleration.

SVM: It is a supervised and linear ML algorithm that is most widely used for image segmentation. SVM has the flexibility of providing solutions to the everyday challenge of data processing and to be used to effectively perform several classification procedures for the distributed detection and classification task.[36] It provides a better result along with other ML techniques. The subset of SVM is SVR (support vector regression), which utilizes the same rules to resolve problems of regression. Data are generally categorized into positive and negative, such that a hyperplane can distinguish the "training classes" precisely.

Training data—Data that helps to explain the trend by focusing on the features Testing data—The model is evaluated to identify the values for unknown data after training on the data, where only attributes are given, and the model can determine whether or not patients are affected. The roc curve assessment found the operational curve of the recipient or the ROC curve. First, the decision function determines the distance from the hyperplane that is divided. For instance, an SVM classifier evaluates hyperplanes that divide space into classification outcomes regions. The distance to the separators is found in this function, given a point.

Maths inside algorithms of solid statistical models and logic, each classification algorithm is composed. Every algorithm must make one decision before the learned model begins to make[38] since the decision tree is one of the classified algorithms. The SVM's rationale is to find the maximal hyperplane that maximizes the training data margin. The biggest challenge in each application in ML is to find the right algorithm. The application is based on training duration, speed, the precision of estimation, number of parameters, number of functions, validation techniques, uncertainty estimates, etc. Thus, in ML, the sum of mathematics is used for justifying the algorithm, such as probability, calculus, linear algebra, complex algorithm, etc.

The svc method is the maximum area under the curve (AUC) and is the most reliable method. Mathematics behind SVM.

The SVM the data in two or more categories by segmenting a similar category utilizing a boundary.

Now discuss how a dataset is described within a space and what is a line equation to further distinguish the same groups, and finally the equation for a distance between a data point and a boundary separating the categories).

A Space Point

Assume that data that allows SVM algorithms to distinguish between patients based on the first analysis of covid 19 characteristics and then classify the unseen data correctly whether or not anyone is influenced by covid 19.

Suppose we know the domain, spectrum, and co-domain definition when defining a function in real space.

When we identify x in an actual space, we understand its domain, and we get range and co-domain by mapping a feature for $y = f(x)$

Initially, thus we are given the data to be algorithmically segregated.

The data generated for dividing/grouping is displayed as a single point in a space where each point has some vector x.

$x \in RD$

Threshold value:

The threshold value for the separation of points into their respective groups is the main separator.

Feature space variable:

The feature space equation is considered as the main separator sequence.

The formula of feature space (for classification) which divides the points can now be read conveniently as:

H: $wT(x) + b =$ zero

Here: b = Feature space interception and predetermining time

Distance Measure:

the data points are interpreted and how a separation segment is designed to fit. However, such a segment distinguishes the data points in the best ways with the minimum mistakes/errors in the misclassification when we were designed to fit the separating linc.

For data points to be counted as the smallest weight, this definition would enable one to know the distance from the data point segregating boundary.

ax + by + c = 0 from the point defined, (x_0, y_0) is given by d of any line.

Consequently, the distance measure of a hyperplane (similarly) is:

wTΦ(x) + b = 0

from a specified point vector Φ(x0):

Where ||w||2 is the Euclidean distance for the length of w given by:

However, not all computational classifying models allow multi-class categorization, algorithms like that of the Logistic Regression and SVM arc optimized for image segmentation.

The purpose of the SVM algorithm was:

If the value is substituted for the positive group in the hyperplane equation for training data that were binary labeled as positive and negative classes, we will obtain a value greater than 0,

wT(Φ(x)) + b > 0

forecasts from the negative class will be pessimistic in the hyperplane's method

wT(Φ(x)) + b < 0.

However, the indications were training evidence, which gives a positive sign for the positive class and a specific value.

If we predict a positive class correctly as positive when evaluating this model on testing results, then two positive signals provide a positive outcome and hence greater than zero. The same is true if the negative category is predicted correctly.

But if the method tends to classify the positive class in negative terms, a plus and a negative are smaller, so less than zero.

For a precise estimate, the result of an expected and accurate labeling must be greater than 0 or less than zero.

$$y_n\left[\omega^T\phi(x)+b\right]=\begin{cases}\geq 0 \text{ if correct}\\ <0 \text{ if incorrect}\end{cases}$$

The AUC was used to compare the trained model's efficiency—the accuracy of the rating related to the accurately categorized percentage of events. The rate of positive cases that were considered positive was defined as sensitivity. Specificity refers to the number of negative instances defined as unfavorable. The calculations indicate that TP, FP, TN, and FN represent real positive, false positives, and true negatives. ROC curves are built by measuring precision and accuracy at all possible confidence score threshold values to describe a true positive outcome. In this analysis, the real data points to the underlying ROC curves. Testing a ROC curve over all precision values involves thorough observation beyond the sample data, resulting in ROC AUC being governed by the field below the extrapolated curve instead of influenced by data points.

The proposed approach substantially supports to reduce the size of the feature image, thus simplifying the function of the SVM classification and increasing its precision. The elimination of the proposed system's computational complexity is another benefit of reducing the number of applications.

7.5 EXPERIMENTAL ANALYSIS

Oracle VM Virtual Box was used as a virtual machine. Orange data mining application was installed on the system for modeling and predictions.

7.5.1 EVALUATION METRICS

Metrics like precision, accuracy, recall, and Area under curve graph were used to assess the proposed ML algorithm's reliability. These metrics are computed based on the TP (true-positive), TN (true-negative), FN (false-negative), and FP (false-positive).

$$\text{Accuracy} = \frac{(\text{TP} + \text{TN})}{(\text{TP} + \text{TN} + \text{FN} + \text{FP})}$$

$$\text{Precision} = \frac{\text{TP}}{(\text{TP} + \text{FP})}$$

The accuracy metric measures the percentage of correctly identify cases relative to the provided dataset. Most ML perform better if the accuracy is higher.

Precision measures the exactness, which is computed as the percentage of positive prediction, which was true positive divided by the number of predicted positives.

The recall is the completeness, which is computed as the percentage of positive that were correctly identified as true positive divided by the number of real positives.

$$\text{Recall} = \frac{\text{TP}}{\text{TP} + \text{FN}}$$

F-measure is calculated, combining precision and recall, which is significant for a test dataset involving an imbalance class.

$$F - \text{measure} = 2\frac{\text{Precision} \times \text{Recall}}{(\text{Precision} + \text{Recall})}$$

The X-ray images were loaded for pre-processing, preparation, and cleaning to avoid data trawling. The Orange application widget was used for the training of the data as well as its validation. Sample of the data used (X-ray) images are shown in Figures 7.3 and 7.4 below.

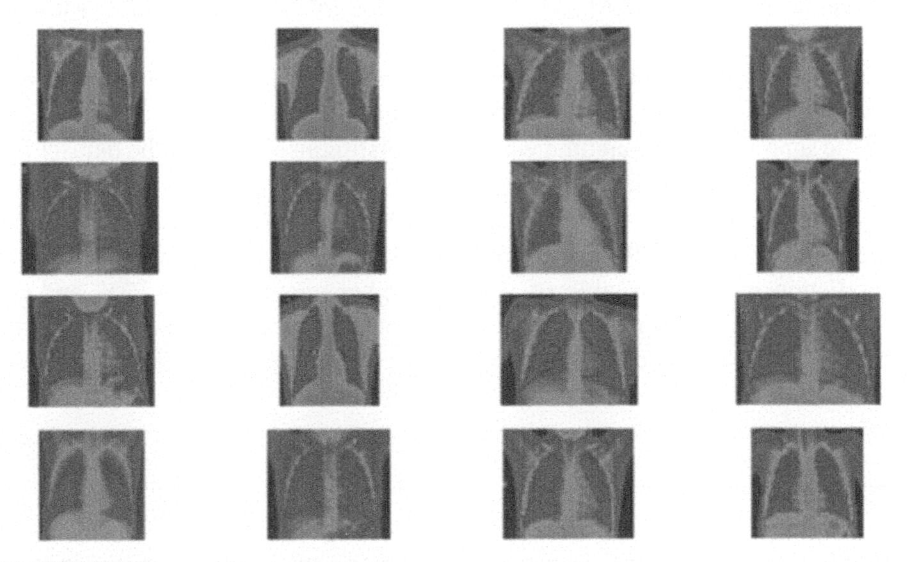

FIGURE 7.3 Normal X-ray images.

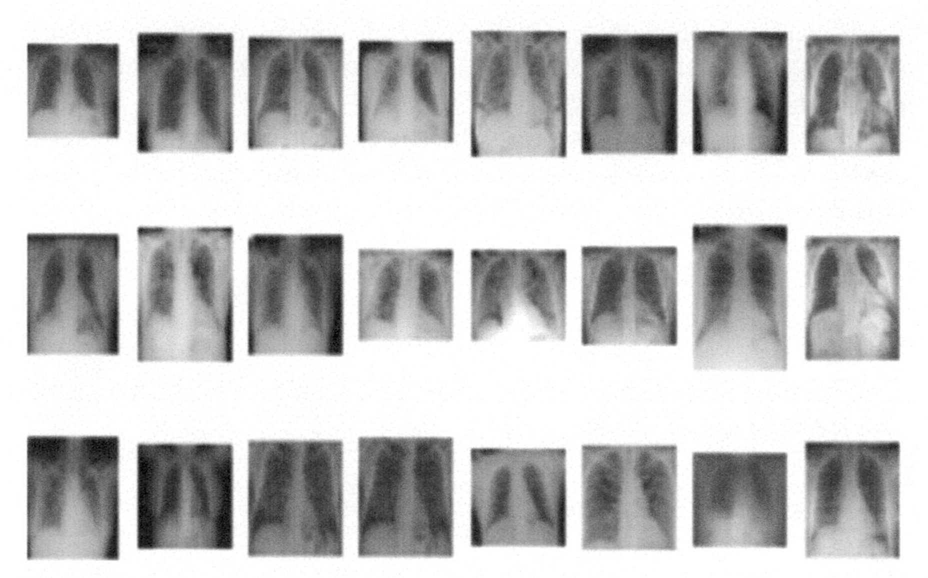

FIGURE 7.4 Abnormal X-ray images.

The model design for diagnosing coronavirus was based on the Support Vector Machine. It was designed to detect coronavirus, by means of X-ray images. The orange application comes with an image analyzer, which is very powerful and can code to suit the kind of patterns needed within the image provided for training.

The detector for COVID-19 was trained and tested on the collected dataset, 70% for training, and 30% as a remainder for testing. The hyper-line of the support vector machine was able to distinguish between the dataset for higher classification accuracy.

The proposed results were very encouraging in the field of diagnosing method in COVID-19. The application ML technique was effective. The prediction model was able to achieve higher accuracy of 97.2% and an F-measure of 94%. Figure 7.5 shows the performance of the Support Vector Machine concerning classification accuracy while Fig 7.6 depicts model loss on the dataset.

7.6 DISCUSSION

In this chapter, the coronavirus model obtains better results when using the Support Vector Machine. Comparing our model accuracy and precision with other models, this proposes coronavirus prediction is best. Receiving 97.7% for accuracy and 96.4% for precision, and the best training process was gained as the training and validation difference became closer. A robust COVID-19 detector was built as the F-measure improved to 0.97. The metrics of AUC were impressive as the model achieved 0.9. Thus, the COVID-19 diagnosis model trained on the X-ray data provides superior performance metrics.

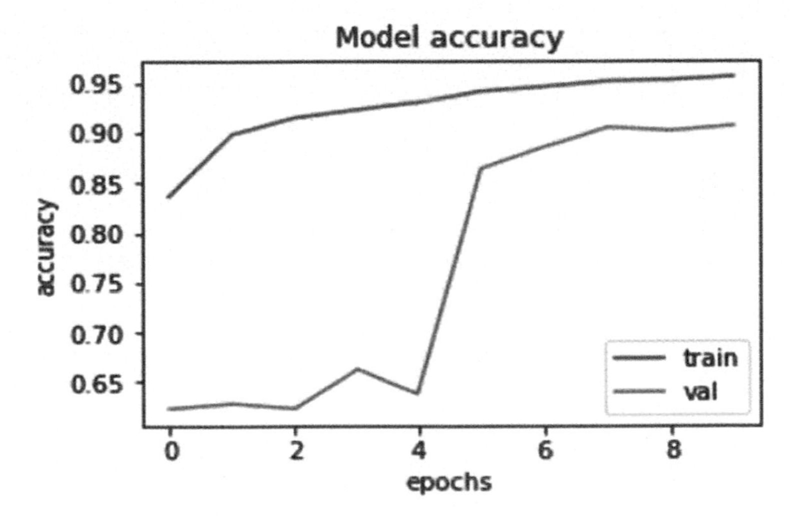

FIGURE 7.5 Performance of SVM model on the dataset.

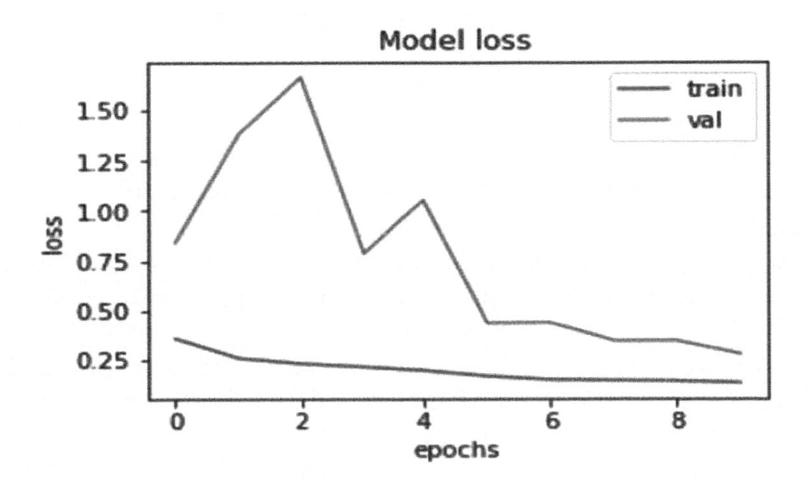

FIGURE 7.6 Model loss on the dataset.

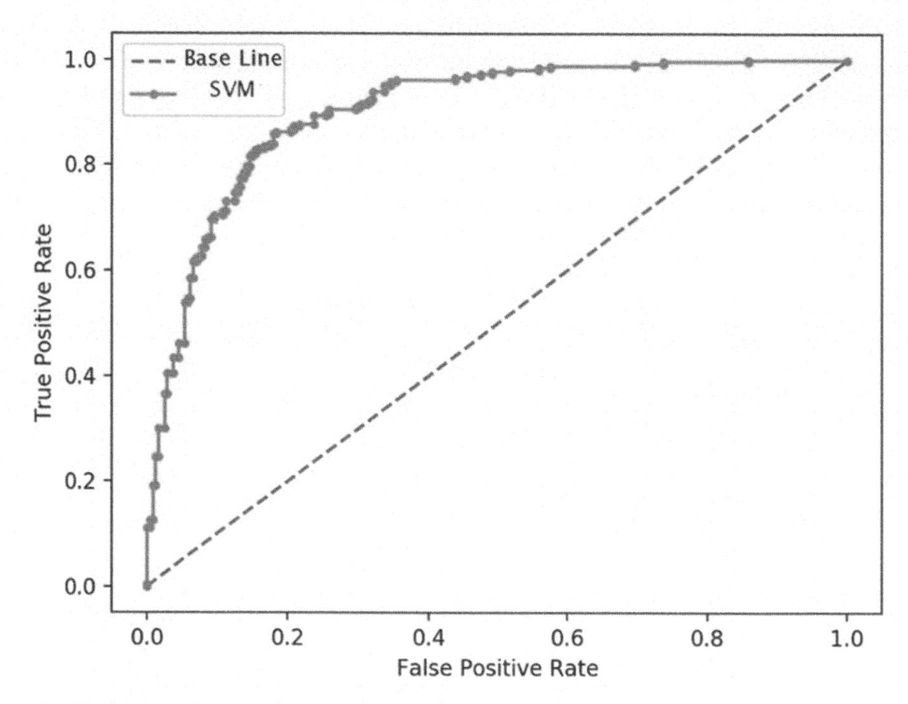

FIGURE 7.7 Area under the curve of SVM.

Using SVM provided highly sensitive, as shown in Figure 7.7, we recommend scientists working on COVID-19 depend on systems with models built using Support Vector Machine.

7.7 CONCLUSION

The proposed work described the classification of positive and negative pandemic evidence from the X-ray image set. Development of a scratch model that distinguishes it from other AUC-related methods. The exponential increase of COVID-19 across the globe and an increasing number of deaths need urgent actions from all departments. In this study, the ML algorithm was utilized to propose a diagnosis system for COVID-19. A diagnosis model using SVM was proposed to detect coronavirus utilizing chest X-ray images. The model was developed utilizing a dataset from the Kaggle database to train and validate the model's effectiveness.

7.8 FUTURE ENHANCEMENT

In the future, we will consider diagnosing COVID-19 in RT-PCR using the VGG-XX version and compare their performance using a high volume of the dataset. Forecasting potential infections will enable authorities to battle the consequences effectively and apply Metaheuristic techniques. To better refine the model. In exchange, refining the model would improve the model's efficiency. If the efficiency improves, the model will create fewer errors, which would be beneficial for COVID 19 predictions with as little error as possible.

KEYWORDS

- **mathematical**
- **model**
- **Coronavirus**
- **curve**
- **SVM**
- **features**

REFERENCES

1. Singh, N. et al. COVID-19 Epidemic Analysis Using Machine Learning and Deep Learning Algorithms, 2020.
2. WHO. Covid-19, 2020.
3. E. Systems. What is Machine Learning. 2020. https://expertsystem.com/machine-learning-definition/.
4. Jiang, F. et al. Review of the Clinical Characteristics of Coronavirus Disease 2019 (COVID-19). *J. Gen. Intern. Med.* 2020, 1–5.
5. Ai, T. et al. Correlation of Chest CT and RT-PCR Testing in Coronavirus Disease 2019 (COVID-19) in China: A Report of 1014 Cases, China, 2020.
6. Tang, Z. et al. Severity Assessment of Coronavirus Disease 2019 (COVID-19) Using Quantitative Features from Chest CT Images. arXiv preprint arXiv:200311988. 2020.
7. Qi, X. et al. Machine Learning-Based CT Radiomics Model for Predicting Hospital Stay in Patients with Pneumonia Associated with SARS-CoV-2 Infection: A Multicenter Study. medRxiv. 2020.
8. Wang, L. et al. COVID-Net: A Tailored Deep Convolutional Neural Network Design for Detection of COVID-19 Cases from Chest Radiography Images. arXiv preprint arXiv:200309871. 2020.

9. W. H. Organization. Laboratory Testing for Coronavirus Disease 2019 (COVID-19) in Suspected Human Cases, 2020.
10. Wang, S. et al. A Deep Learning Algorithm Using CT Images to Screen for Corona Virus Disease (COVID-19). medRxiv, 2020.
11. Alazab, M. et al. Survey in Smartphone Malware Analysis Techniques. *New Threat. Countermeas. Digit. Crime Cyber Terror*. 2015, *6*, 105–130.
12. Mesleh, A. Feature sub-set Selection Metrics for Arabic Text Classification. *Pattern Recognit. Lett.* 2011, *32*, 1922–1929.
13. Wynants, L. et al. Systematic Review and Critical Appraisal of Prediction Models for Diagnosis and Prognosis of COVID-19 Infection, 2020.
14. Panovska-Griffiths, J. Can Mathematical Modelling Solve the Current Covid-19 Crisis? *BMC Public Health* 2020, *20* (1), 551.
15. Giordano, G. et al. Modelling the COVID-19 Epidemic and Implementation of Population-wide Interventions in Italy. *Nat. Med.* 2020, 26, 855–860.
16. Nandal, U. Mathematical Modeling the Emergence and Spread of New Pathogens: Insight for SARS- CoV- and Other Similar Viruses. *Pharma R&D Today* 2020.
17. Jewell, N. et al. Predictive Mathematical Models of the COVID-19 Pandemic Underlying Principles and Value of Projections. *JAMA* 2020, 323, 1893.
18. Santosh, K. AI-Driven Tools for Coronavirus Outbreak: Need of Active Learning and CrossPopulation Train/Test Models on Multitudinal/Multimodal Data. *J. Med. Syst.* 2020, *44*, 1–5.
19. Mesleh, A. Support Vector Machine Text Classifier for Arabic Articles. *VDM* 2010.
20. Hemdan, E. E.-D.; Shouman, M. A.; Karar, M. E. A Framework of Deep Learning Classifiers to Diagnose COVID-19 in X-Ray Images. arXiv Prepr., 2020, *2003*, 11055.
21. Yang, Z.; Zeng, Z.; Wang, K. et al. Modified SEIR and AI Prediction of the Epidemics Trend of COVID-19 in China under Public Health Interventions. *J. Thorac. Dis.* 2020, *12*, 165–174.
22. Wu. J. T. et al. Nowcasting and Forecasting the Potential Domestic and International Spread of the 2019-nCoV Outbreak Originating in Wuhan, China: A Modeling Study. *Lancet* 2020, *395*, 689–697.
23. Tang, B. et al. Estimation of the Transmission Risk of 2019-nCoV and Its Implication for Public Health Interventions. *J. Clin. Med.* 2020, *9*, 462.
24. Imai, N. et al. Report 3: Transmissibility of 2019-nCoV, published online January 25, 2020.
25. Li, R. et al. Substantial Undocumented Infection Facilitates the Rapid Dissemination of Novel Coronavirus (SARS-CoV2). *Science* 2020, *368*, 489–493.
26. Leung, K. et al. First-wave COVID-19 Transmissibility and Severity in China Outside Hubei after Control Measures, and Second-wave Scenario Planning: A Modeling Impact Assessment. *Lancet* 2020, *395*, 1382–1393.
27. Sethy, P. K. et al. Detection of Coronavirus Disease (COVID-19) Based on Deep Features, 2020.
28. Debnath, S. et al. Machine Learning to Assist Clinical Decision-Making during the COVID-19 Pandemic. *Bio Electron. Med.* 2020, *6*, 14.
29. Pandey, G. et al. SEIR and Regression Model Based COVID-19 Outbreak Predictions in India, 2020.
30. Rachna, C. Difference between X-ray and CT Scan, 2020.
31. Alazab, M. et al. COVID-19 Prediction and Detection Using Deep Learning, June 2020.

32. Hassanien, A. E. et al. Automatic X-ray COVID-19 Lung Image Classification System based on Multi-Level Thresholding and Support Vector Machine, medRxiv, 2020.
33. LeCun, Y. et al. Deep Learning. *Nature* 2015, *521*, 436–444.
34. Kumar, P. et al. Forecasting the Dynamics of COVID-19 Pandemic in Top 15 Countries in April 2020 through ARIMA Model with Machine Learning Approach, medRxiv, 2020.
35. Fu, M. et al. Deep Learning-Based Recognizing COVID-19 and Other Common Infectious Diseases of the Lung by Chest CT Scan Images, medRxiv, 2020.
36. Huang, C.-J.; Chen, Y.-H.; Ma, Y.; Kuo, P.-H. Multiple-Input Deep Convolutional Neural Network Model for COVID-19 Forecasting in China, medRxiv, 2020.
37. Awad, M. et al. *Support Vector Machines for Classification BT—Efficient Learning Machines: Theories, Concepts, and Applications for Engineers and System Designers*; Apress: Berkeley, CA, 2015; pp 39–66.
38. Dutta, N. et al. In *Mathematical Models of Classification Algorithm of Machine Learning*, International Meeting on Advanced Technologies in Energy and Electrical Engineering 2018. http://doi.org/10.5339/qproc.2019.imat3e2018.3.

CHAPTER 8

Detection and Classification of Knee Osteoarthritis Using Texture Descriptor Algorithms

ANJANI HEGDE*, RISHMA MARY GEORGE, and H. D. RANJITH

Department of Electronics and Communication Engineering, Mangalore Institute of Technology and Engineering, Moodabidri, India

Corresponding author. E-mail: anjani7978@gmail.com

ABSTRACT

The frequently observed disorder in the population is arthritis, tenderness, and swelling in the joint. The primary symptoms of arthritis are stiffness and joint pain. Osteoarthritis (OA) is one of the categories in arthritis, degradation of knee joint cartilage that leads to numbness and loss of movement. Due to overstress on the knees also leads to damage of the cartilage. OA is common in old age people. Cartilage is the slippery tissue that helps to smooth the frictionless movement of joint. OA is the condition where the protective layer of the cartilage damages in the knee, the whole rupture leads to severe pain, and loss of motion, and joint losses of its normal shape. Early detection of OA helps to repair and regenerate the knee cartilage. Once the cartilage wears down completely, then it is not possible to regenerate.

The detection of cartilage in knee MRI is a principal task. The cartilage is a thin layer between the femur and tibia. Cartilage damage leads to friction between the bones. Low-contrast of the cartilage denotes biochemical properties of it, and cartilage fluid content widens the joint space width.

Intelligent Interactive Multimedia Systems for e-Healthcare Applications. Shaveta Malik, PhD Amit Kumar Tyagi, PhD (Eds.)

The automatic texture analysis involves two methods cartilage segmentation and texture analysis. Knee cartilage is a thinnest and low contrast region that is difficult to extract. The preprocessing of the input MRI image improves the visuality of the cartilage region. The preprocessing involves filtering, contrast enhancement, and thresholding. The binary image generated after preprocessing differentiates between background and bone. The thresholded image is converted to an edge-detected image. In the cartilage segmentation method, the mask is used to extract the cartilage region that is in between the femur and tibia. The mask is a rectangular box, an edge detection method used to identify the edges of the bones. Low-contrast cartilage is located between the boundaries of two bones extracted from the input MRI image. The microtexture descriptor algorithm is applied to the cartilage to find the continuity. In the microtexture descriptor, each pixel value is replaced by the value obtained by taking a weighted average of eight neighboring pixels. The automatic segmentation of cartilage helps to identify the condition of OA. The cartilage texture was analyzed by the microtexture descriptor algorithm and joint space width was measured. Based on the result obtained by the width and the texture the condition was categorized as normal, mild, moderate, and severe. The width of the cartilage is computed and matched with the standard thickness values defined by the researcher.

8.1 INTRODUCTION

Worldwide most people are suffering from osteoarthritis (OA). It is a great challenge in the health system. The Joint, where two bone ends with slippery protective tissue cartilage, comes together. OA occurs when cushions between two ends of the bone wear down gradually due to stress on legs, utmost injury because of an accident, wearing down of cartilage tissue worsen over time. Slippery, firm tissue cartilage, helps to frictionless joint movement. OA occurs mostly in oldster, may also occur in adults. OA is also popularly called wear and tear arthritis or degenerative arthritis or degenerative joint disease.

The OA can be classified as two main categories: primary and secondary. Primary OA , not only caused by aging, is a chronic degenerative disease. As a person's age increases, the water content in the cartilage reduces, thus weakening the strength, allowing it to degenerate. Primary OA exact reasons are unpredictable Secondary OA is unusual pain across the joint with post-traumatic causes or abnormal articular cartilage. Secondary OA due to stress on the knee, injury, long time squatting, diabetes, obesity, family history, etc. OA is not possible to cure, but symptoms are minimized by excise, alternate medicine, canes, or braces like physical aids, healthy diet. Symptoms of OA are stiffness, pain, grating sensation when a bone moves, unbearable pain during bone movement, swelling.

The factors responsible for OA are as follows:

Age: As the age increases, the capacity of cartilage healing itself decreases.

Heredity: OA has been linked with some particular genes. The disorders such as bowlegged, knock-kneed, double-jointed also have high chances of OA.

Obesity: Because of more bodyweight, more stress on the knee leads to the degeneration of cartilage.

Knee Injury: Knee injury occurs during accidents or sports that lead to OA.

Overuse: Repetitive stress, walking, standing for a longer time, over excise, repeated kneeling, heavy lifting all increase the risk factor of OA.

Depending upon the degree of rupture doctor decides the treatment method:

Change of lifestyle: If the degradation is very less, the doctor may recommend losing weight, switching from stressed excise to easy excise.

Therapy: Some physical therapy strengthens the muscles and increases the range of motion.

Medication: Early stage of degeneration cured, also the pain can be reduced by medicines suggested by the physician.

8.2 STRUCTURE OF KNEE

The very complex and massive joint structure is the knee joint. The thigh-bone (femur) and shinbone (tibia) join together, forming knee structure. The bone runs along with the tibia known as the fibula and the kneecap (patella) is present in the knee joint. The knee joint is like a hinge-type joint. Like a door hinge, the same kind of movement happens in the knee. Along with it also allows rotational movements.

The four main bones of the knee are:

Femur

This is one of the longest bones in the body. Femur ends with a round knob known as condyles. At knee joint femur ends covered by hyaline cartilage also is known as articular cartilage.

Tibia

It is the bone connecting the knee and ankle. two flatten surfaces are covering the top portion of the tibia. It is covered with the articular cartilage. Menisci are the two shock-absorbing C-shaped cartilage.

The knee cap or patella

It is the triangle-shaped semi-flatten bone bends as the knee bends. The major task of this is to improve the force generation by the quadriceps muscles. If the patella is damaged then it is not possible to pull the tibia and straighten the knee. Kneecap also protects the joint from trauma.

Fibula

It is the thin bone that runs alongside the tibia. It is situated in the lower position of the leg connecting the knee to the ankle. The tibia carries major weight and the fibula helps to carry a small portion of the weight. For muscle such as the biceps, it acts as an attachment.

The cartilage of the knee

There are two types of fibrocartilage and hyaline cartilage. Fibrocartilages are giving extra strength and stability to the knee which is strong and rubbery. Hyaline cartilage is slippery and flexible. Since hyaline cartilage is flexible it is helpful for shock absorbance. Synovial fluid is the oil lubricant within a joint that makes cartilage more slippery. So the bones can move smoothly without any pain. But cartilage does not have a blood supply and is more difficult to repair.

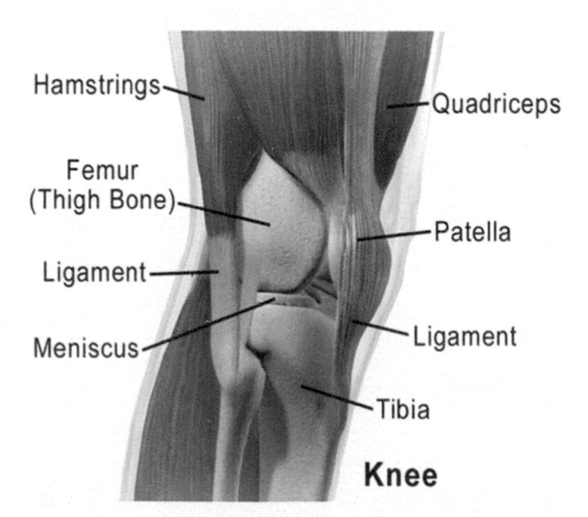

FIGURE 8.1 Physical structure of the knee.[12]

Source: Blausen.com staff (Ref. 12). https://creativecommons.org/licenses/by/3.0/

8.3 LITERATURE SURVEY

The automatic segmentation of biomedical images uses many methods. According to the survey few methods are explained in this section. Automatic texture-based segmentation uses *k*-means clusters and co-occurrence features. Features are extracted from a set of 256 gray-level information. Feature extraction aims to segment image regions regarding textural homogeneity. In this method, $n \times n$ sliding window is used for co-occurrence calculation, helpful for feature extraction. For every input pixel pi, the window is positioned such that the first box of the window matches the position of pi. For co-occurrence and feature, the calculation uses a gray level co-occurrence indexed list (GLCIL) algorithm.[3]

Cartilage segmentation accuracy improved in steerable feature-based design with a modified Layered Optimal Graph Image Segmentation of Multiple Objects and Surfaces (LOGISMOS). The object and multiple surfaces interacted mutually with the help of the LOGISMOS optimized segmentation method. Algorithm models the individual graphs and n-D graphs associated with the cost. The first step is to determine the volume of interest (VOI) along with bone and cartilage. Ada Boost Classifiers are trained using 3D Haar features by making use of VOI. This method generated errors while analyzing femur and tibia along with a long cartilage plate called a long surface positioning. The error may occur for two reasons one may be VOI segmented may not be of clinical usage or the algorithm fails to extract the interesting region automatically. For individual subregions generated error will be less compared to long cartilage plate regions which were attributed to the edges of the cartilage. In Ref. [6] the author explained the segmentation of menisci. The OA also affects the integrity of ligaments and menisci rather than joint cartilage. OA symptoms have been found that menisci damage and cartilage loss. This uses the shape creation model to segment menisci. The author believes that parameter optimization and a large dataset to train may give accurate segmentation of menisci automatically.

To segment cartilages and bones, a 3-dimensional statistical shape and thickness model is used developed. The shape models with each meniscus parameterized to have 10,242 vertices. In[7] also author explained the segmentation of the menisci. In this paper, Law's texture algorithm is used to segment femur and tibia bones. In the beginning, the author selected or pre-segmented three regions one region is a bone region, and the rest two regions are not bone regions. In each region author selected ten random 5×5 values to create a database. The five Law's vectors are used to create 14 Law's masks for texture analysis. Divide the whole image into 5 × 5 non-overlapping regions and convolve this region with Law's masks repeated the procedure for the whole image. Then convolve the database 5 × 5 values with the Law's mask and find the Euclidian distance between those two convolved values, select the best of the image blocks repeating the same thing for the not bone region, and superposition of all the images gives the best results. The graph cut methods are used to separate bone from not bone part. But this method takes too much time to run the program in MATLAB.

In[8] the biomedical, morphological, and mechanical features of the cartilage are used to detect OA. The detection is based on the fluid content and the thickness of the knee cartilage. Fluid, proteoglycan, and collagen together give mechanical support to cartilage. The unbalance of these damages the articular cartilage. This method involves segmentation, thickness measurement, and quantitative MR parameter measurement. The developed system using a novel algorithm measures the thickness and fluid content in the cartilage. hence the severity of OA can be detected using the model. In[9] compares the manually segmented and automatically segmented cartilage by examining volume and area estimates. For the segmentation, the author uses the voxel classification method. The classifier uses the principle of the kNN-classifier. The interscan reproducibility gives the best results for medial tibial cartilage. Even though low field MRI gives less accuracy this method is used to train and evaluate low field MRI images. To overcome the problem of obtaining the criterion function required for multiclass classification can be solved by the binary classifier.

In[10] author explained the texture algorithm for the articular cartilage. But the cartilage is selected manually from the MR images. The MTD algorithm is used for texture analysis. MTD uses local neighbors for the analysis. 3 × 3 neighbors are used for the MTD value calculation. the degeneration cartilage indicates the state of OA. The art of segmenting is done with the help of a 2-D semi-automated algorithm. But these algorithms require more time and supervision by the experts, so the automatic segmentation of cartilage is the aim in the medical field. The proposed approach in[10] initiated the segmentation of the cartilage automatically by generating a patella cartilage deformable model by finding the spatial relationship between the bone and cartilage. It uses an optimized statistical shape model to build the 3D model of the patella cartilage.

In[11] the author uses active shape models to segment articular cartilage. For the analysis, Photon Emission Computed Tomography images are used. For this process many datasets are manually segmented, resulting in unorganized point clouds. the closest point algorithm is used to establish a correlation between different surfaces. In the author proposed a new technique for segmentation, which is the simple method that works effectively on the knee MRI but the author made one assumption that cartilage is the present middle of the knee. The cartilage thickness will reduce as the cartilage degenerates and joint space width increases. The measure of cartilage thickness will decide the stage of arthritis. For all medical image applications, there is not a specified algorithm. A more generalized algorithm can be applied to a variety of data. By taking prior knowledge of any algorithm, it is possible to achieve better efficiency and accuracy in segmentation.

On average 71 millions people are having a problem with OA, it is a damaging degenerative disease creating mechanical abnormalities. Knee OA is growing common among older people, obese and women. The percentage of people affected by OA increases with age value. The early detection of KOA could aware people retard the progression of the ailment. So the automatic texture analysis method is used to find the texture of the cartilage region, which uses vertical and horizontal histogram plots for segmentation and the MTD algorithm used for texture analysis.

8.4 PROPOSED WORK

On the basis of the literature survey, for the early detection of OA automatic texture analysis are explained as follows:

The automatic texture analysis consists of two parts those are:

1. Automatic cartilage segmentation.
2. Texture analysis using micro-texture descriptors.

The OA disease also can be characterized by the fact of morphological and biochemical degradation of cartilage or tissue. Morphological degradation and biochemical degeneration reveal the condition of OA such as fibrillation and cartilage thinning. Biochemical changes in the articular cartilage happen in advance of morphological degradation. By testing the biochemical features of cartilage may be useful; in diagnosing the disease.

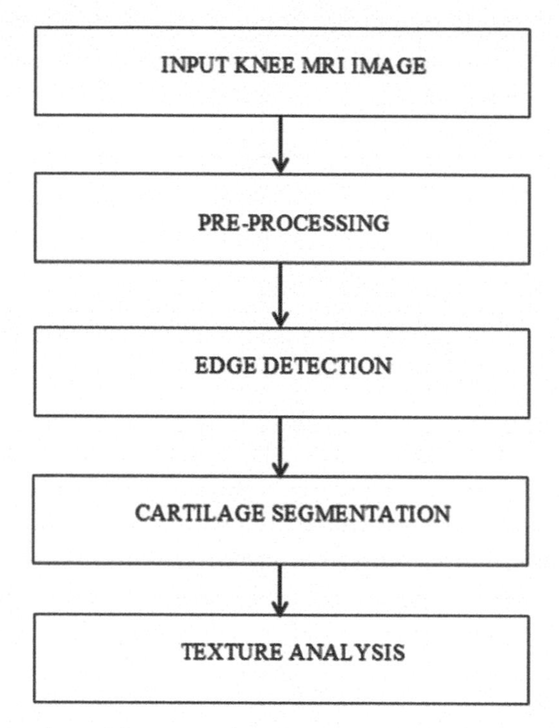

FIGURE 8.2 Flow chart of the proposed work.

In Figure 8.2 pre-processing stage includes adaptive histogram equalization, filtering, and contrast enhancement. The filter removes the noise present in the image. For this analysis T2 map sagittal view MR images of the knee were given as the input with an echo time of 88 ms, repetition time of 3800 ms. The input image set can be divided into three categories normal knee, OA, and unresolved OA. The proposed algorithm automatically segments the articular cartilage and the texture algorithm defines the condition of OA, or the segmentation of the cartilage, the vertical and horizontal projection of thresholded images is used. The thresholded image obtained using some in-built functions the below figure shows the thresholded image of the knee.

FIGURE 8.3 Expected cartilage region.

The thresholded image contains only two types of values that are zero and one. For the detection of the cartilage region, the column-wise and row-wise sum are calculated separately (Fig. 8.4).

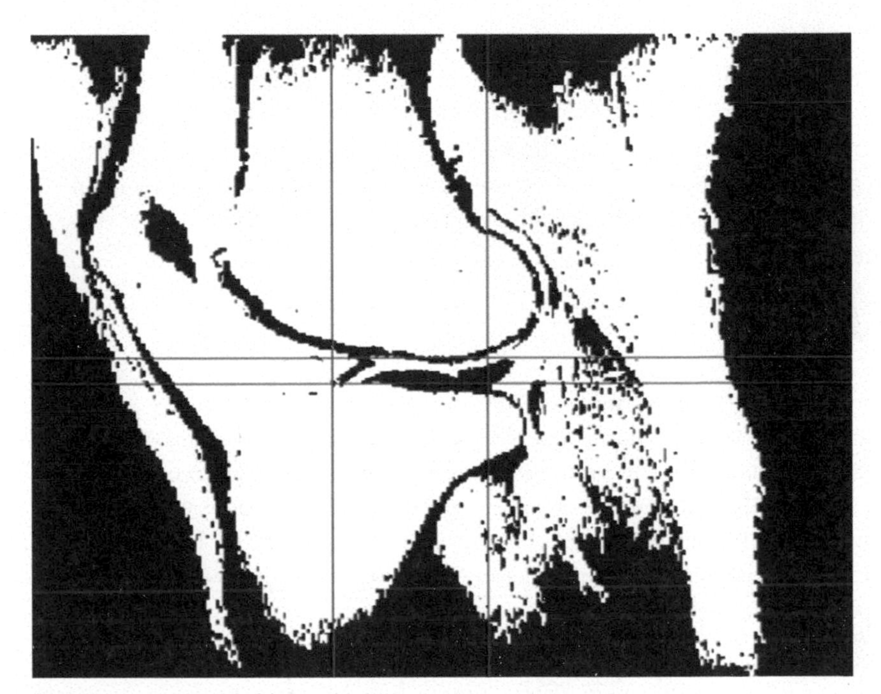

FIGURE 8.4 Thresholded image highlighted the cartilage region.

The column-wise sum gives two vertical lines in which cartilage is expected and the row-wise sum gives the two horizontal lines in which cartilage regions are present. Figure 8.3 shows the expected cartilage region. The region between inter-sections of the blue lines is cartilage which is a boat-shaped thin region that can be cropped from the original image. The MTD algorithm is then applied to this for texture analysis.

First-row-wise sum is calculated, we will get an array such that the number of elements in an array is equal to the number of rows in an image. The region with minimum value represents the cartilage region since it is present between the femur and tibia and also cartilage is the low contrast region so the binary image has minimum value regions. The MATLAB contains in-built functions for generating a thresholded image and for the thresholded value.

8.5 MICROTEXTURE DESCRIPTOR

In radiological images variation of texture intensity in local gray level pixel modeled using the Micro Texture Descriptor algorithm. In MTD operation nonparametric gray level, invariant texture characterization is accepted, also spatial structure of an image is summarized. To extract local texture variation MTD is one of the simple and effective algorithms. The MTD makes use of 3 × 3 neighborhood to find the value of the descriptor by utilizing the local gray intensity levels. In the first step, the central pixel compares its intensity with its eight neighboring pixels. In the resulting 3 × 3 window it substitutes the values either 0 or 1, defined in eq 8.1.

$$d_k = \begin{cases} 1 & if\ p_k \geq p_c \\ 0 & if\ p_i < p_c \end{cases} \tag{8.1}$$

The pattern in an MTD texture description is indicated by a set of nine elements P = $\{p_c, p_0, p_1 \ldots . p_7\}$ where p denotes the intensity level of the pixel. In this work, the input knee MRI image is reduced to eight bits by the descriptor algorithm so a total of 256 ($2^8 = 256$) MTD codes can be generated. The spatial binary pattern of 3 × 3 pixels is described by MTD code. In the MTD algorithm p_c represents the central pixel and p_k represents neighboring pixels where k=0,1,2,3...7.

$$\text{MTD values central pixel} = \sum_{k=0}^{7} d_k\, x 2^k \tag{8.2}$$

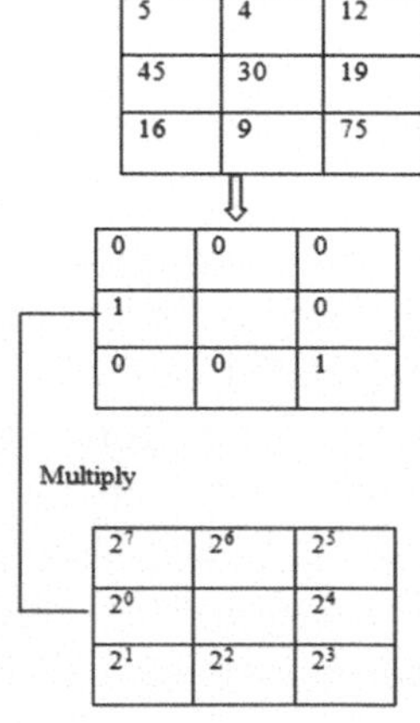

FIGURE 8.5 MTD value generation.

The MTD value for any pixel can be calculated by comparing the central pixel p_c with neighboring pixel p_k. If the central pixel value is equal to or greater than the neighboring pixel value pi then in the binary window that position is written as 1. This process is similar to hard thresholding. If the neighboring pixel p_k intensity is greater than or equal to that of central pixel p_c then in the binary window that position is written as 0. The resulting binary patterns are addressed as d_k. In the next step, the d_k is multiplied with 2^k then the sum of $d_k \times 2^k$ gives MTD a new value for the central pixel p_k. The same procedure is followed for all the pixel values in the entire MR image.

MTD is used to extract texture using local binary value patterns. The thresholding of neighboring pixels with center pixels gives the binary pattern that has only two values they are 1 and 0. The MTD is the weighted average of the local neighboring pixels which helps to find the degenerative part of the cartilage. In OA the biochemical property of the cartilage will change, that is fluid content and other biological compositions unbalance, etc. This affects the MR imaging pixel values so MTD helps to detect this change. An MTD is a single integer value that can take any value from 0 to 255, which encodes information about local microstructure around any pixels. A single MTD code can be generated by a central pixel and its neighbor. The frequency of MTD codes in the image can be described by the histogram process. The representation of the histogram offers the creation of the MTD feature vector thus indicating information of image micro-texture. Radiological sparkle noise and motion blur artifacts become texture sensitive by generating MTD code by hard thresholding the neighboring pixel.

$$d_k = \begin{cases} 0 & \text{if } p_k < p_c - T \\ \dfrac{T - p_k + p_c}{2T} & \text{if } p_c - T \le p_k < p_c + T \\ 1 & \text{if } p_k \ge p_c + T \end{cases} \tag{8.3}$$

To form a robust MTD operator to overcome imperfection and uncertainty, characterization with fuzzy logic may be helpful. This helps to improve the discrimination power of the MTD operator. The fuzzy logic reproduces or imitates the human decision-making process. Fuzzy MTD can minimize the uncertainty in data description. In fuzzy MTD process transfers the input variable to the fuzzy variable.

8.6 RESULT AND DISCUSSION

The Input MR Images has the following specification:

Siemens Magnetom Verio 1.5T MRI scanner.
Scanning Sequence: spin echo (SE).
View: Sagittal.
Slice Thickness: 3 mm.
Repetition Time (TR): 3800 m sec
Echo Time (TE): 88 m sec
Image matrix size: 256 × 256

The input image is if the form of a Dicom image. The cartilage, since it is the low contrast region pre-processing is a very important stage to improve the properties of the cartilage region. The pre-processing stage contains contrast enhancement, adaptive histogram, filtering. The mask is used to extract femoral-tibial cartilage. The pink color region in the output image indicates the healthy cartilage region which is formed by the continuity of the cartilage texture, the yellow color indicates the dissimilarity of the cartilage region. Figure 8.5 shows the cartilage of a normal person.

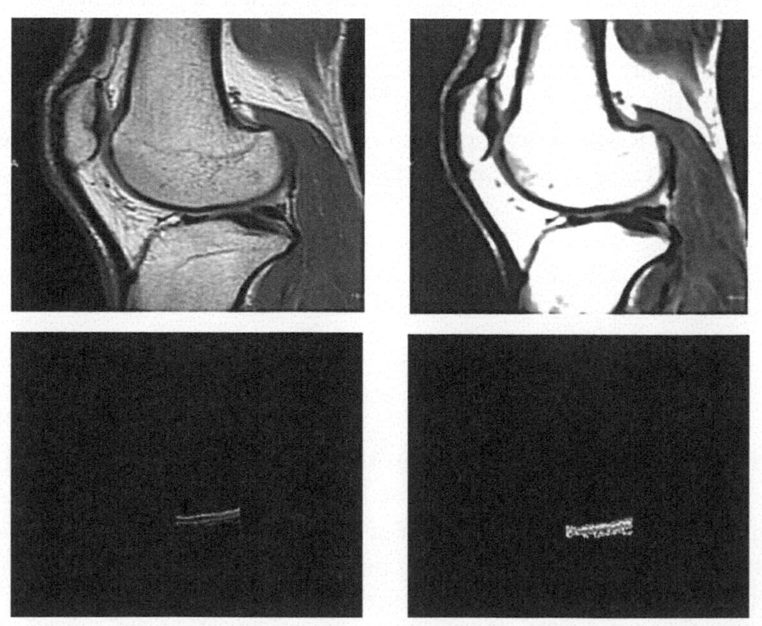

FIGURE 8.6 (a) Normal Knee image, (b) preprocessed input image, (c) extracted cartilage from the normal knee image, and (d) texture description algorithm on the normal cartilage.

Figure 8.6 indicates the normal knee structure. Figure 8.6(d) has a less yellow color portion that gives the indication of more fluid present in the cartilage.

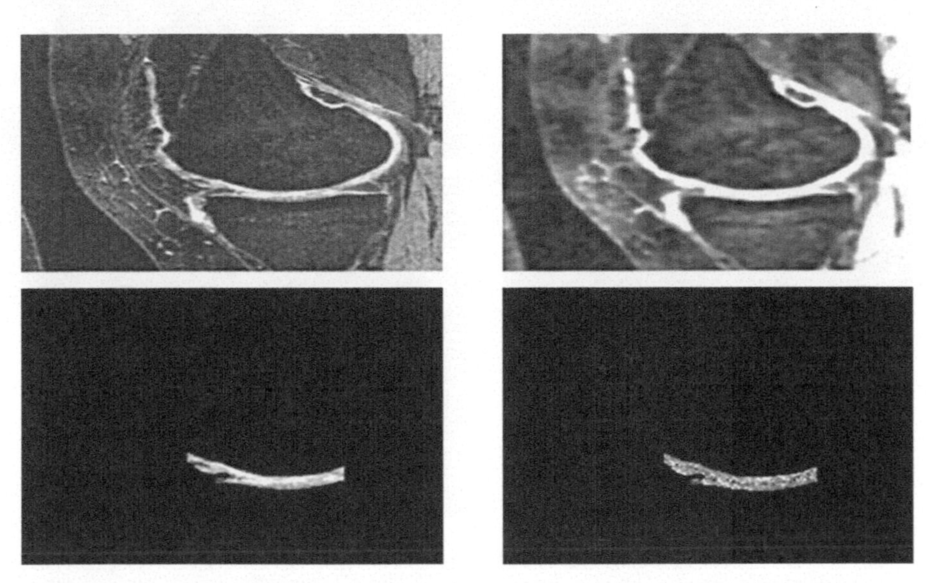

FIGURE 8.7 (a) Knee image, (b) preprocessed input image, (c) extracted cartilage from the input knee image, and (d) texture description algorithm on the doubtful KOA cartilage.

It is observed in the output cartilage for the normal knee joint the yellow-colored portion is less and in the OA case, there will be more yellow regions in the output cartilage region. This is because of non-similar pixel values in the neighbors of the pixel. In Normal cartilage, the pixel values are almost similar. Because of OA the fluid content in the cartilage region will increase so the intensity value of knee cartilage may change.

Figure 8.7 indicates the doubtful KOA type of images. It is difficult to take a decision by looking at the colors present in 8.7(d). Also, the thickness is in between normal knee cartilage and OA-affected cartilage.

Figure 8.8 is the OA effected image because of the less fluid content and more space between femur and tibia, the yellow color indicates the dissimilarity in the texture representation of the cartilage.

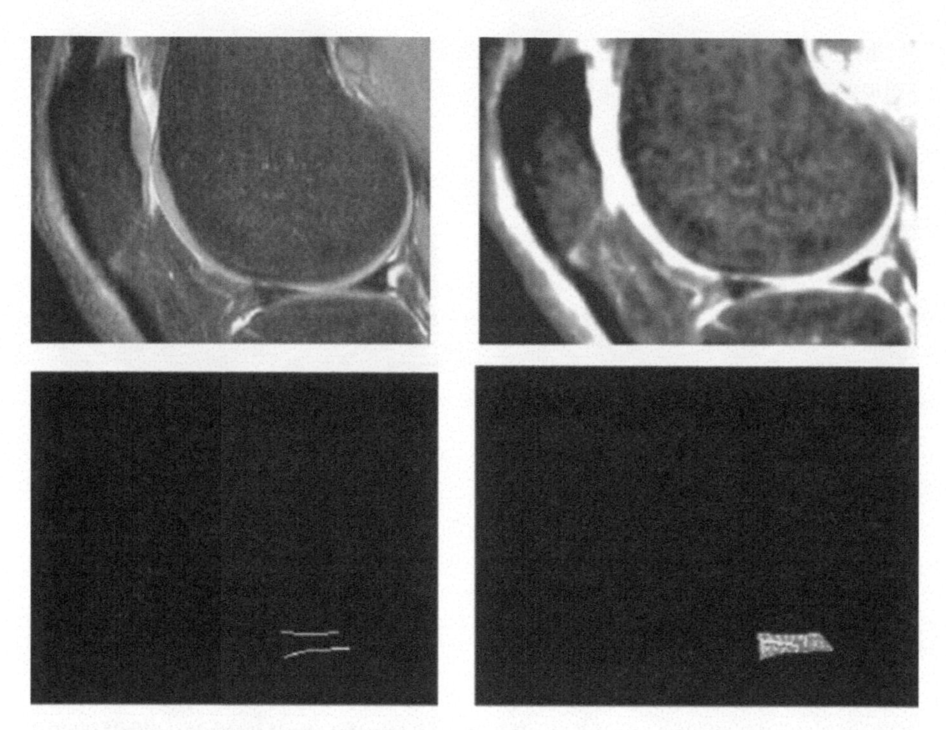

FIGURE 8.8 (a) Knee image, (b) preprocessed input image (c) boundary to extract cartilage from knee image (d) texture description algorithm on the KOA cartilage.

8.7 CONCLUSION

The cartilage is the thinnest tissue in between the bone detection of the that automatically is the challenging task. The automatic segmentation of the cartilage is performed by the thresholding and adoptive histogram methods. The automatic algorithm gives boundaries to the crop cartilage region. The biochemical properties like water content, protein, and mechanical properties like thickness can be measured using MTD algorithms. The fuzzy MTD improves the efficiency and accuracy of the texture analysis. The MTD is a simple computational algorithm that helps to predict OA. For future work different views of the knee, MRI considered and the other algorithm can be applied to the cartilage region to give an accurate result.

KEYWORDS

- **osteoarthritis**
- **microtexture descriptor algorithms**
- **knee cartilage**
- **MRI**
- **knee joint**
- **image processing**

REFERENCES

1. Braun, H. J.; Gold, G. E. Diagnosis of Osteoarthritis: Imaging. *Bone* **2012**, *51* (2), 278–288.
2. Kubakaddi, S.; Ravikumar, K. M.; Harini, D. G. . Measurement of Cartilage Thickness for Early Detection of Knee Osteoarthritis (KOA). In *2013 IEEE Point-of-Care Healthcare Technologies (PHT)*; IEEE, 2013.
3. Mallikarjunaswamy, M. S.; Holi, M. S. . Knee Joint Cartilage Visualization and Quantification in Normal and Osteoarthritis. In *2010 International Conference on Systems in Medicine and Biology*; IEEE, 2010.
4. Kashyap, S.; Yin, Y.; Sonka, M. Automated Analysis of Cartilage Morphology. In *2013 IEEE 10th International Symposium on Biomedical Imaging*; IEEE, 2013.
5. Togni, S.; Appendino, G. Curcumin and Joint Health: From Traditional Knowledge to Clinical Validation. *Bioactive Food as Dietary Interventions for Arthritis and Related Inflammatory Diseases: Bioactive Food in Chronic Disease States*; 2012; p 67.
6. Chetty, G.; Scarvell, J.; Mitra, S. Fuzzy Texture Descriptors for Early Diagnosis of Osteoarthritis. *2013 IEEE International Conference on Fuzzy Systems (FUZZ-IEEE)*; IEEE, 2013.
7. http://www.ivysportsmed.com/en/knee-pain/knee-joint-functionTuceryan, M.; Jain, A. K. Texture Analysis, The-Handbook of-Pattern Rec. and Computer Vision, 1998.
8. Srinivasan, G. N.; Shobha, G. Statistical Texture Analysis. *Proc. World Acad. Sci. Eng. Technol.* **2008,** 36.
9. Bastos, L. de O.; Liatsis, P.; Conci, A. Automatic Texture Segmentation Based on k-means Clustering and Efficient Calculation of Co-occurrence Features. In *2008 15th International Conference on Systems, Signals and Image Processing*; IEEE, 2008.
10. Kumar, D. et al. Development of a Non-Invasive Diagnostic Tool for Early Detection of Knee Osteoarhritis. In *2011 National Postgraduate Conference*; IEEE, 2011.
11. Fripp, J. et al. Automatic Initialization of 3D Deformable Models for Cartilage Segmentation. In *Digital Image Computing: Techniques and Applications (DICTA'05)*; IEEE, 2005.
12. Blausen.com staff (2014). "Medical gallery of Blausen Medical 2014". *WikiJournal of Medicine 1* (2). DOI:10.15347/wjm/2014.010. ISSN 2002-4436
13. Dalvi, R. et al. Multi-contrast MR for Enhanced Bone Imaging and Segmentation. In *2007 29th Annual International Conference of the IEEE Engineering in Medicine and Biology Society*; IEEE, 2007.

14. Anjani, S. K. Automatic Texture Analysis of Cartilage for Early Detection of Osteoarthritis. *Int. J. Eng. Comput. Sci.* **2014**, *3* (5).
15. Cope, P. J. et al. Models of Osteoarthritis: The Good, the Bad and the Promising. *Osteoarthritis Cartilage* **2019**, *27* (2), 230–239.
16. Kundu, S. et al. Enabling Early Detection of Osteoarthritis from Presymptomatic Cartilage Texture Maps via Transport-Based Learning. *Proc. Nat.l Acad. Sci.* **2020**, *117* (40), 24709–24719.

CHAPTER 9

Sensor Cloud-Based Theoretical Machine Learning Models for Predicting Pandemic Diseases

PRASHANT SANGULAGI[1*] and ASHOK V SUTAGUNDAR[2]

[1]*Bheemanna Khandre Institute of Technology, Bhalki Karnataka 585328, India (Affiliated to Visvesvaraya Technological University, Belagavi Karnataka, India)*

[2]*Basaveshwar Engineering College, Bagalkot Karnataka 587102, India*

[*]*Corresponding author. E-mail: psangulgi@gmail.com*

ABSTRACT

Many pandemic diseases spread across the country or the whole world and affect the people as well as take their lives, for example, Cholera, severe acute respiratory syndrome, H1N1, Ebola, and Coronavirus disease 2019, etc. Research shows that the risk of pandemics has expanded during the past few decades owing to enhanced world travel and migration, urbanization, land use modifications, and greater natural environmental destruction. The increasing possibility of such a pandemic combined with the threat's predicted spread and death tolls increases the need to avoid this global epidemic before it happens. The only strategy to prevent a large-scale pandemic is to avoid the transmission of such an epidemic to and across living organisms. Nevertheless, mitigation allows the public health community to reliably anticipate what the suspect pathogen is to have the appropriate solutions to either deter the transmission of the novel infection to humans' altogether

Intelligent Interactive Multimedia Systems for e-Healthcare Applications. Shaveta Malik, PhD
Amit Kumar Tyagi, PhD (Eds.)

or control the disease as a community epidemic. This chapter presents the theoretical models to predict the pandemic diseases before they spread in the community or the whole world. Sensor cloud paradigm is used to continuously monitor the patient activity and sensed information is saved into a cloud server where various machine learning (ML) algorithms are used to find the type of disease and provide the appropriate medication to the affected person. Advanced theoretical ML models are proposed to get the information about the infection caused to the human body before it becomes a pandemic to the county or world.

9.1 INTRODUCTION

World Health Organization (WHO) describes pandemics as the widespread transmission of an infectious epidemic, but this term was granted to other historical diseases, such as the Black Death, focused mostly on the severity of the illness than global impact. Many pandemics arise as a consequence of contact with a novel bacterial virus, development, or reappearance.[1] Irrespective of the features of pandemics, they also appear to have significant effects on both the affected nations as well as the whole planet.[2] Such diseases have taken millions of lives, causing significant economic casualties, and interrupting various countries' growth and development. The rising probability of pandemics combined with the threat's predicted distribution and casualty figures generates the need to avoid the humanitarian crisis before it happens. The only approach to avoid a large-scale pandemic is to avoid the transmission of such an epidemic to and across human societies. However, prevention efforts the global public health society to accurately estimate what the cause of this problem virus is and provide the right information to either reduce the spread of the novel pathogen to humans entirely or comprise the infection as a regional epidemic. This chapter highlights the prediction algorithms required for detecting the type of virus//bacteria available and causing the human body, for example, severe acute respiratory syndrome (SARS), H1N1, Cholera, Ebola, Coronavirus disease 2019 (COVID-19).[3] Several major medical advances were used to diagnose individuals with communicable diseases, fears about a potentially deadly pandemic would not be overlooked. The pandemic of 1918 was not the first large-scale global epidemic in history that destroyed millions of citizens, indicating that equally serious outbreaks might arise again (presently COVID-19).[4] In fact, pandemics had broader social effects than simply the amount of diseases and fatalities.[5] Communicable diseases pandemics had already made their mark all across history by impacting the economy, society, and political framework of various countries and world organizations, changing the universe we currently know.

9.1.1 SENSOR CLOUD

Sensor cloud (SC) is a new trending paradigm in the present context where it combines two conventional technologies, wireless sensor network (WSN), and cloud computing (CC).[6] The information generated by the sensor nodes is processed and stored in the CC for better utilization of the services and combination can be used for a wide variety of applications. SC combines various networks with different sensing technologies and cloud infrastructure framework by enabling for cross-disciplinary systems that can span organizational types.[7] A SC collects and processes information from several sensor networks enables information sharing on big-scale and collaborates the applications on cloud among users. It integrates several networks with several sensing applications and CC platform by allowing applications to be cross-disciplinary that may span over organizational ranges.[8] SC enables users to easily gather, access, processing, visualizing, and analyzing, storing, sharing, and searching a large number of sensor data from several types of applications. These vast amounts of data are stored, processed, analyzed, and visualized by using the computational information technology and storage resources of the cloud.

WSN's technology innovation motivates various real-life technologies such as target detection, military surveillance, remote monitoring, inevitable surveillance, and a variety of other applications.[9] Some WSNs, though, are centralized to unique users. On either side, the information from single sensor networks can indeed be shared among different applications such as weather forecasting and remote monitoring. SC could play an important role in some of these situations in delivering the Sensors-as-a-Service software when fulfilling various device criteria by creating virtual sensors on a CC platform.

Architecture of SC as shown in Figure 9.1 comprises of physical WSN which intern has sensing nodes to sense the information and send it to the sink node (SN). It also consists of a gateway to pass the collected information to the SC server. At SC server, the information is analyzed, computed, stored, and given access to the requested users. The users can request for the information or they can access the stored information based on their requirements. Sometimes the information saved might be having emergency actions to be taken in that case administrator shall be given rights to take appropriate action. The processed information can be modified or changed according to the system requirements.

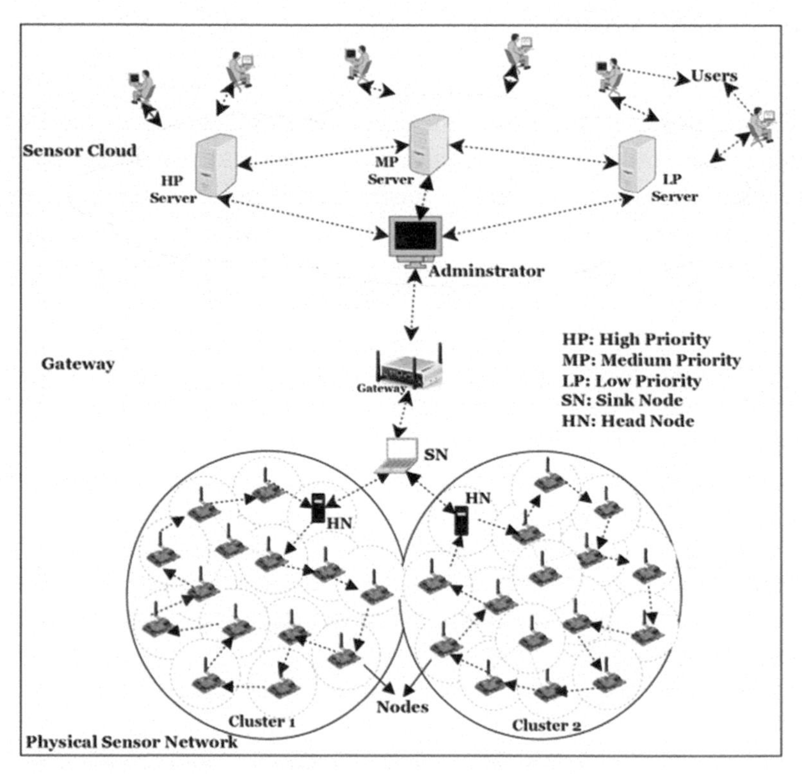

FIGURE 9.1 Sensor cloud architecture.

9.1.2 MACHINE LEARNING

Machine Learning (ML) is a branch of artificial intelligence (AI) and becoming popular for many applications especially healthcare application.[10,11] The purpose of ML commonly is to characterize the functionality of information and match that information into frameworks that can be acknowledged and implemented by users. While ML is a discipline under computer science, it varies from conventional computational methods. In conventional programming, algorithms are collections of directly coded instructions used by computers to quantify or come up with solutions. In skilled tutors, ML methods have been used to gain new information regarding describing their expertise and discover different methods of instruction.[12] They keep improving learning by watching regularly how well the students understand and generalize subjects or their laws. They utilize prior experiences to guide progressive accumulation, encourage mentors to adjust to different situations, and imply new information. The ML techniques rather allow for computers to train on available data and use statistical analysis to produce common characteristics in a limited range. Because of this, ML helps computers in modeling techniques from sample dataset to optimize decision-making processes derived from data input variables. The ML algorithms are broadly classified into three types such as supervised, non-Supervised, and reinforcement learning (RL).[13] It can also be categorized in terms of three factors such as classification, regression, and clustering. ML is a field that continues to expand. Over this, when people deal with ML methodologies, there have been some things to take into account or determine the impact of ML processes.

9.1.3 RECENT EPIDEMICS AND METHODOLOGY OF FINDING

Old techniques for the estimation of communicable diseases contain primarily differential equation forecasting models and time series forecasting mechanisms focused on statistics and random processes. The differential equation forecasting models are to create a differential equation that may represent the important features of bacterial infections as per world population patterns, infection frequency, and transmitting policies inside the community. Using qualitative and quantitative study and computational modeling of system mechanisms, the emergence mechanism of diseases is shown, the transmitting rules are identified, the transition and growth patterns are anticipated, the effects and major issues of the spread of disease are examined, and the appropriate preventive and control approaches are checked. Popular models for determining differential equations for communicable disease mechanics include general differential methods that explicitly represent the relation between the individuals' immediate rate of increase across each containment and its time among all containers. A popular assessment method when taking into consideration age distribution is the partial differential method. Delay differential system is indeed a sort of method of separating which happens whenever the scene framework is termed.[14] The classical differential equation forecasting model predicts that the cumulative number of citizens in a given environment is a fixed that would trigger the normal propagation mechanism of communicable diseases, define the evolutionary relationship between various nodes within the network with a period, and show the ultimate rule of communication of data. Yet, across time, the population is shifting. In regards to food, energy, and staying, there is still being a sort of contact with several other communities. The relation among persons is arbitrary, and the variation among persons being distributed is overlooked, thereby restricting the model's field of implementation.

The various enduring impacts on domestic and foreign communities that previous pandemics have had indicate several directions that large-scale epidemics can affect civilization. Although people exist in a "new" world, discrimination, prejudice, and racial prejudice occupy their minds. Pandemic diseases cannot distinguish such variations given the prevalence of classifying individuals depending on one's characteristics. The emergence in such a communicable disease thus has the power to fully rework very complexities in our modern culture. Without selecting and choosing who catches an illness does not imply that a pandemic can be faced similarly by all. Even modern technical methods may not eradicate the possibility of a large-scale pandemic that might destroy millions, provided advancements in fighting contagious diseases, especially when confronted with epidemic crises. A fresh challenge emerges with every move forward the public health system is taking in the fight against communicable diseases. While antibiotic production has helped reduce the risk of bacteria-caused bacterial pneumonia and diseases, these microbes are developing tolerance to several drugs widely used as medication. The emergence in antimicrobial resistance, partially prompted by the overuse and abuse in medications that destroy bacteria, has produced bacterial species.[15] Consequently, antibiotic-resistant bacteria are a growing global concern, triggering more lives year by year. Thereby, the growth of these microorganisms raises the chances of severe infection and deaths from viral infections such as plague and cholera, and secondary infections are sometimes seen by flu epidemics.

A theoretical framework for communicable disease pandemics is a good support for combining data from contextual and action modeling methods alongside the information in planning records to offer detailed recommendations on potential reaction choices. Planning is crucial to preventing an unexpected communicable disease pandemic and ultimately devastating effects on culture, but that's far from straightforward.[16] Decisions should be taken in quickly evolving, unpredictable circumstances, with little previous expertise throughout a pandemic. In a variety of nations even vaccination, a method used to eliminate both viruses and bacteria, had already lost momentum. Following an effective vaccination program led to the eradication of smallpox in 1980, the public health establishment assumed that vaccines were the secret to preventing all infectious diseases. Smallpox gave itself extermination not just because of its wide appeal and a brief moment of implantation but rather because people have been the only hub of the viral infection. Such variola viral features distinguish smallpox from other pathogens, such as polio and measles, which endure despite current eradication efforts.

The H1N1 pandemic of 1918 was believed to have triggered hundreds of millions of deaths globally. It is promising that the medications and vaccinations accessible to us currently would hopefully minimize the effects of a potential pandemic outbreak, but with cities around the world progressively linked by air transport, we are likely to experience a pathogen susceptible to quickly spreading throughout the world. The 2009 H1N1 pandemic, a virus believed to be little transmissible than the 1918 pandemic,[17] spreading to 74 nations by just four months. Computational and analytical simulations are essential instruments for preparing and reacting to pandemics. Even though it is doubtful that it would be able to anticipate exactly how or when the forthcoming pandemic would occur[1] after a possible epidemic has been detected, systems have tremendous potential to increase the efficiency of their responses. Individuals could be used to generate the observational research to improve spatial awareness, to determine the future trajectory of the disease outbreak and its associated socioeconomic effects and to implement a strategy for prevention. Such pathogenic traits may remain unclear at the start of a pandemic, and would, therefore, remain identified when they arise, because epidemics with very well-characterized diseases can vary enough in such steps to generate confusion as to the appropriate solution. When the knowledge of a pandemic's potential effects increases, decision representatives would also use those analytics to better determine the overall action plans, protection mechanisms to enforce, and when to apply them.[18] Considering the reliance of mitigation strategies and decision-making on spatial awareness measures, the collection of pertinent information at an epidemic as soon as feasible was established as a goal for monitoring and real-time information analytical techniques.[19] Unlike state and international health service agencies, the ultimate objective of only one publicly financed corporation is worldwide public safety against contagious diseases. Apparently, the WHO is the most government-supported global health body with representatives representing all continents. The corporation gives useful information and assistance via its international headquarters to every sovereign nation. Following developments in various community health awareness and innovation, over the last seven decades, the WHO has continued to develop, extending its roles and services to assist with global safety programs. As far as disease outbreaks are concerned, the WHO has been a leading determinant behind the attempts to eliminate smallpox, polio, and measles among several other pathogens, supplying both technical support and resources for global vaccines against such diseases. Most notably, the WHO has become the key corporation for mobilizing foreign funds and capital to tackle public health issues worldwide, particularly in places with smaller economies.

In this chapter, we develop the theoretical ML models to detect pandemics before they damage the human being. A decision model for disease transmission pandemics is an effective mechanism for incorporating evidence from contextual and intervention analytical techniques, along with the data in policy statements, to provide reliable suggestions on probable possible responses. Various additional features are added to ML algorithms for preparing the accurate model for detecting diseases. Multiple scenarios and conditions are to be considered to check whether the person is having flu or not. Various datasets are considered to train the machine and it helps to give the results accurately on time. The theoretical ML models considered here to predict the pandemic are neural network (NN), fuzzy logic (FL), and random forest (RF). All these ML algorithms are playing an important role in the computational world to maximize the output with greater accuracy. In SC, nodes are implanted on the human body or placed near to them to monitor the activities of the infected person. The sensor sends the information to the head node (HN) or SN in a timely manner and submitted to the cloud server. At the cloud server, the information is tested, analyzed and the decision is taken using ML algorithms. The results are stored in the cloud database and it can be shared with the concerned patient/hospital facility for further actions.

The chapter is organized as follows: Section 9.1 presents an introduction to SC, like the architecture of SC, working of SC, advantages of SC, applications of SC, introduction to ML, and recent pandemics. Section 9.2 describes the various ML algorithms. Section 9.3 depicts the proposed theoretical ML models for predicting pandemic disease in the present context. Section 9.4 depicts the prediction and discussion. Section 9.5 provides an open discussion regarding this work. Finally, Section 9.6 concludes the chapter in brief.

9.2 MACHINE LEARNING APPROACH

In general, ML is about discovering more valuable knowledge for particular issues from vast quantities of information utilizing its algorithm pattern. The ML comprises a wide variety of topics, including medication, engineering, statistical data, psychology, and so on.[20] For example, the NN can predict any high-dimensional, non-linear optimum representation among outputs and inputs by copying the computing process of the nervous system of the human biological neuron. The conventional statistical approach is not as successful when dealing with dynamic data relationships which do not provide reliable findings as the NN. ML is an information processing methodology that trains algorithms to do what happens easily to people and animals: know through practice. ML algorithms utilize statistical techniques to "read" knowledge from the data immediately without depending as a template on a fixed approximation. When the number of observations necessary for training improves, the algorithms adaptively enhance their efficiency. The sophisticated method of ML is deep learning. ML methods allow experimental knowledge of systems. ML relates to the capability of a framework to obtain and integrate information via large-scale predictions and also to broaden itself from gaining knowledge instead of programming this with that understanding. ML is a scientific discipline which has generated multivariate analysis-computational hypotheses about effective learning, constructed supervised learning which is frequently used during large enterprises, and floated.[21] ML methods are used in a wide variety of systems to address numerous interesting issues examples like healthcare, data analyst, image processing, robotics, mining, text analysis, and industry The advantages of the ML are, affective computing, improved decision making and prediction, quicker storage, accuracy, easy information planning, less costly, and dynamic large information analysis.[22]

ML is composed of three distinct forms, that is, supervised learning (regression and classification methods are classified in these datasets), unsupervised learning (not classified in these datasets and methods such as dimension reduction and clustering have been used) and RL (method in which system learns within each intervention) for the creation of ML solutions.[23] Often a majority of ML analysts and business geeks utilize supervised and unsupervised type ML. RL is also potent and complicated to implement for difficulties. The detailed explanation of all three types is described below.

(a) Supervised Learning

Supervised learning[24] is a form of ML that is used to train data from the datasets labeled. This helps one to forecast the performance for information to come or go unnoticed. Supervised ML constructs a model mostly in the face of disturbances that make accurate predictions based on observation. The algorithm requires a defined set of inputs and defined statistics (output) reactions and trains a prototype to create satisfactory predictions about future responses about the received information. Using supervised learning users recognize the information individuals have been expecting for the outcome. Supervised learning builds prediction models through the use of classification and regression methodologies. It is the most effective method of training NN and decision-making trees. All of these strategies rely heavily on the details provided by the pre-defined classifications. In the context of NN, the classification is often used to assess the error code and then modify the network to reduce it, and the classifications have been used in the decision trees to decide the features have the most knowledge which can be utilized to resolve the classification challenges. These both are looked at even more specifics, and in the meantime, it must be sufficient to conclude that several of these illustrations succeed in possessing some oversight in the shape of predefined classifications.[25] The working of supervised learning is as depicted in Figure 9.2.

Supervised learning builds predictive methods through the use of classification and regression strategies. Classification methods foresee distinct reactions. The steps involved in this type of learning process are, preparing data, steps to train, assessment or review phase, and deployment of production. Popular classification methods include support vector machine (SVM), enhanced and bagged decision trees, quadratic, k-nearest neighbor, Naïve Bayes, discriminant analysis, logistic regression, and NNs. Regression method predicts continuous replies. Popular regression models include linear models, nonlinear models, regularization, stepwise regression, enhanced and bagged decision trees, NNs, and adaptive neuro-fuzzy learning. When we are about to train a model to generate a forecast, use supervised learning, for instance, the potential value of variables, like temperature or market price, or grouping. Define car makers from camera video clips, for instance.

(b) Unsupervised Learning

Unsupervised learning looks a little tougher; the primary objective is to get the device to understand more about something we do not tell us how to do it! Currently, there are two methods of unsupervised instruction.[26] The first approach is to educate the agent not by providing categorizations directly, but by utilizing some type of incentive program to demonstrate progress. Remember that this form of training should ideally fit within the context of the judgment issue, as the aim is not to generate a distinction, but to decide things that optimize benefits. This strategy validates well to the physical world, where individuals may be praised for performing some acts and disciplined for just doing otherwise. Unsupervised learning has created numerous achievements, like world-champion caliber backgammon systems and even devices able to drive vehicles. This can be a useful method whenever there is a simple way of assigning attributes to decisions. The working of unsupervised learning is depicted in Figure 9.3.

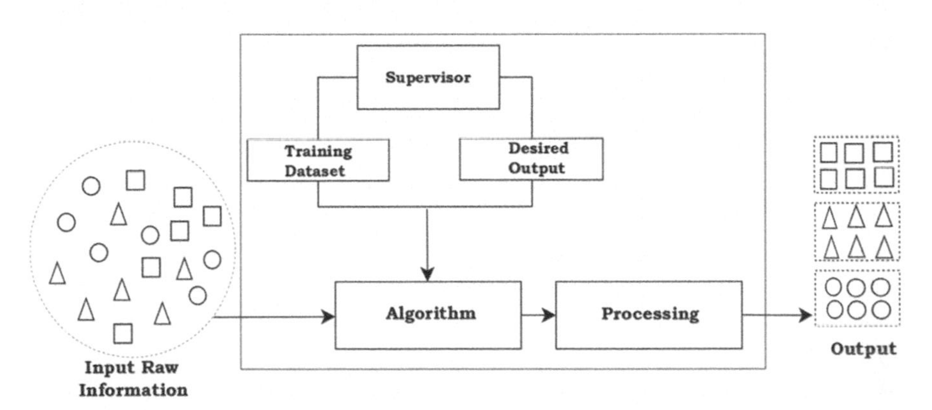

FIGURE 9.2 Supervised learning.

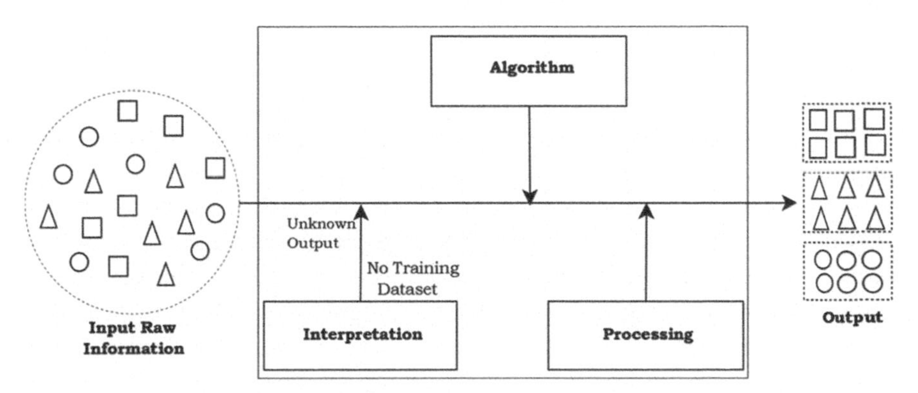

FIGURE 9.3 Unsupervised learning.

Whenever there is a need to discover someone's information and would like to train a model to determine a suitable internal view, select unsupervised learning, like dividing the population in clusters. Unsupervised learning techniques are developed to extract raw data from structures. An optimization problem that is generally greatly reduced to indicate optimum parameters labeling the obscured pattern in the data, describes the effectiveness of a configuration. Reliable and trustworthy estimation needs an assurance that derived configurations are representative of the source of the information, that is, identical configurations may be derived from a second collection of the same source for information. Shortage of reliability through statistics and the evidence on ML is regarded as over performing.

Clustering is one of the types of unsupervised learning and it is popular in its category. It is being used to identify patterns or gatherings in information for qualitative information analysis. Its applications like, genetic diversity assessment, business study, and entity detection software. Clustering may be helpful where there are sufficient data to create clusters and particularly where additional information regarding cluster participants may be used to provide added results regarding data dependency. Interpretation of the cluster is not always a particular methodology, however, the further to be fixed. Multiple algorithms, which vary considerably in one's definitions of what makes a cluster as well as how to discover them effectively could even obtain all these. Common cluster theories involve classes of limited gaps among participants of the cluster, large features of data space, periods, or unique empirical functions. Consequently, clustering could be developed as an issue of the multi-objective optimization process. The required clustering algorithm and variable configurations are based on the particular sample population and the tests expected to be used.

(c) Reinforcement Learning

Reinforcement learning is a form of ML that allows the learning model to analyze the scenario and learn the optimal behavior based on attempting to optimize certain sort of accumulated compensation.[27] A few of RL unique qualities are the learning which recognizes the atmosphere, chooses and takes specific activities, and receives benefits in payback. The module knows the plan or strategy that over term which maximizes its returns. The example for it is a technology that utilizes deep RL to recognize a machine from a single box in a production facility and place it in a tray. These are learned by the robot/agent through a benefits-based learning process that also encourages it to take the necessary decision. RL varies from supervised learning with not having to show labeled input/output sets, as well as not trying to directly rectify suboptimal behavior. The objective is on just achieving a balance among exploration and manipulation. An RL configuration is usually comprised of two parts, one agent and one environment as shown in Figure 9.4.

RL is all about sequential decisions. Simply developers may assume that the outcome depending on the present input status and the next input is depending on the outcome of the previous input. RL's applications are manufacturing automation and processing of information. RL greatly increases efficiency, enhances behavior, and defies minimum requirements.

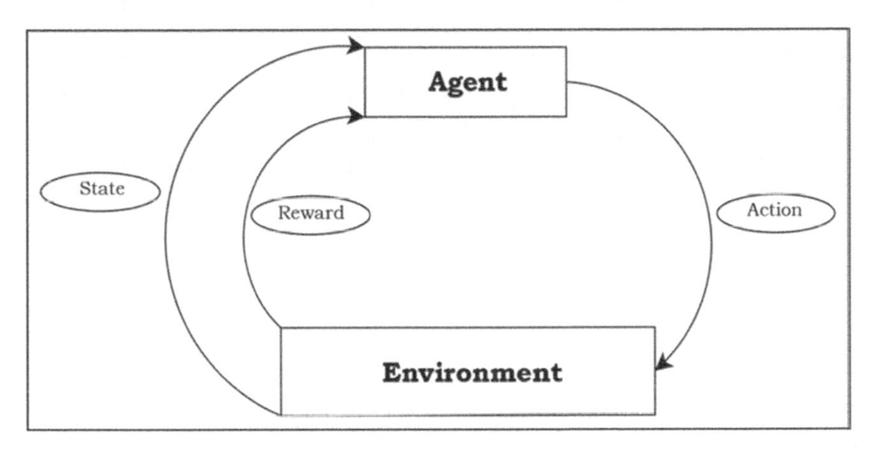

FIGURE 9.4 Reinforcement learning.

9.3 PROPOSED THEORATICAL MODELS

This section describes the proposed theoretical ML models for predicting pandemic diseases. The pandemic diseases have been considered here namely, cholera, COVID-19, Ebola, and H1N1. All three have quite similar characteristics and they have taken many lives when they are exposed all over the world.

This proposed work gives more accurate results about the causes using ML algorithms. The sensor nodes are used to sense the health condition of the infected person and send the sensed information to the sensor cloud server (SCS) where ML algorithms are applied to the input details to detect and recognize the type of cause attacked to the infected person and it also suggests the related medication to cure the cause. The proposed pandemic prediction model consisting of sensor network, gateway, and SC server is as shown in Figure 9.5. The sensor nodes are implanted on the human body to monitor the status of the infected person. Sensors collect the information like fever, blood pressure, immunity level, throat status, blood contamination, and diabetic level. This information is sent to SCS via a gateway. At the SCS, all the information fetched by the sensor are stored and sent to a model for analyzing the type of cause and which virus has affected the person. The results are analyzed with other models for conformation and finally, the results are displayed based on the input received from the patient.

Cholera: Cholera is an infectious illness that is commonly spread by contaminated water.[28] Cholera induces extreme vomiting and diarrhea. Left unaddressed, cholera could be dangerous within hours, including in people who were previously well.

Standard management of solid waste and water effectively prevented cholera in developing nations. Yet there is already cholera in Europe, in Southeast Asia, and Haiti. If starvation, violence or environmental disasters cause humans to live in cramped environments lacking sufficient hygiene, the likelihood of a cholera outbreak is strongest.

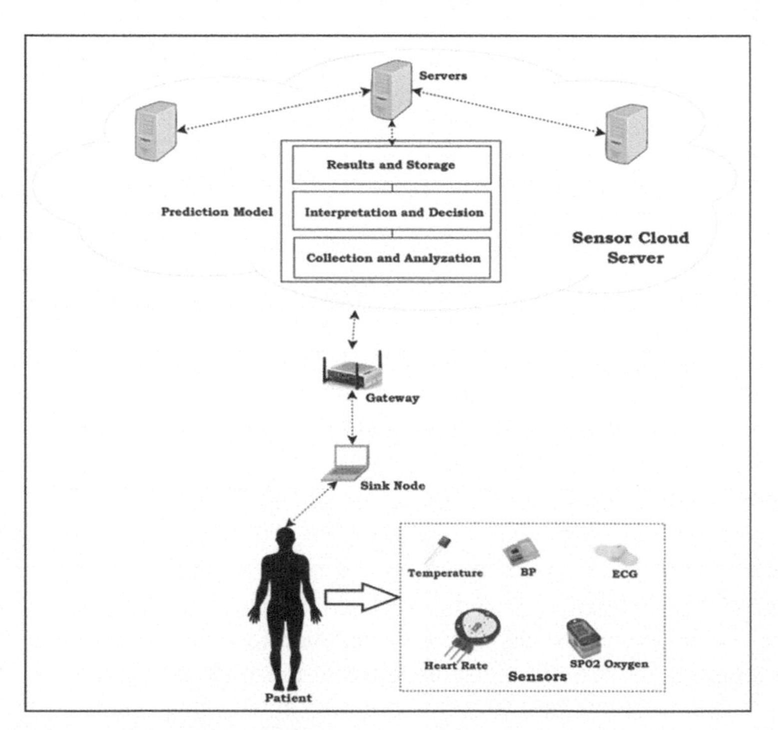

FIGURE 9.5 Proposed pandemic prediction model.

Cholera is simple to handle. An easy and affordable rehydration solution may avoid death from extreme dehydration. Many people are susceptible to the cholera bacterium do not get sick and also do not realize they got contaminated. Even once they have lost 7–14 days with cholera in the stool, individuals may also harm someone by water contamination.

Many forms of cholera triggering signs involve mild to severe diarrhea and it is also difficult to ascertain other than diarrhea induced by many other things. Some show quite severe Cholera signs and effects, typically in a few days after exposure.

COVID-19: Corona virus is another viral infection that may cause diseases such as colds, SARS, and respiratory disease.[29] A modern corona virus was reported in 2019 as the source of an epidemic of the infection that occurred in China.[30] The novel COVID-19 leaves a significant worldwide socio-economic impact because of the simplicity of transmitting the infection, mainly by droplets from the nose through an infected patient when he/she sneezes or coughs, highly urbanized nations have to be on a considerably high warning.

The infection is also classified as corona virus 2 (SARS-CoV-2), a serious respiratory infection disease.[31] The infection which it induces is named the 2019 corona virus disease (COVID-19). WHO declared the spread of COVID-19, a disease outbreak in 2020. COVID-19 early symptoms can show up 2–14 days after infection. This phase it is named the duration of incubation following exposure and before developing symptoms. Common causes and indications could include fever, runny nose, and fatigue. The extent of the effects of COVID-19 can differ from quite slight to serious. Many persons can still have certain signs, while other people do not have any signs at all. Many individuals may develop worse signs, such as worse breathing and pneumonia smallness, around a week after problems started. Elderly individuals have a greater chance of severe COVID-19 infection,[32] and the frequency rises with aging. Individuals who have known underlying medical problems may often be at increased risk.

Ebola: Ebola virus in humans and animals is a significant, sometimes deadly situation. Ebola is among the infectious hemorrhagic diseases triggered by a Filoviridae group infection, the genus Ebola virus.[33]

Ebola's death level differ bowing to pressure. The virus spreads through close interaction with contaminated livestock or a person's saliva, bodily fluids, and tissues. Severely ill patients need intensive care. Ebola virus disease also involves the sudden appearance of fatigue, severe exhaustion, abdominal discomfort, fatigue, and runny nose. Ebola continues to propagate progressively across family and associates, since they become subjected to bacterial secretions as they care about an infected person. The infection eventually travels to humans by close interaction with blood, bodily fluids, and membranes. Ebola, therefore, travels to others by close interaction with a patient's body fluids whose ill with and even deceased with Ebola. This may happen anytime a human encounters certain infectious body fluids (or things which are polluted with them), and the infection gets stuck in the eyes, nose, or mouth by torn mucosal membranes. Persons could get the infection via sexual relations with others who are infected with Ebola, even though after Ebola recovery. Upon healing from the disease, the infection may remain in other body fluids, such as semen. The period of Ebola exposure to symptoms presentation varies from 2 to 21 days. Ebola is a potentially life-threatening viral infection, which is often devastating. Healing from Ebola is based on adequate healthcare and the immune system from the individual. Reports suggest that Ebola virus exposure patients have antibodies that remain active in the bloodstream for up to 10 years post-treatment. Victims are believed to provide some defensive antibodies against the Ebola form they were sickened with.

H1N1: Technologically, the phrase "swine flu" applies to pig illness. Pigs sometimes spread viral infections to humans, mostly to pig owners and pediatricians. More commonly those with swine flu spread the virus to others.[34]

Researchers analyzed a unique type of flu virus identified as H1N1 mostly in 2009. Apparently, this virus is a mixture of pig, birds, and person viruses. In the influenza season 2009–2010, H1N1 triggered the severe respiratory illness widely known as swine flu. The WHO deemed the influenza triggered by H1N1 to be a worldwide pandemic since too many citizens across the globe were ill that year.

Flu virus H1N1 is infectious. The H1N1 flu virus is transmitted from one to another, and the virus is quickly transmitted between humans. This is thought to have the same expansion as normal influenza viruses. Patients infected with the H1N1 influenza virus could even contaminate someone else from 1st day before symptoms occur and up to 7 days more than upon illness. Whenever the infection reaches the organism via the eyes, nose, and mouth, infection is transmitted from individual to individual. Nausea and snoring expel the pathogens into the atmosphere, where someone can exhale them in. Even the infection will stay on flat floors such as door handles, all time money (ATM) keys, and surfaces.

New technologies are required to generate, maintain, and interpret broad data on the increasing network of people with the disease, clinical records, social activities and merge information with medical, clinical, genetic, and public health studies.[35] Different information sets, which include texts, E-mail interactions, social networking, and journal posts, may be of tremendous benefit in evaluating the development of group activity infection. Keeping this information together with ML and AI, investigators will predict how and when the infection is expected to propagate and alert certain areas to suit the preparations needed. Travel background of infectious subjects may be automatically monitored, to research observational associations with the infection transmission.

The different sensors like temperature, electro cardiogram (ECG), motion, BP, and arterial oxygen are implanted on the human body to check the parameters like temperature, stress, sleep, activity, respiration rate, heart rhythm, blood oxygen level, and blood pressure.[36,37] Along with sensory information, it is necessary to get the other physical situation of the infected person to provide accurate results of the disease caused. The models depicted below collects the information about the patient using sensors and physical situation. The collected information is sent to a predicting model to decide what kind of disease the patient is suffering from and also the model traces the person's recent travel history using the global positioning system (GPS) which was attached to affected persons bodies. It is necessary to check whether the person is affected by any pandemic disease or not because sometimes the person might be having a common fever along with cold and might not have any influenza. The model receives the important information from the data received by the person who is suffering from the mentioned symptoms and predicts what has caused him/her. The model also gives necessary precautions as well as the recent history of the person who is affected who finds the contacts and makes them aware of the disease and also avoiding the further spreading of it. The currently existing models are like SI model, SIS (Susceptible – Infectious – Susceptible) model, SIR (Susceptible – Infectious – Recovered) model, SEIR (Susceptible – Exposed – Infectious – Recovered) model, Autoregressive-Integrated Moving Average model (ARIMA), Exponential Smoothing method (ES), Markov Chain (MC) method, Grey Model (GM), some mathematical models include logistic, Gompertz, and Bertalanffy models.[14,32,38]

(a) Fuzzy Logic-Based Pandemic Disease Prediction

The ML models are constructed to provide a strong early forecast of the newly diagnosed cases and the times whenever the epidemic can stop. We suggest a platform to implement such models on SC data-centers to provide fail-safe computing and efficient information review. The healthcare facilities constantly give their optimistic report of patients in an SC-based system. The FL-based pandemic disease prediction is as shown in Figure 9.6. The sensory information from the patient is collected using wireless body area network technology along with non-sensory information like physical characters and behaviors which are non-sensible. Both information are sent to the fuzzy model where numerous membership functions with their rules have been defined.[39,40] Using an If-then argument usually reflects the knowledge in a fuzzy rule-based method.[41]

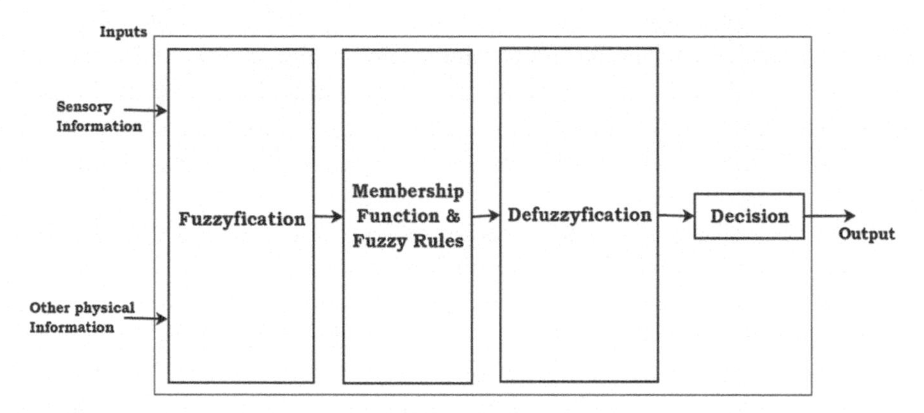

FIGURE 9.6 Fuzzy logic-based pandemic prediction.

FL contains two components: the predecessor part is the specific conditions defined as the inner variables, and the corresponding part that represents the outer variables. The decision-making process has faced several obstacles in engaging with real-world issues. The standard crisp theory is focused on conditional thinking, where everything is marked as either Yes (True) or No (False). FL can consider taking a value from 0 to 1 and all these attributes are degrees of complexity and individual adaptive decision-making styles. 0 is fake elements and 1 is real numbers. Fuzzy set is a type of description of data appropriate for notions that could not be described which rely on the experiences. They are indeed a crisp sets addition. A crisp set is a conventional way whereby complete membership or section membership is permitted for each feature. Now, the fuzzy membership function acts on the input information and provides the decision result which tells what has been caused to the patient. All the inputs have two-three subordinates (Ex: Temperature: High, Medium and Low & SpO2: Above 90 and Below 90) and among them, each subordinate have a different result and helps better in predicting what type of disease has caused to the patient. If the decision result says the patient has been attacked by some pandemic diseases then his/her recent travel history is traced to avoid further spreading of the disease. The patient always has an active GPS tracker with them to track their movements. The symptoms of various pandemic diseases are listed in Table 9.1. Based on the symptoms, membership functions are created and applied to the incoming input information. After fuzzification and defuzzification, the decision result is taken forward and sent to the SC database.

TABLE 9.1 Symptoms of Various Pandemic Diseases.

Symptoms of H1N1	Symptoms of cholera
• Cold • Cough • Sore throat • Runny nose • Body aches • Fatigue • Fever (Sometimes)	• Diarrhea • Dehydration • High heart rate • Loss of skin elasticity • Low BP • Thirst • Vomiting
Symptoms of Ebola • Fever • Body aches • Fatigue • Diarrhea • Vomiting • Stomach pain	Symptoms of COVID-19 • Fever • Cough • Sore throat • Increases resting heart rate • Increases respiratory rate • Decrease in sleep • Decrease in blood oxygen level

Example Scenario for Predicting COVID-19 Using Fuzzy Logic

The following example shows the prediction of COVID-19, ECG, heart rate, temperature, and SpO2 sensors are implanted on the patient body to continuously monitor their status. Along with sensory information, some physical characteristics are also taken into consideration like weakness, cough, sore throat, and body response to check whether the patient is affected by COVID-19 or not. The prediction method is as follows,

Begin

1. Collect the details about the patient (both sensory and non-sensory)
2. Check temperature (high, medium, low),
   If (temperature is high) then
   Go to next step else
   Go to step 8

3. Check cough status (common cough and dry cough)
   If (dry cough) then
   Go to next step else
   Go to step 8
4. Check oxygen level (below 90 and above 90)
   If (oxygen level below 90) then
   Go to next step else
   Go to step 8
5. Check respiration rate (high, medium, low)
   If (respiration rate is high) then
   Go to next step else
   Go to step 8
6. Check body status (weakness and normal)
   If (weakness is present) then
   Go to next step else
   Go to step 8
7. The patient has covid-19 and needs immediate medication
8. The patient has mild/low symptoms of covid-19 and needs home quarantine.

End.

For mild or low symptoms, it is not necessary to have medical treatment but mild symptom patients are required to stay in home at-least for 7 days as per the regulations given by the WHO. The positive tested patients need immediate medical treatment and their travel history must be traced to get the information about which all have come in contact with the affected person to avoid further spreading of the pathogen.

(b) Random Forest-Based Pandemic Disease Prediction

Random forest is much like the bootstrapping method for a decision tree model. The RF provides even more accurate forecasts in many cases than the standard regression analysis. RF is a guided algorithm used for classification as well as for regression. After all, this is mainly used with issues of classification. Since forests are made up of plants so more plants mean better woods. RF algorithms create decision trees on combinations of knowledge and then get the estimate from each and eventually select the best result by voting. A decision tree is the base of RF and an evolving sequence. Typically, the decision tree is being used in data mining techniques. A decision tree is a collection of yes/no questions concerning our knowledge which ultimately leads to an intended group. This is an understandable paradigm as it makes classifications the same as we are doing; until we make a judgment, we pose a set of questions regarding the knowledge they provide. RF running flow is as seen in Figure 9.7.

RF function is as follows : (1) Sorting of the RF from the specified set, (2) Creation of a decision tree for each test gathering predictive outcomes for each decision tree, (3) Polls for predictions; and (4) Predicting the outcomes with most poll findings is the final predictive outcome.

The advantages of RF are: (1) RF removes the effects of over-fitting by averaging the results of various decision trees, (2) RF functions nicely for a large information set than a single decision tree, (3) RF has lesser variance than a single decision tree, (4) RFs are flexible and exact, and (5) RF retains their accuracy. The limitations of RF are (1) Uncertainty always remains, (2) Development method takes longer than decision tree, (3) Estimation capital criteria are greater, and (4) Prediction mechanism is a time-intensive one. The Table 9.2 shows example scenario of the COVID-19 in which input values received from the different patients are stored and they are checked for detecting the type of disease caused to the patient.

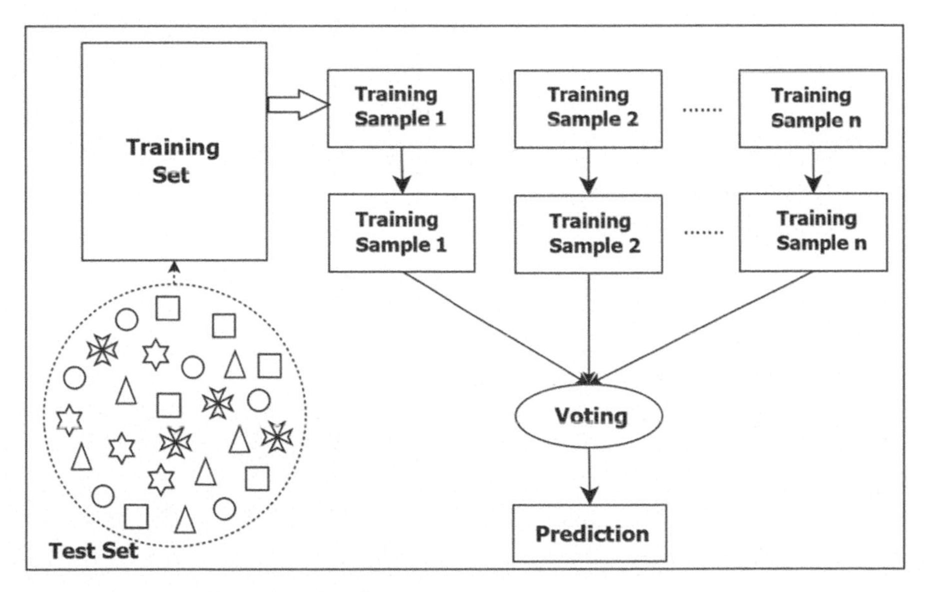

FIGURE 9.7 Working of random forest.

TABLE 9.2 Patient Input Information for COVID-19 Prediction.

Patient ID	Temperature (°C)	Cough (Yes or No)	SpO2 level (below 90 or above 90)	Heart rate (high or normal)	Weakness (yes or no)
1	38	Yes	Above 90	High	Yes
2	37	No	Above 90	Normal	Yes
3	37.5	Yes	Above 90	Normal	No
4	38.4	Yes	Below 90	High	Yes
5	39	Yes	Below 90	High	Yes
6	36.5	No	Above 90	Normal	No
7	36	No	Above 90	Normal	No
8	35.4	No	Above 90	Normal	No
9	36.9	No	Above 90	Normal	Yes
10	38.1	Yes	Below 90	High	Yes

The algorithm for predicting the pandemic disease using RF is given below. The information obtained from sensors and physical characteristics and sent to the model for predicting the pandemic disease. The information is trained before the samples are testing RF. The decision tree is formed based on the samples received and final result is obtained after classifying the raw input.

Nomenclature: I = Information, C = Clusters, IS = Same Information, HP = High Priority, MP = Medium Priority, LP = Low Priority, I_{HS} = High Sensitive Information, I_{MS} = Medium Sensitive Information, I_{LS} = Low Sensitive Information, R = Result.
Begin

1. SC server accepts I $\forall$ C
2. is I valid?
   If yes, go to next step
   Else accepts next I.
3. Send I to RF
   3.1 Consider $I_1 \in I$ sample and apply to decision tree
   3.2 Store outcome
   3.3 Consider next sample $\in$ I & repeat
   3.4 All samples finished?
   If yes, go to step 4
   Else repeat until last sample selected.

4. Obtain R from all outcomes of voting scheme.
5. Check R for I
   5.1 If I∈ I_{HS} then keep R in HP
   Else
   5.2 If I∈ I_{MS} then keep R in MP
   Else
   5.3 If I∈ I_{LS} then keep R in LP
   Go to step 9
6. Send another selected samples into a decision tree and go to step 5
7. Repeat step 5 for all samples
8. Are all samples are analyzed and classified?
   If yes then go to next step
   Else repeat step 5 until all samples are analyzed and classified.
9. Stop

End

(c) Programming Languages

This section explains some of the popular programming languages used for ML.

Python is a powerful object-oriented programmable language, with high-level programming capability.[42] It is trendy these days thanks to its simple-to-understand syntax and functionality capabilities. Python's acceptance may be attributed to the improved production of newly accessible deep learning models for that language, particularly TensorFlow, PyTorch, and Keras. As a language including understandable syntax and the potential to use this as a scripting tool, Python shows to be efficient and transparent in both pre-processing as well as directly working from the information. The scikit-learn ML library is developed on the basis of many current Python packages which might already be common to python programmers, notably NumPy, SciPy, and Matplotlib.

Java is commonly used for corporate computing, and is typically used by programmers of front-end software applications who often focus on enterprise-level ML. Java-ML is often a series of algorithms for ML and data-mining that strives to be an easy access and extendable API with both software developers and researchers.[43] The interfaces are made minimal with each implementation sort, and the implementations obey the corresponding interface specifically. Hence, it is simple to compare various classification models or clustering methods, and it is easy to incorporate evolutionary methods. The algorithm formulations are explicitly labeled, are well recorded, and can, therefore, be using it as a benchmark. Typically, it was not the first option for some of those fresh to the program who would like to know regarding ML, but to contribute to ML it is preferred by somebody with experience in Java. As far as ML applications in the enterprise are concerned, Java appears to be seen more toward computer security than Python, allowing use across malicious threat and vulnerability management situations. Java's ML library is Deeplearning4j, an open-source and shared deep-learning framework developed for both Java and Scala; MALLET (ML for LanguagE Toolkit) enables text-based ML applications, involving natural language analysis, topic modeling, information classification, and clustering; and Weka, a list of ML algorithms to be used.

R is also an open-source framework mostly used for quantitative computation.[44] In recent times, this had gained increasing popularity, and is preferred by most in research. R also is not usually used during production systems in manufacturing, but has been growing in commercial processes due to higher focus in information science. Common ML packages in R involve caret for building predictive designs, RF for classification and prediction, and e1071 for quantitative reasoning theoretical implementations.

C++ library being used for different uses such as large math tasks. C++ might render the ML systems very effective and quicker.[45] Despite the success of python, there are many fields whereby C++ surpasses over Python. For one aspect, C++ does have the benefit of becoming a strongly typed code because at compilation there would not have type mistakes occurring. The efficiency title even belongs to C++, as C++ makes runtime code more lightweight and quicker. C++ is the predominant language for ML and AI in computer or robotic application areas. In ML applications, integrated computer hardware designers and electronic technicians are always more likely to consider C++ or C because of the language skills and control strategy. Many ML libraries that could use on C++ provide the flexible mlpack, the broad variety of ML algorithms provided by Dlib, and the modular and open-source Shark.

9.4 PREDICTION AND DISCUSSION

Through systematically combining outcomes from situational and appropriate arrangements to pandemic response strategy, epidemics management capacities can be enhanced. Prediction of pandemic disease is a major concern in the present context. Finding the affected person and knowing their recent travel and his/her contacts is a mandatory task to avoid the further spreading of the pandemic disease. There are more ways to predict the disease caused to a particular person but accuracy is not achievable. With proper prediction models and proper information about the infected person, it is easy to find the total infected persons and also their criticalness. The theoretical prediction model also helps to find the total infected persons in a day/week/month as well as total death and total recovery. In our proposed method, we have developed two prediction models to check the epidemic caused to a particular person. A framework built using this method would guarantee that judgments have access to one of the most accurate, reliable information at functionally important moments in time. Using the total tests, the rate of cases can be calculated as,

$$\text{Case rate} = 100 \times \left(\frac{\text{Predicted total cases}}{\text{Total tests}} \right) \tag{9.1}$$

Also using the predicted total deaths, the expected death rate can be calculated as,

$$\text{Deat rate} = 100 \times \left(\frac{\text{Predicted total deaths}}{\text{Predicted total cases}} \right) \tag{9.2}$$

Total recovery rate can be calculated using the equation,

$$\text{Recovery rate} = 100 \times \left(\frac{\text{Total recovered}}{\text{Total predicted cases}} \right) \tag{9.3}$$

The prediction models defined in this chapter are fuzzy-based prediction models and RF-based prediction model. Both models are working well in predicting the type of disease that has been caused to the person who is suffering from illness. Both models give accurate results based on the input information considered (sensory as well as physical characteristics). The models work fine in finding many pandemic diseases like H1N1, Ebola, Cholera, and COVID-19 type of pandemics. One should get proper characteristics of pandemic diseases then it's very easy for the ML to predict the pandemic among infected patients and proper treatment can be given within a short time and also avoiding spreading of disease among common people.

The identification of pathogens and viral constituents through sensor/biosensors is a comparatively recent technique and it is still in its development for several of the methods explored. Existing approaches to virus tracking are time-consuming and costly, and the need for effective viral biosensors through quick detectors is facilitating the development. Due to inappropriateness in the field and the need for highly qualified workers, certain approaches are limited. Many have shortcomings concerning the composition of the sensor surface, repeatability, sustainability, and required processing of samples. To expand this extensive library of key steps into functional diagnostics, further research is required. There is a generalized basis for methods for detecting linking viral analyses to particular bio-receptors, but it needs improvement and further miniaturization in some instances. It is unavoidable that the methods mentioned will develop alongside emerging nanotechnologies with the emergence of metallic nanoparticles with optical and electronic resistivity and nanoparticle-based assays. In addition, in order to improve the efficiency, specificity and amplification of viral detectors, bio-receptor architecture and immobilization strategies are advancing. Sensors are becoming lighter and much more compact, and if less or more different innovations are used, prices will decrease.

The transmission of the infection can be tracked with effective and up-to-date monitoring mechanisms. When the agencies have details on the transmission of the virus, it is possible to make strategic decision, namely shutting down priority areas and blocking trouble spots. Growing steps for testing in nearby cities could minimize the negative consequences of the transmission of this infection with only rigorous and scheduled research. The SC platforms along ML can be used by government agencies to implement such systems, feeding information from such monitoring sensors and forecasting the number of instances in the coming days in almost real time. In addition, if it periodically updates the datasets and uses other demographic variables in the prediction architecture, such as population growth, climates, and age structure, for the last predicted case, it can render increasingly reliable and precise projections. This helps the officials to carry out the lock-down while holding a watch on the post-lockdown spike in cases. To prevent the spread of the epidemic, traveling and community activities have been limited worldwide. The new cases and fatalities could differ dramatically from the currently expected patterns as lock-downs are relaxed. Other variables will also impact the spread in the future, such as infection mutations. Continuous work is also needed to ensure that reliable forecasts are rendered and that appropriate steps can be taken.

The proposed work aimed to provide efficient model to predict the pathogen and protect the human being before it spreads and take more causalities. The models defined are efficient to detect the any pandemic diseases and also avoids its spreading.

9.5 AN OPEN DISCUSSION

The world has witnessed many pandemic diseases from past many centuries and till today the world is suffering from their effects (recently COVID-19). The causalities happened due to these pandemics are huge and affecting the mankind. Earlier due to non-availability of technology, the spreading of such pandemics was unrecognizable. But, now the scenario has changed and we have developed many advanced technology to recognize the acceleration of such epidemic. The spreading of disease among society can be restricted by including the prediction formula to detect the effected persons and stopping them before they spread through the society. Many tools are available to detect the affected person and simultaneously the required remedial can be given to cure the cause or spreading of same. The theoretical ML models considered here to predict the pandemic, which are FL and RF. These ML algorithms are playing an important role in computational world to recognize the infected person with their learning capability and produce the results with greater accuracy. The sensor nodes are implanted on the human body to continuously monitor the activity and sensed output is sent to SC server where ML algorithms are applied and results are obtained. The results are stored in the cloud database and it can be shared with the concerned patient/hospital facility for further actions. The methodology is most needed in-order to avoid the spreading of the pandemic which may harm the mankind.

9.6 CONCLUSION

In developing an accurate prediction for the probability of disease or illness, many variables may be recognized as crucial ingredients. The predictability including its population density and sequential period are paramount among others. Predictions must be related to intervention measures to enable them to be changed based on such behavior. Thus authorities responsible for preventing an epidemic may examine the efficacy of predictions, the efficiency of the reactions, and the amount of effort needed for an accelerated epidemic. This chapter has presented the theoretical models for predicting epidemic diseases like H1N1, Ebola, Cholera, and COVID-19 before they spread across the community and harm humanity. The SC has been utilized for gathering the necessary information from the patient to check whether a patient is affected by any pandemic disease or not. FL- and RF-based ML models are used to predict the epidemic disease and give results of the concerned before it spreads. The ML available in the SC server predicts the disease and shares the results to concern for further necessary actions. The models presented here give more accurate results and are very quick.

KEYWORDS

- **machine learning**
- **fuzzy logic**
- **sensor cloud**
- **random forest**
- **pandemic**
- **COVID-19**

REFERENCES

1. Holmes, E. C.; Rambaut, A.; Andersen, K. G. Pandemics: Spend on Surveillance, Not Prediction. *Nature* **2018,** *558* (7709), 180–182.
2. Mills, C. E.; Robins, J. M.; Lipsitch, M. Transmissibility of 1918 Pandemic Influenza. *Nature* **2004,** *432* (7019), 904–906.
3. Myers, M. F. et al. Forecasting Disease Risk for Increased Epidemic Preparedness in Public Health. *Adv. Parasitol.* **2000,** *47*, 309–330.
4. Tuli, S. et al. Predicting the Growth and Trend of COVID-19 Pandemic Using Machine Learning and Cloud Computing. *Internet Things* **2020,** 1–16.
5. DiNardo, A. Predicting Pandemics: Past Disease Outbreaks and What They Teach Us about Preparing for the Next Pandemic. *Senior Capstone Projects* **2019,** *917*, 1–82.
6. Alamri, A. et al. A Survey on Sensor-Cloud: Architecture, Applications, and Approaches. *Int. J. Distrib. Sensor Netw.* **2013,** *9* (2), 1–18.
7. Dwivedi, R. K.; Kumar, R. Sensor Cloud: Integrating Wireless Sensor Networks with Cloud Computing. In *2018 5th IEEE Uttar Pradesh Section International Conference on Electrical, Electronics and Computer Engineering*; IEEE, 2018; pp 1–6.
8. Sindhanaiselvan, K.; Mekala, T. A Survey on Sensor Cloud: Architecture and Applications. *Int. J. P2P Netw. Trends Technol.* **2014,** *6*, 49–53.
9. Kurata, N. et al. Actual Application of Ubiquitous Structural Monitoring System Using Wireless Sensor Networks. In *Proceedings of the 14th World Conference on Earthquake Engineering*, 2008; pp 1–9.
10. Choi, S.; Lee, J.; Kang, M. G.; Min, H.; Chang, Y. S.; Yoon, S. Large-scale Machine Learning of Media Outlets for Understanding Public Reactions to Nation-Wide Viral Infection Outbreaks. *Methods* **2017,** *129*, 50–59.
11. Yu, K. H.; Beam, A. L.; Kohane, I. S. Artificial Intelligence in Healthcare. *Nat. Biomed. Eng.* **2018,** *2* (10), 719–731.

12. Mannila, H. Data Mining: Machine Learning, Statistics, and Databases. In *Proceedings of 8th International Conference on Scientific and Statistical Data Base Management*; IEEE, 1996; pp 2–9.
13. Lison, P. An Introduction to Machine Learning. *Language Technology Group (LTG)* **2015,** *1* (35).
14. Wang, L. D. *Research on Epidemic Dynamics Model and Control Strategy*; Shanghai University, 2005.
15. Tenover, F. C. Mechanisms of Antimicrobial Resistance in Bacteria. *Am. J. Med.* **2006,** *119* (6), 3–10.
16. Sands, P.; Mundaca-Shah, C.; Dzau, V. J. The Neglected Dimension of Global Security—A Framework for Countering Infectious-Disease Crises. *N. Engl. J. Med.* **2016,** 1281–1287.
17. Mills, C. E.; Robins, J. M.; Lipsitch, M. Transmissibility of 1918 Pandemic Influenza. *Nature* **2004,** *432*, 904.
18. Lipsitch, M.; Finelli, L.; Heffernan, R. T.; Leung, G. M.; Redd, S. C. Improving the Evidence Base for Decision Making during a Pandemic: The Example of 2009 Influenza A/H1N1. *Biosecur. Bioterror* **2011,** *9* (2), 189–115.
19. Lipsitch, M.; Santillana, M. Enhancing Situational Awareness to Prevent Infectious Disease Outbreaks from Becoming Catastrophic. In *Global Catastrophic Biological Risks: Current Topics in Microbiology and Immunology*; Inglesby, T., Adalja, A., Eds.; Springer: Berlin, Heidelberg, 2019.
20. Huang, P. *Research and Implementation of Prediction Model for Class B Infectious Diseases Based on Machine Learning*; University of Electronic Science and Technology, 2019.
21. Woolf, B. P. Machine Learning. *Build. Intell. Interact. Tutors* **2009,** 221–297.
22. Albahri, A. S.; Hamid, R. A. Role of Biological Data Mining and Machine Learning Techniques in Detecting and Diagnosing the Novel Coronavirus (COVID-19): A Systematic Review. *J. Med. Syst.* **2020,** *44* (7).
23. Ayodele, T. O. Types of Machine Learning Algorithms. *New Adv. Mach. Learning* **2010,** *3*, 19–48.
24. Amit Kumar Tyagi, Poonam Chahal, "Artificial Intelligence and Machine Learning Algorithms", Book: Challenges and Applications for Implementing Machine Learning in Computer Vision, IGI Global, 2020.DOI: 10.4018/978-1-7998-0182-5.ch008
25. Al Hasan, M. et al. Link Prediction Using Supervised Learning. *SDM06: Workshop Link Analy. Counter-Terror. Secur.* **2006,** *30*.
26. Ghahramani, Z. Unsupervised Learning. In *Summer School on Machine Learning*; Springer: Berlin, Heidelberg, 2003.
27. Sutton, R. S.; Barto, A. G. *Reinforcement Learning: An Introduction*. MIT Press, 2018.
28. Colwell, R. R. Global Climate and Infectious Disease: The Cholera Paradigm. *Science* **1996,** 2025–2031.
29. Seshadri, D. R. et al. Wearable Sensors for COVID-19: A Call to Action to Harness Our Digital Infrastructure for Remote Patient Monitoring and Virtual Assessments. *Front. Dig. Health* **2020,** *2*, 1–8.
30. Wang, C.; Horby, P. W.; Hayden, F. G.; Gao, G. F. A Novel Coronavirus Outbreak of Global Health Concern. *Lancet* **2020,** 470–473.

31. Maier, B. F.; Brockmann, D. Effective Containment Explains Sub-Exponential Growth in Confirmed Cases of Recent Covid-19 Outbreak in Mainland China. *Science* **2020,** *368* (6492), 742–746.
32. Jia, L.; Li, K.; Jiang, Y.; Guo, X. Prediction and Analysis of Coronavirus Disease 2019. *arXiv preprint arXiv: 2003.05447*, 2020.
33. Petrosova, A. et al. Development of a Highly Sensitive, Field Operable Biosensor for Serological Studies of Ebola Virus in Central Africa. *Sens. Actuat. B Chem.* **2007,** *122* (2), 578–586.
34. Yang, S.; Wu, J.; Ding, C.; Cui, Y.; Zhou, Y.; Li, Y. Epidemiological Features of and Changes in Incidence of Infectious Diseases in China in the First Decade after the SARS Outbreak: An Observational Trend Study. *Lancet Infect. Dis.* **2017,** *17* (7), 716–725.
35. Zhao, S.; Lin, Q.; Ran, J.; Musa, S. S.; Yang, G.; Wang, W.; Lou, Y.; Gao, D.; Yang, L.; He, D. et al. Preliminary Estimation of the Basic Reproduction Number of Novel Coronavirus (2019-ncov) in China, from 2019 to 2020: A Data-Driven Analysis in the Early Phase of the Outbreak. *Int. J. Infect. Dis.* **2020,** *92*, 214–217.
36. Dinis, H.; Mendes, P. M. Recent Advances on Implantable Wireless Sensor Networks. *Wireless Sensor Networks-Insights and Innovations*; Intech Open, 2017.
37. Darwish, A.; Hassanien, A. E. Wearable and Implantable Wireless Sensor Network Solutions for Healthcare Monitoring. *Sensors* **2011,** *11* (6), 5561–5595.
38. Li, Z. L.; Zhang, L. M. Mathematical Model of SARS Prediction and Its Research Progress. *J. Math. Med.* **2004,** *17* (6), 481–484.
39. Mei, S. et al. Individual Decision Making Can Drive Epidemics: A Fuzzy Cognitive Map Study. *IEEE Trans. Fuzzy Syst.* **2014,** *22* (2), 264–273.
40. Arji, G. et al. Fuzzy Logic Approach for Infectious Disease Diagnosis: A Methodical Evaluation, Literature and Classification. *Biocybernetics Biomed. Eng.* **2019,** *39* (4), 937–955.
41. Ross, T. J. *Fuzzy Logic with Engineering Applications*; John Wiley & Sons, 2005.
42. Raschka, S. *Python Machine Learning*; Packt Publishing Ltd, 2015.
43. Witten, I. H.; Frank, E. Data Mining: Practical Machine Learning Tools and Techniques with Java Implementations. *ACM Sigmod. Record* **2002,** *31* (1), 76–77.
44. Bischl, B. et al. MLR: Machine Learning in R. *J. Mach. Learn. Res.* **2016,** *17* (1), 5938–5942.
45. Curtin, R. R. et al. MLPACK: A Scalable C++ Machine Learning Library. *J. Mach. Learn. Res.* **2013,** 801–805.

PART 3

Applications for Intelligent and Automated Healthcare Systems

CHAPTER 10

Applications of Machine Learning Algorithms in Fetal ECG Enhancement for E- Healthcare

YOJANA SHARMA[1], SHASHWATI RAY[2], and OM PRAKASH YADAV[3*]

[1]*Department Electronics and Communication Engineering, Bhilai Institute of Technology, Durg 491001, Chhattisgarh, India*

[2]*Department Electrical Engineering, Bhilai Institute of Technology, Durg 491001, Chhattisgarh, India*

[3]*Department Electrical and Electronics Engineering, PES Institute of Technology, Shivamogga 577204, Karnataka, India*

[*]*Corresponding author. E-mail: omprakashelex@gmail.com*

ABSTRACT

The fetal ECG (FECG) records the electrophysiological activity of the fetal heart and is considered as a useful tool for detecting fetal abnormalities during fetus growth. However, the recorded signals are often contaminated by various artifacts during acquisition. The vernix caseosa causes electrical isolation, which actually shrinks the signal amplitude, resulting in a poor signal-to-noise ratio of the FECG signal. Even the 3D morphology of FECG is quite challenging as the recorded FECG depends upon the position and shape of the uterus, fetal size, and presentation. The domain of FECG processing is attractive to many researchers as there is a lack of trustworthy exemplary databases with proficient interpretations and standard assessment tools for assessing the algorithms. Additionally, due to the lack of a

Intelligent Interactive Multimedia Systems for e-Healthcare Applications. Shaveta Malik, PhD Amit Kumar Tyagi, PhD (Eds.)

sufficient number of cardiologist/radiologist and costly FECG recording device, this domain is gaining popularity in e-healthcare services. Machine learning (ML) algorithm, because of its adaptive learning approaches, find vast applications in FECG analysis. Algorithms like support vector machine (SVM), K-nearest neighbor (KNN), and Bayesian network are usually utilized for the detection of FECG from multichannel abdominal leads, whereas random forest algorithm helps in FECG acquisition. ML algorithms are also utilized for the classification of FECG signals using decision trees, linear and nonlinear regression models, principal component analysis (PCA) with Gaussian mixture models, and linear discriminant analysis. ML algorithms have been also used as a diagnosis tool to predict preterm births, fetal hypoxia, fetal heart rate, uterine contractions, and other chromosomal aneuploidies. Deep learning algorithms which are subsets of ML are gaining more popularity in this domain and are used for FECG signal enhancement, classification, and diagnosis purpose. In this chapter, firstly, we introduce FECG, their extraction methods, available databases for e-healthcare, and assessment tools. Secondly, different types and existing ML algorithms for FECG signals are explored. Finally, the outcomes of ML algorithms on fetal ECG enhancement for e-healthcare are presented.

10.1 INTRODUCTION

Advances in modern medical devices provide useful information for better healthcare services. The healthcare industry is major source of revenue and employment. In India, it is growing rapidly because of its extension of coverage, services, and additional investments by both public and private sectors. Heavily populated countries like India is planning to spend 2.5% of the country's GDP by 2025 for providing better health services to the citizens. Healthcare market in India is proposed to reach 372 Billion USD by 2022 to provide better health services. In continuation to that information, Indian Government has launched World's largest funded healthcare scheme, Ayushman Bharat on 23 September 2018.[2,39]

Healthcare services and healthcare industry mainly depend upon the quality, value, and result they are offering. Technology-driven smart healthcare is no longer a flying flight, as we know that medical devices connected to the Internet can simultaneously maintain the healthcare system and prevent the population from falling. At the same time, however, large amounts of data are generated and require a special type of observation that takes time. Development of advanced analytical tools and machine learning (ML) algorithms has done remarkable job to interpret voluminous amount of data even up to significant level of accuracy. ML significantly reduces the efforts by saving time. Also, it is considered as a cost-effective method as it can perform multiple tasks, such as feature extraction, preprocessing, analysis, optimization, and data fitting simultaneously. Thus, the results obtained from ML algorithms are found to be more accurate and validated which can further be used by organizations to develop statistical models even for real-time operations. Thus, these tools would ease the task of physicians to diagnose the diseases and they would also help in their research. Since, the ML models can analyze the medical data in multidimensional way, it helps physicians to improve the healthcare services. These tools also give patient satisfaction and provide an impartial opinion to improve performance, reliability, and precision.[22,27]

Since, developed algorithms of ML concepts are to be used in healthcare services over human beings, care must be taken to ensure its effectiveness. Also, these algorithms must not mislead the diagnosis. Thus, a systematic procedure for ML models is required to be practiced on routine basis in clinics. However, the ML algorithms produce more accurate results over large dataset. These algorithms may also be applied to small datasets.

To improve the effectiveness of treatment options, ML algorithms are now integrated with real-time data available from different healthcare systems. The available applications of ML for healthcare services include identification of disease and diagnosis, manufacturing of drugs, smart health records, diagnosis of medical imaging, personalized medicine and clinical trials, etc.[4] Existing models that revolutionize the medicine industry include QUOTIENT HEALTH which is used to reduce the cost of electronic medical record, KENSCI to predict illness, and it also offers treatment options, CIOX HEALTH is the health management system, PATHAI is used for pathological purpose, InnerEye to detect tumors, BETA BIONICS to manage sugar levels. Details about these healthcare devices may be obtained from web.[1]

Although healthcare industries based on ML models are providing better services to both patients and physicians, hardware models for automatic analysis of FECG signals are still untouched.

10.2 FETAL ELECTROCARDIOGRAPHY

India's healthcare system features a universal healthcare system in which a large number of public hospitals and primary health centers (PHCs) have been developed across India. Although there is a significant difference between rural and urban healthcare systems, the National Rural Health Initiative sought to address the gap and aimed at delivering efficient healthcare to rural communities as well.[47]

The Neonates Death Rate, which is considered to be a crude sign of the overall healthcare services and systems available for neonates in India, is 33 neonates deaths per thousand alive births for the year 2017. This figure is much lower, that is, 129 neonates deaths per thousand alive births in the year 1971.[21] Although Neonates Death Rate has declined, but it cannot be ignored and special thrust for healthcare service to neonates needs to be enhanced. According to the reports of WHO 2012, the Congenital Heart Anomalies, that is, heart defects in neonates was approximately 1 in 33 neonates which resulted in almost 3.2 million birth defects-related disabilities per year.[45] Therefore fetal heart rate monitoring is acute and can predict fetal asphyxia early in acidosis evolution.

Through this chapter, we bring attention of readers towards e-healthcare services for fetus heart analysis, especially over FECG which is often used to obtain significant information, such as heart rate, waveform, fetal development, fetal maturity, congenital heart disease, etc.

Among the most prevalent birth defects are heart defects. In the development of neonates, cardiac diagnosis is invaluable since it provides an opportunity to handle the baby's condition during the stages of antepartum and intrapartum. It is only from the antepartum stage, one can actually detect the FECG signal.

The heart is the very first organ which is formed in the fetus, even its development at the early stages of pregnancy is significant. The most important time is after fertilization, that is, from 3 to 7 weeks. The heart is believed to start beating by life's 3rd week. After 7th–9th week the fetal heart can be externally controlled. After 20th week, heart sound can be recorded easily without external support.[25,52]

FECG is a graphical representation of electrical activity of fetus heart and it is characterized by a series of waves viz., P, Q, R, S, T and sometimes U waves.[31,55] An irregular FECG can mean that the fetus does not get sufficient oxygen nor may suffer from other problems. Often an irregular pattern may lead to an emergency or caesarean delivery.[32]

FECG measurement can be done thoroughly in two different ways: invasive (direct) method and noninvasive (indirect) method.[7] Figure 10.1 shows the direct and indirect method of measurement of FECG signals.

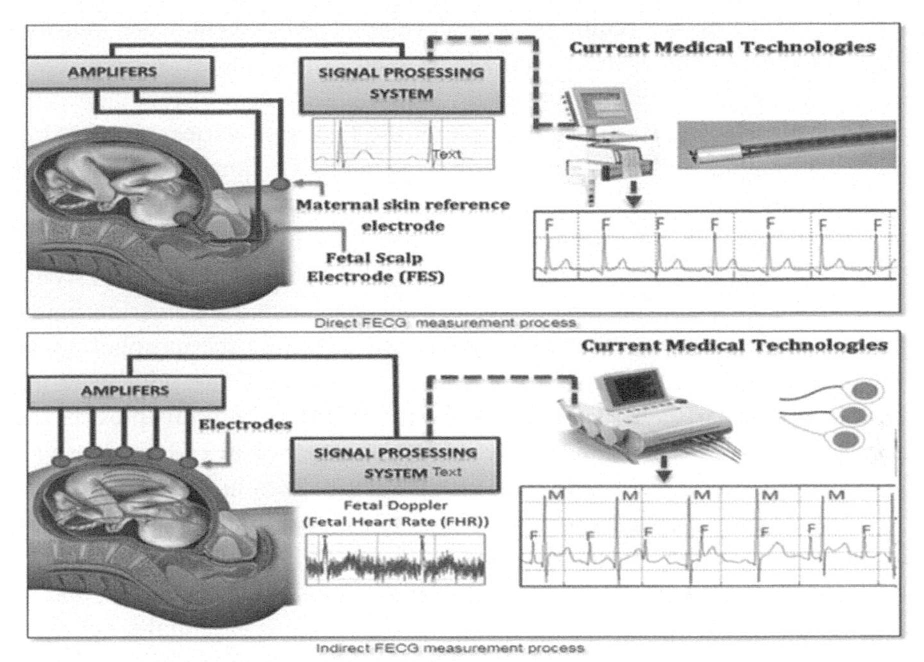

FIGURE 10.1 Direct and indirect measurement processes involved in EFCG determination.

Indirect way (noninvasive) of FECG measurement is the extraction of FECG recorded on the mother's abdomen. Abdomen signal includes MECG (mother ECG), FECG and noise signal. The MECG signal is 5–20 times larger in amplitude and always superimposes FECG. Common noise signals affecting FECG signals are powerline interference which is due to power supply, muscular noise is due to the movement of muscles at the time of recording, electrodes noise is due to electrodes placed on abdomen. Baseline noise is a shift in baseline drift and noise due to recording system. The indirect FECG measurement technique is the preferred method[3,51] because it is safe as compared with direct method. However, the quality of the FECG signal is poor as it is superimposed with noises of variable frequency and magnitude. The electrode should, in direct way, move through the mother's abdomen and reach the womb to meet the head of the fetus. Although the method offers pure FECG signal (with little interference), this can cause problems for both mother and the fetus, and the method is intrusive unfortunately. Popular direct methods include scalp electrode (spiral, clip or suction) method.[36,59] The present healthcare systems would perform signal processing task successfully to some extent. However, systems to analyze the FECG signal completely are required to be developed. Also, healthcare industry should focus over measurement of FECG even from home.

Morphologically, adults and fetus have similar pattern except their relative magnitude. The relative amplitude of the fetal also varies considerably during gestation and also after birth.[15] The FECG signals detected and extracted from the maternal abdomen typically give an undesirable signal-to-noise ratio (SNR) resulting in poor detection of the fetal heart rate.[17]

FECG signals are highly nonstationary signals due to their time-varying statistical character. The various types of noises, such as, powerline interference maternal noise, electrode contact, etc. affecting the morphology of FECG are explained in Clifford et al.[19] These noises along with maternal ECG and other artifacts degrade the quality of FECG and thus prevent the correct analysis and diagnosis. The FECG strength of FECG signals with respect to mother's ECG and other artifacts can be seen in Figure 10.2.

It is also important to extract the FECG from the polluted signal to obtain accurate fetal status information and to detect anomalies that will further help ensure fetal well-being, to confirm if the fetus is alive or dead, and to assess the presence of twin pregnancies. Therefore, it is essential for clinical diagnosis as well as for biomedical applications to extract the FECG signals from noisy surroundings.

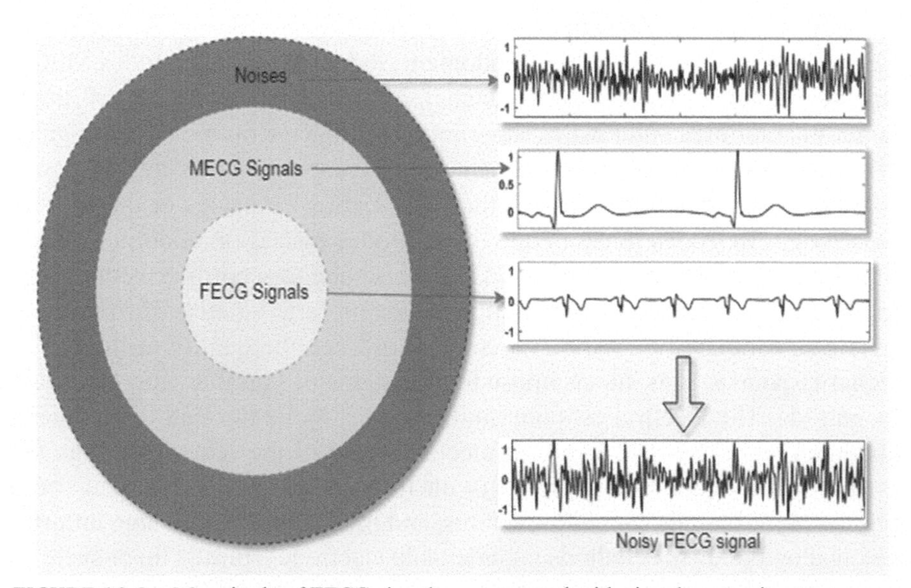

FIGURE 10.2 Magnitude of FECG signals as compared with signal contaminants.

The amount of data generated by existing FECG machines is also enormous. In critical cases, analyzing this big data is too tedious for physicians. ML algorithms can be effectively and efficiently utilized to extract clinically important features from FECG signals. So, advanced healthcare systems are required to extract FECG characteristics from noisy FECG signals. Also, systems are required to be developed that can be easily accessible even to remote locations for the measurement of FECG signals. Even systems that can automatically determine the fetus status are required to be developed.

10.2.1 ONLINE AVAILABLE FECG DATABASE

The primary requirement to work in the field of signal analysis is the collection of data over which algorithms are to be tested. Also, selection of suitable data plays very important role for validation of the algorithms, that is, if the algorithm is to be developed for disease identification, then different data are required, whereas to reduce the artifacts from the signal, a different set of data is required. Accordingly, depending upon the objective, suitable data are to be collected. Also, length of data, sampling rate, resolution, noise levels, and data format play equally important role for algorithm validation. Although the different algorithms asserted their robustness for FECG analysis, these algorithms were not evaluated quantitatively as there is a lack of popular databases with adept elucidations and methodology for evaluating the algorithms. Also, one has to depend upon online database as it is difficult to get signals from hospitals/clinics because of either their privacy policy/ government rules.

The available FECG data for research in this domain may be obtained from:

1. The FECGSYNDB is a broad database of simulated adult and noninvasive fetal ECG (NI-FECG) signals, providing a comprehensive resource that allows for reproducible field research. The data are created through the simulator FECGSYN.[8] Individual signals are recorded from 34 channels, with addition of five different levels artifacts, for seven different cases with 5 times repetition. The dataset is suitable for signal restoration algorithms, fetal heart rate measurement, and knowing the effect of uterine contractions.
2. One set of The Abdominal and Direct Fetal Electrocardiogram Database contains signals, four maternal abdomen signal and one direct signal recorded from the fetal head during 39 to 42 weeks of

gestation.[37] The signals are recorded at a sampling rate of 1 KHz with 16-bit resolution. The dataset is mainly utilized for the extraction of FECG signal from the abdominal maternal ECG.

3. The Noninvasive Fetal ECG Arrhythmia Database contains noninvasive set of normal and abnormal FECG signals. These signals are sampled either at 1 KHz or 500 Hz. The dataset is suitable to determine the fetal heart rate of regular and irregular cardiac rhythm.[14]
4. The Abdominal Indirect Fetal Electrocardiogram recording consists of a set of 55 multichannel recordings. These recordings are recorded from a single patient between 21 and 40 weeks of pregnancy. The duration of these records is variable and these recordings are done at an interval of 2 weeks. These records may be extremely helpful for testing the signal separation algorithms.[28]
5. The DAISY database, which is a part of SISTA (Signals, Identification, System Theory, and Automation) database from the department of Electrical Engineering of Belgium's Katholieke Universiteit Leuven data, has abdominal FECG signals consisting of eight channels, out of which three are thoracic, four are abdominal, and one reference channel. The duration of individual channel is 10 s. The sampling rate used is 250 Hz.[20] These data are suitable for FECG extraction algorithm.

ML algorithms may be implemented on any of the data set. However, to derive any conclusion for any algorithm, either the same data set or different may be used. Also, duration of the samples, sampling rate, level of noise added, type of noise added, and bit resolution must be included for any concrete decision.

10.2.2 PERFORMANCE ASSESSMENT TOOLS FOR FECG EXTRACTION METHODS

The efficiency of any algorithm depends upon the way the data are utilized. It also depends upon the objective of the algorithm. Here, we present the standard performance assessment tools for the measurement of quality of the extracted FECG from composite signals consisting of mother, fetus, and noises.

We can evaluate the output of an FECG extraction algorithm by estimating the similarity of the restored/reconstructed signals with those original/actual signals.

Let us consider x(n) and y(n) as discrete samples of original and reconstructed samples of FECG of length N.

Mean Absolute Error (MAE): This parameter provides mean of absolute difference between the original and reconstructed FECG signals.[16]

$$MAE = \frac{1}{N}\sum_{n=1}^{N}\left|x(n) - y(n)\right| \quad (10.1)$$

The Root Mean Square Error (RMSE): This assessment tool provides mean of the square errors between the original and reconstructed FECG signal. This tool is useful when large errors are particularly undesirable.[60]

$$RMSE = \sqrt{\frac{\sum_{n=1}^{N}\left|x(n) - y(n)\right|^2}{N}} \quad (10.2)$$

The Percentage Root Mean Square Difference (PRD): This assessment tool is used to find out the deformations in the reconstructed FECG signal. This is calculated as the sum of square errors. And this is always expressed in (%).[16]

$$PRD(\%) = 100\sqrt{\frac{\sum_{n=1}^{N}\left|x(n) - y(n)\right|^2}{\sum_{n=1}^{N} y(n)^2}} \quad (10.3)$$

The Normalized PRD (PRDN) is the normalized version of PRD.[16]

$$PRDN(\%) = 100\sqrt{\frac{\sum_{n=1}^{N}\left|x(n) - y(n)\right|^2}{\sum_{n=1}^{N}\left[y(n) - \bar{y}(n)\right]^2}} \quad (10.4)$$

where $\bar{y}(n)$ is the norm of the reconstructed FECG signal. PRDN is generally chosen to remove the baseline that is additional to FECG signals for storage purpose.

Correlation Coefficient (CC) measures the degree to which two signals are associated in terms of their morphology.[18]

$$CC = \frac{N(\Sigma xy) - (\Sigma x)(\Sigma y)}{\sqrt{\left[N\Sigma x^2 - (\Sigma x^2)\right]\left[N\Sigma y^2 - (\Sigma y^2)\right]}} \quad (10.5)$$

Signal-to-Noise Ratio (SNR) assesses the signal over noise, and it is measured in the root mean square sense.[16] The noise present in any signal may be evaluated at two different stages, that is, before processing SNR*o* of original noisy signal and after processing (SNR*r* of reconstructed signal).

$$SNR_o\left(dB\right)=10log_{10}\left(\frac{\sum_{i=1}^{N}\left[y(n)\right]^2}{\sum_{i=1}^{N}\left[y(n)-y'(n)\right]^2}\right) \tag{10.6}$$

$$SNR_r\left(dB\right)=10log_{10}\left(\frac{\sum_{i=1}^{N}\left[y(n)\right]^2}{\sum_{i=1}^{N}\left[y(n)-x(n)\right]^2}\right) \tag{10.7}$$

$$SNR_{im}\left(dB\right)=SNR_o\left(dB\right)-SNR_r\left(dB\right) \tag{10.8}$$

The performance assessment parameter MAD, RMSD, PRD, and PRDN are negatively oriented performance tool parameters and thus lesser value is desired. It means reconstructed signals having lesser value are more close to the original signal. The CC and SNR are positively oriented tools and are expected to be high. Algorithms producing high SNRim are more capable of extracting FECG signals from the maternal ECG signals or from noisy signals. The CC near to 1 indicates closer resemblance between the original FECG and the reconstructed signals. Equation 10.1 to 10.8 provides a clear understanding of above defined parameters in detail.

These performance parameters are mostly found in the literature to assess ML algorithms developed for the extraction of restoration of FECG signals.

10.3 DEVELOPMENTS IN MACHINE LEARNING

ML is the most common branch of AI that predicts the future by automatically learning and improving from experience, that is, ML algorithms use input data and use it for learning and training to analyze data behavior to autonomously predict the future.[5,46] Figure 10.3 shows the progress in ML techniques and intelligence offered by the algorithms.

Data are the central backbone of ML systems, and has thus opened up a huge opportunity for healthcare. The ML algorithms have given another dimension to the way we perceive information because of its ability to learn efficiently, systematically and quickly from data. This property of ML algorithms makes them an evolving tool for e-healthcare. All ML models are accessed through accuracy and if the accuracy is within the acceptable limits, the ML model is applied, else the model is trained again and again with the training data with different model parameters. Optimization of the models is also required to produce accurate results.[23,30,50]

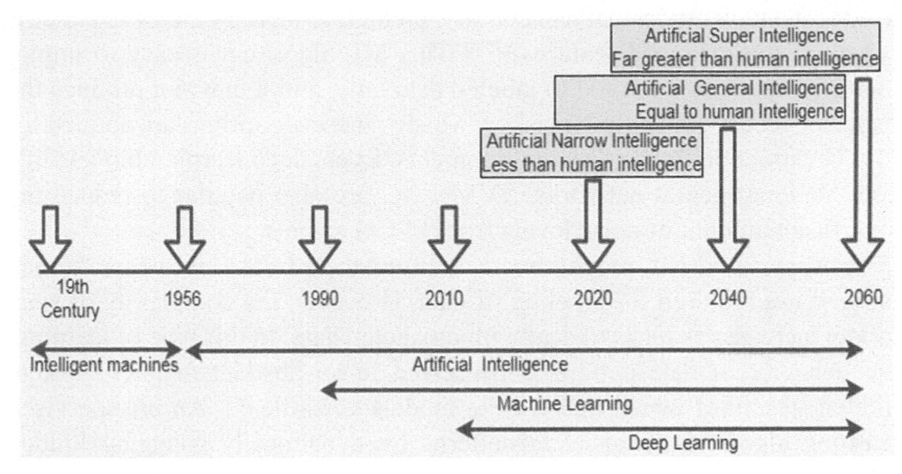

FIGURE 10.3 Progress and developments in machine learning algorithms.

A ML algorithm categorizes data input into two groups, that is, data labeled and data unlabeled. The labeled data have both inputs and corresponding outputs and so a great deal of effort is needed to mark the data. Unlike labeled data, unlabeled data has only inputs, and thus need more complex solutions.[65] Normally, the extraction of FECG signals from MECG signals are mainly done on labeled data. However, few ML algorithms also utilize unlabeled data to FECG restoration purpose. Determination of twin pregnancy may be done using labeled data. Cardiac defect can also be easily identified by unlabeled data. In short, labeled data are preferred for classification task.

10.3.1 POPULAR MACHINE LEARNING ALGORITHMS

Supervised ML algorithms are optimized for data labeling and thus accurate data labeling is necessary to achieve the correct results. The ML model is fed in supervised learning for working with the input data, that is, training data. This training data provides necessary information such as the problem and the solution required associated with the data pattern.[5] This trained model then can be utilized for unknown data. The benefit of this algorithm is that it will continue to improve discovering new relationships in every training.[40] Popular supervised ML models include linear regression, back propagation neural network, decision trees, support vector machine (SVM), deep learning, or a more sophisticated combination of methods. Supervised learning mainly used to develop model that can classify, predict, or identify the EEG patterns based on the extracted features.[5,12,43] This ML algorithm is easy to implement, however, it is limited to labeled data only, and it may not produce the expected results for complex tasks.[30] Mostly, these algorithms are applied for FECG extraction from noisy environments. Even, deep learning models like convolutional neural networks, SVMs, etc. are also popular to reduce the significant amount of noise levels from FECG signals.

Unsupervised ML models work with unlabeled data and hence human labor is not required for labeling of data. However, the complexity of such model increases as it has to deal with unlabeled data. In this type of learning, the behavior of data pattern is perceived in an abstract manner through hidden structures which make these models versatile.[12,46] An unsupervised learning algorithm learns data patterns by dynamically changing hidden structures and thus finds more applications in clustering/classification algorithms.[12,30,62,] Unsupervised ML algorithms are usually deployed for clustering and association. Popular unsupervised algorithms existing in literature include clustering, K-means clustering, KNN, mixture models, singular value decomposition, and self-organizing maps. Advantage associated with unsupervised learning is that it can even identify hidden pattern in FECG signals and it can be used for real-time applications. These algorithms are used to find the fetus heart rate and defects in fetus heart.

Reinforcement learning (RL) is reward–punishment-based training of models to make a sequence of decisions based on a trial and error method. Desirable outputs are endorsed, and undesirable outputs are discouraged.[5,12] This type of ML is required where we are not sure about the result of any action. Two widely used reinforcement ML models are Markov Decision Process and Q learning. RL algorithms are categorized as: model-free, that is, without model and model-based, that is, with model. Current action is based on some trial and error experience in model-free learning, whereas in model-based learning, current action is dependent upon the previously learned lessons.

The deep learning models are based on advanced artificial neural networks. This deep model consists of multiple hidden layers and hierarchical architecture that can learn more precisely as compared with artificial neural networks and thus provides better results. Each layer in deep learning models transforms input data from input layer to successive layer in nonlinear fashion. Deep learning models accept large amount of unlabeled data and trains the models in layer-wise fashion. Several architectures, such as recurrent neural networks, deep neural networks (DNN), convolutionary neural networks

(CNN), and deep belief networks are used in deep learning. Several existing applications of these models are found in speech and speaker recognition systems, computer vision, natural language processing, and design of drugs, bioinformatics, and medical image processing. The deep learning model requires huge processing power and huge amount of data, which is easily available these days. Since, these algorithms accept huge amount of data, these data must be presented in a systematic manner to obtain the desired results.[29,49]

The deep RL combines the features of both the deep and RL algorithms. Currently, deep RL algorithms such as Q-learning are coupled with deep learning to construct a powerful model of deep reinforcement.[33] The deep RL has already been successfully implemented in healthcare industry for identification and behavior of drugs, disease and pattern identification, entertainment industry for development of video games, finance and robotics industry. Even deep RL is capable of solving many unsolvable problems of neural networks. The deep RL finds applications in various domains of healthcare, including dynamic treatment of chronic diseases, automatic analysis of structure and unstructured clinical data, and for the treatment of critical cases.[9,61] ML classification methods are shown in Figure 10.4.

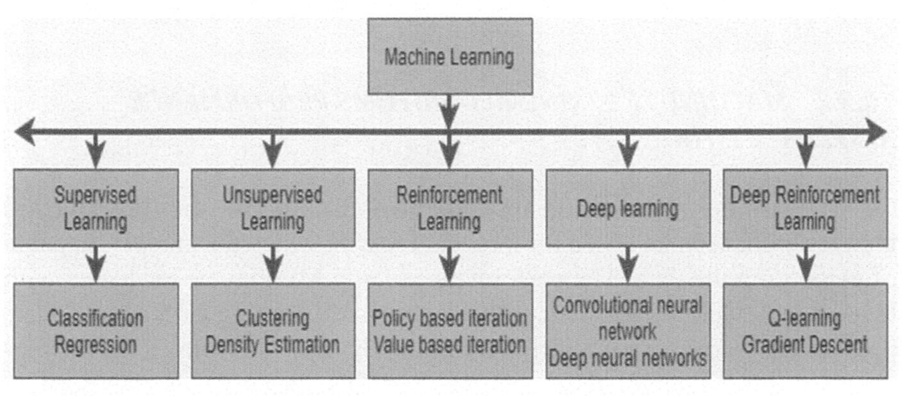

FIGURE 10.4 Classification of machine learning algorithms.

ML algorithms help to analyze vast amounts of data. While they generally provide faster, more reliable results to spot fruitful opportunities or hazardous threats, it may also take additional time and data to properly train it. The combination of ML with AI and cognitive technology will make it even more successful in processing large quantities of information associated with biomedical signals, thus helps the healthcare industry to come up with reliable solutions. Applications of ML in healthcare are presented in Figure 10.5.

This chapter mainly focuses over ML algorithms for the enhancement of FECG signals so that it could support e-healthcare monitoring, which in turn can help in improving the prenatal diagnoses of fetal abnormalities.

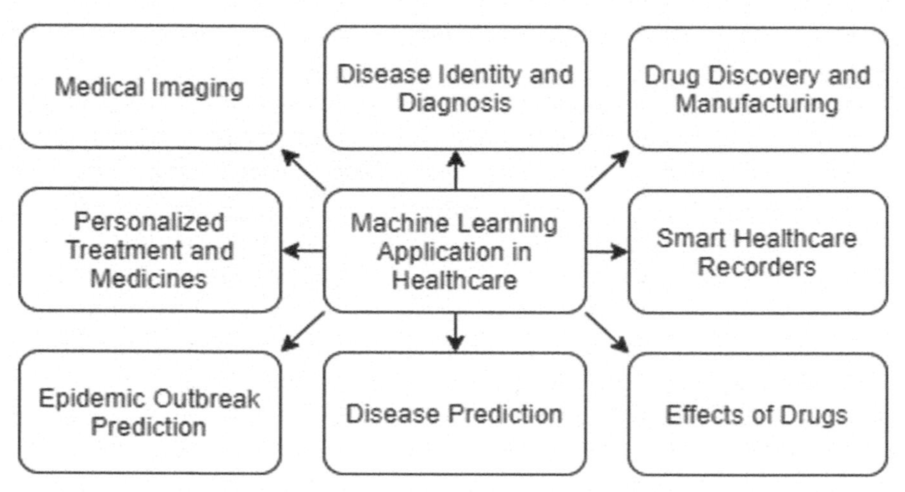

FIGURE 10.5 Applications of machine learning algorithms in healthcare.

10.3.2 MACHINE LEARNING ALGORITHMS PERFORMANCE ASSESSMENT TOOLS

Once a classifier algorithm is applied to any data, it classifies the data and then the model can be tested for its accuracy with which the model is capable of classifying the data. The performance of any ML may be evaluated with respect to a table, that is, a confusion matrix consisting of true and false predictions. A confusion matrix has generally four attributes, that is, true-positive, true-negative, false-positive, and false-negative. A true-positive (TP) predicts a positive class whereas a true-negative (TN) correctly predicts a negative class. A false-positive (FN) incorrectly predicts the positive class whereas a false-negative (FN) incorrectly predicts the negative class.[35,56]

Accuracy is defined as the ratio that indicates true predictions about the classifier model.

$$Accuracy = \frac{TP + NP}{TP + TN + FP + FN} \tag{10.9}$$

Misclassification is refers to how often the predictions are wrong. It is also equivalent to 1 minus accuracy.

$$Misclassification = \frac{FP + FN}{TP + TN + FP + FN} \quad (10.10)$$

Precision represents the proportion of true-positives which were actually true.

$$Precision = \frac{TP}{TP + FP} \quad (10.11)$$

Recall evaluates the proportion of actual positives that were identified correctly.

$$Recall = \frac{TP}{TP + FN} \quad (10.12)$$

F-Measure balances both the precision and recall in one number.

$$F - Measure = \frac{2 * Precision * Recall}{Precision + Recall} \quad (10.13)$$

The parameters, viz., accuracy, precision, and recall are positive-oriented parameters, and the model which is producing higher value will be considered better. Care should be taken while improving precision as precision and recall are inversely proportional. Depending upon the requirements, F-measures are set. In some cases, high value of F-measure is expected and in some cases, lower value is preferred. There is always a trade-off between (Equations 10.9–10.13) precision and recall.[35]

10.4 MACHINE LEARNING ALGORITHMS ON FETAL ELECTROCARDIOGRAM SIGNALS

Traditional AI algorithms based on the neural networks were used to record and analyze the FECG signals in real time. Efficient adaptive noise cancellation and adaptive signal enhancement based on artificial neural network for the extraction of FECG signals from abdominal ECG were presented in Selvan and Srinivasan.[53] However, these algorithms were based on labeled data. Also, selection of parameters for these algorithms was based on hit and trial method.

Amin developed an adaptive linear neural network model to extract FECG from abdominal signal for easy monitoring of fetus status in Amin et al..[6] The input to the model is maternal ECG and the target assigned is the abdominal signal having FECG and maternal ECG. The model resemble maternal signal as closely as possible to abdominal signal, thus only predict the maternal ECG from the composite signal. The FECG signal is then obtained by subtracting the maternal ECG from the composite signal. The model training parameters particularly the learning rate, momentum and initial weights are adjusted until the efficient results are obtained. According to their results, the FECG estimation model performs betters at high learning rate, low momentum, and small initial weights. In[13] a neural network based on fuzzy sets is developed to detect QRS complexes from ambulatory recording of single-lead maternal abdominal signal. In the proposed model, the FECG signal is initially extracted from the composite signal using autocorrelation method, and then fuzzy rule-based sets are used for the extraction of FECG signal by utilizing time domain characteristics of FECG signal. Results obtained for noise-free ambulatory recordings are found to be satisfactory.

Fang and Pei[42] utilized the concept of deep learning to restore FECG signals from multichannel abdominal ECG signals by utilizing the time-frequency features of the abdominal ECG signals through short time Fourier transform. The feature maps are then given to 2D Convolutional Neural Network deep learning model having three stages, that is, two convolutional layers, two pooling layers, and one fully connected layer. Inference accuracy reported for the model for large dataset is 90% which is better than the k-nearest neighbor classifiers.

A residual convolutional encoder–decoder network is proposed to restore FECG from the single-channel maternal ECG collected from the mother's abdomen. The deep learning architecture has five convolutional–deconvolutional blocks consisting of convolutional and deconvolutional layers. These layers are then followed by an activation, that is, tanh layer and one fully connected layer. To increase the learning capacity of the model, additional interconnections were provided. The model developed then is presented with the signal channel AECG which extracts features and estimates the FECG signal.[64] Jia implemented an adaptive linear neural network which is trained based on the input and target values for extraction. The error signal produced by the network is only the fetal ECG. Training rule adopts the network weights according to W-H learning rule. The results indicate that the algorithm can extract the FECG even when small amount of data are presented and the outcome is higher than the adaptive filter technique which is ensured by visual comparison.[38]

Assaleh proposed a method to separate the FECG by utilizing a thoracic and an abdominal ECG recording. To map nonlinearly the signals of thoracic ECG and abdominal ECG, the polynomial networks technique is utilized. Then, FECG is separated by deducting the aligned thoracic ECG from the abdominal ECG signal. The qualitative nature of the FECG extracted by this approach exceeds the results obtained from other methods like independent component analysis.[11]

Swarnalatha and Prasad provided a method for maternal ECG cancellation using adaptive neuro fuzzy inference system (ANFIS) and wavelets with three different approaches, namely, ANFIS alone, the wavelets followed by the ANFIS and the ANFIS followed by the wavelets. In wavelet processing, either preprocessing or post-processing technique involves the decomposition and reconstruction process using Coiflets since it decreases the noise components and afford high-resolution output and also it is related to the shape of fetal ECG. The performance is assessed through signal-to-noise ratio and CCs. ANFIS followed by wavelet post-processing approach produced best extraction compared with other two approaches.[54] Assaleh also utilized ANFIS to extract FECG signal from the abdominal ECG of mother correlating maternal ECG in nonlinear sense with the components of maternal ECG present in the abdominal ECG signal. The extracted maternal signals are then cancelled from the composite signal then the residual is the extracted FECG signal.[10]

FECG extraction using adaptive neuro-fuzzy logic technique by cancelling the maternal ECG signal from abdominal has been presented in the works of Vijila et al.[57] Instead of using the conventional filters, adaptive filters based on neuro-fuzzy logic which has the capability of self-adjustments according to the environments were developed. When the noisy signal having major component as mothers ECG is passed through the filter, it suppresses the noise and filters the FECG signals. The advantage of the proposed filter is that it does not require prior knowledge of signal or noise characteristics.

Zhong et al.[63] developed deep learning models to recognize QRS complexes of FECG signals. Initially, the features of noninvasive FECG signals from Cardiology Challenge database were extracted and then normalized. The normalized features were then applied to convolutional neural network designed to detect QRS complexes. The deep learning parameters, viz., precision, recall, F-measure, and accuracy were used to assess the performance of the model. The proposed model reported 75.33% precision, 80.54% recall,

77.85% F-measure score which is better than the performance obtained by K-nearest neighbor, Naive Bayes, and SVM-based classifier. Deep learning model presented in Muduli et al.[48] is used to recover FECG signals in a faster way by enhancing the computational speed in a telemonitoring system. The model consists of a stacked denoising autoencoder (SDAE) which is used for nonlinear mapping. The training algorithm utilized is off-line mini-batch gradient descent-based backpropagation which enhances the training speed.

Zhong proposed a residual convolutional encoder–decoder network (RCED-Net) model to extract FECG signals from single-channel AECG recording. The convolutional networks are designed to extract features of the time series data by themselves. And these characteristics, convolutional networks used to extract FECG features at input stage which further outputs the estimate of the FECG signals from the abdominal signals.[64]

Invasive monitoring of the FCCG always presents a potential risk of fetal infection. Advancement in FECG measurement devices and extraction algorithms in has allowed FECG to monitor the mother's abdominal ECG from the early stages of pregnancy. The extraction algorithms, however, include the reference maternal ECG as well as heavy feature drafting makes FECG tracking out-of-clinics not yet feasible in everyday life. In Vo et al.'s paper,[58] this issue has been solved, where a pure end to end deep learning model is built to remove the FECG QRS complexes. In addition, the model has the e-residual network architecture (ResNet) that adopts the novel 1-D octave convolution (OctConv) for learning multiple temporal frequency functions, leading to decrease in memory and computational costs. Essentially, the model is able to show the role of regions which are more dominant in the detection process. The data from the PhysioNet 2013 Challenge with named QRS complex annotations were used in the original form to test the proposed model, and the data were then updated with Gaussian and motion noise, imitating real-world scenarios. The model can obtain an F1 score of 91.1% while saving more than 50% computational costs with less than 2% loss of output, demonstrating the efficiency of the model.[58]

Currently, FECG research is mainly focused on the control of the fetus which is based solely on the heart rate. Research using the FECG structure for cardiac-anomaly populations is usually not conducted. The foremost reason behind this is the unavailability of standard FECG measuring devices. Jagannath et al.[34] proposed a hybrid deep learning model, that is, Bayesian deep learning (BDL) consisting of a Bayesian filter and a deep learning model for the cancellation of maternal ECG and other high frequency nonlinear noises from the composite signal to provide high quality noninvasive fECG signal.

The results obtained by the proposed model shows a high improvement in signal-to-noise ratio, that is, 25.76 dB and 40.23 dB for input SNR of 15 dB and −30 dB, respectively. Also, PRD reported is 0.32 which is very low and is considered good for signal restoration problems.

Monitoring of the FECG has also become important owing to the relatively larger growth in the number of cardiac patients. Muduli et al.[48] proposed a model that could help cardiologist to diagnose more and more patients. The objective of their research is to develop a model that can be used to restore FECG signals in a faster way so that it can be easily available to physician. For the same, a deep learning model architecture incorporating a nonlinear mapping using a SDAE is developed. The encoding of the raw nonsparse FECG data takes place through a DNN at the transmitter side. After pre-training, the mini-batch gradient-based backpropagation algorithm will further fine tune the entire deep SDAE. Although SDAE preparation is typically time-consuming, due to the one-time off-line training phase, this does not affect the results. Reconstruction of the FECG in real time is easier because of a few multiplications of the matrix-vector at the receiver end. The simulations conducted using normal noninvasive FECG databases show positive outcomes with regard to the consistency of the reconstruction.

A fully convolutional deep learning model is presented to detect QRS complexes in FECG signals to determine the fetal heart rate, to identify FECG segment characteristics, and to differentiate between normal and abnormal heart beats. To assess the performance of the proposed model, 685 echocardiograms signals from fetuses from 18 to 24 weeks were used. The best findings were obtained with a sensitivity and accuracy of 100% and 90%, respectively in the diagnosis of hypo plastic left heart syndrome versus normality. While the findings sound positive, one of the key drawbacks of this analysis is that only two congenital heart disorders were examined and the deep learning system was equipped only with images from 1 US computer without taking into account the heterogeneity.[26]

Eleni and Rik in 2020 developed a convolutional encoder–decoder deep learning model to suppress maternal ECG present in a mutichannel FECG. The proposed model has symmetric skip layer connections and it performs learning in end to end mappings to restore the FECG signals. The proposed model is evaluated on SNR and an improvement of 9.5 dB is observed for fetal ECG signals with input SNR ranging from −20 to 20 dB. Additionally, the presented model is capable of preserving beat-to-beat morphological characteristics of FECG signals.[24]

Convolutional neural networks were also utilized to automatically detect irregular heartbeats in FECG signals. In their model, the morphology and rhythm of heartbeats were combined into a two-dimensional feature vector, which was further used by the convolutional neural network to detect beats. When the proposed model was tested on the MIT-BIH arrhythmia database, wherein authors have reported the average accuracy of 99.1% for five heartbeats and 97% for eight heartbeat categories. The sensitivity and positive predictive rate reported are suitable for the e-home health monitoring of cardiovascular disease.[41]

An echo state recurrent neural network (ESN) is proposed in Lukoˇseviˇcius and Marozas[44] that combines the FECG characteristics to cancel noise from FECG signals. The process starts by filtering and normalization of FECG signals, and then followed by the detection of maternal QRS complexes by ESN trained by supervised ML, and finally the averaged maternal ECG obtained is discarded.

10.5 CONCLUSIONS

Healthcare industries provide services to both patients and physicians and are playing an essential part in countries economy. Due to large number of diseases, it is really impossible to provide better services to patients particularly in countries which are densely populated. Since, healthcare services and healthcare industry has to maintain the quality, value, and results, it must be cautiously permitted and routinely monitored. ML algorithms because of their popularity, are now gaining importance in healthcare services and industries. In this context, so many healthcare devices were launched by healthcare industries and these devices are now offering better services to society. FECG signal which measures electrical activity of fetus heart is very important signal for fetus related issues. However, these signals are always associated with noises during recordings which hamper the correct diagnosis. Also, these signals generate voluminous amount of data. In chapter, FECG signals, noises affecting FECG characteristics, available database and performance tools to evaluate FECG-based algorithms are presented. ML algorithms and tools are available to assess ML algorithms are also explored. Finally, ML algorithms available for FECG extraction methods are presented that would provide sufficient literature to implement these algorithms for e-healthcare.

KEYWORDS

- **fetal ECG (FECG)**
- **Mother ECG (MECG)**
- **artifacts**
- **machine learning**
- **deep learning**
- **confusion matrix**

REFERENCES

1. Madhav A. V. S., Tyagi A. K. (2022) The World with Future Technologies (Post-COVID-19): Open Issues, Challenges, and the Road Ahead. In: Tyagi A. K., Abraham A., Kaklauskas A. (eds) Intelligent Interactive Multimedia Systems for e-Healthcare Applications. Springer, Singapore. https://doi.org/10.1007/978-981-16-6542-4_22.
2. Healthcare Industry in India, 2020b. https://www.ibef.org/industry/healthcare-India.
3. Adam, J. The Future of Fetal Monitoring. *Rev. Obst. Gynecol.* **2012,** *5* (3–4), e132–e136.
4. Ahmed, Z.; Mohamed, K.; Zeeshan, S.; Dong, X. Artificial Intelligence with Multi-Functional Machine Learning Platform Development for Better Healthcare and Precision Medicine. Database, 2020.
5. Alpaydin, E. *Introduction to Machine Learning*. MIT Press, 2020.
6. Amin, M.; Mamun, M.; Hashim, F., Husain, H. Separation of Fetal Electrocardiography (ECG) from Composite Using Adaptive Linear Neural Network for Fetal Monitoring. *Int. J. Phys. Sci.* **2011,** *6* (24), 5871–5876.
7. Anandanatarajan, R. *Biomedical Instrumentation and Measurements*. PHI Learning Pvt. Ltd, 2011.
8. Andreotti, F.; Behar, J.; Zaunseder, S.; Oster, J.; Clifford, G. D. Anopen-source Framework for Stress-Testing Non-Invasive Foetal ECG Extraction Algorithms. *Physiol. Meas.* **2016,** *37* (5), 627.
9. Arulkumaran, K.; Deisenroth, M. P.; Brundage, M.; Bharath, A. A. Deep reinforcement Learning: A Brief Survey. *IEEE Signal Process. Mag.* **2017,** *34* (6), 26–38.
10. Assaleh, K. Extraction of Fetal Electrocardiogram Using Adaptive Neuro fuzzy inference Systems. *IEEE Trans. Biomed. Eng.* **2006,** *54* (1), 59–68.
11. Assaleh, K.; Al-Nashash, H. A Novel Technique for the Extraction of Fetal ECG Using Polynomial Networks. *IEEE Trans. Biomed. Eng.* **2005,** *52* (6), 1148–1152.
12. Ayodele, T. O. Types of Machine Learning Algorithms. *New Adv. Mach. Learn.* **2010,** *3*, 19–48.
13. Azad, K. A. K. Fetal QRS Complex Detection from Abdominal ECG: A Fuzzy approach. *In Proceedings of IEEE Cordic Signal Processing Symposium*; Kolmarden, Sweden, 2000; pp 275–278.
14. Behar, J. A.; Bonnemains, L.; Shulgin, V.; Oster, J.; Ostras, O., Lakhno, I. Noninvasive Fetal Electrocardiography for the Detection of Fetal Arrhythmias. *Prenatal Diag.* **2019,** *39* (3), 178–187.

15. Behrman, R. E.; Butler, A. S. et al. Measurement of Fetal and Infant Maturity. In *Preterm Birth: Causes, Consequences, and Prevention*; National Academies Press, US, 2007.
16. Blanco-Velasco, M.; Cruz-Rold′an, F.; Godino-Llorente, J. I.; Blanco-Velasco, J.; Armiens-Aparicio, C.; L′opez-Ferreras, F. On the Use of PRD and CR Parameters for ECG Compression. *Med. Eng. Phys.* **2005,** *27* (9), 798–802.
17. Catalano, J. T. *Guide to ECG Analysis*; Lippincott Williams & Wilkins, 2002.
18. Chiu, C.-C.; Lin, T.-H.; Liau, B.-Y. Using Correlation Coefficient in ECG Waveform for Arrhythmia Detection. *Biomed. Eng.: Appl. Basis Commun.* **2005,** *17* (03), 147–152.
19. Clifford, G. D.; Silva, I.; Behar, J.; Moody, G. B. Non-invasive Fetal ECG Analysis. *Physiol. Meas.* **2014,** *35* (8), 15–21.
20. De Moor, B.; De Gersem, P.; De Schutter, B.; Favoreel, W. Daisy: Adatabase for Identification of Systems. *Journal A* **1997,** *38*, 4–5.
21. Division, V. S. *SRS Bulletin Sample Registration System*, 2019.
22. Dua, S.; Acharya, U. R.; Dua, P. Machine Learning in Healthcare Informatics 2014, *56*.
23. Flach, P. A. On the State of the Art in Machine Learning: A Personal Review. *Artif. Intell.* **2001,** *131* (1–2), 199–222.
24. Fotiadou, E.; Vullings, R. Multi-channel Fetal ECG Denoising with Deep Convolutional Neural Networks. *Front. Pediat.* **2020,** *8*, 508.
25. Freeman, R. K.; Garite, T. J.; Nageotte, M. P.; Miller, L. A. *Fetal Heart Rate Monitoring*; Lippincott Williams & Wilkins, 2012.
26. Garcia-Canadilla, P.; Sanchez-Martinez, S.; Crispi, F.; Bijnens, B. Machine Learning in Fetal Cardiology: What to Expect. *Fetal Diag. Ther.* **2020,** *47* (5), 363–372.
27. Ghassemi, M.; Naumann, T.; Schulam, P.; Beam, A. L.; Ranganath, R. Opportunities in Machine Learning for Healthcare, 2018. *arXiv preprint arXiv:1806.00388.*
28. Goldberger, A. L.; Amaral, L. A.; Glass, L.; Hausdorff, J. M.; Ivanov, P. C.; Mark, R. G.; Mietus, J. E.; Moody, G. B.; Peng, C. K.; Stanley, H. E. Physiobank, Physiotoolkit, and Physionet: Components of a New Research Resource for Complex Physiologic Signals. *Circulation* **2000,** *101* (23), e215–e220.
29. Goodfellow, I.; Bengio, Y.; Courville, A. *Deep Learning*; MIT Press, 2016.
30. Gutierrez, D. D. *Machine Learning and Data Science: An Introduction to Statistical Learning Methods with R*; Technics Publications, 2015.
31. Haghjoo, M.; Khorgami, M. Electrocardiography: Basic Knowledge with Focus on Fetal and Pediatric ECG. In *Congenital Heart Disease in Pediatric and Adult Patients*, **2017;** pp 245–278.
32. Hasan, M. A.; Reaz, M.; Ibrahimy, M.; Hussain, M.; Uddin, J. Detection and Processing Techniques of FECG Signal for Fetal Monitoring. *Biol. Procedures Online* **2009,** *11* (1), 263.
33. Henderson, P.; Islam, R.; Bachman, P.; Pineau, J.; Precup, D.; Meger, D. Deep Reinforcement Learning that Matters, 2017. arXiv preprint arXiv:1709.06560.
34. Jagannath, D.; Raveena Judie Dolly, D.; Peter, J. D. A Novel Bayesian Deep Learning Methodology for Enhanced Foetal Cardiac Signal Mining. *J. Exp. Theoret. Artif. Intell.* **2019,** *31* (2), 215–224.
35. Javaid, A.; Niyaz, Q.; Sun, W.; Alam, M. A Deep Learning Approach for Network Intrusion Detection System. In *Proceedings of the 9th EAI International Conference on Bio-inspired Information and Communications Technologies (formerly BIONETICS)*, 2016; pp 21–26.

36. Jezewski, J.; Horoba, K.; Matonia, A.; Gacek, A.; Bernys, M. A New Approach to Cardiotocographic Fetal Monitoring Based on Analysis of Bioelectrical signals. *Proc. 25th Annu. Int. Conf. Eng. Med. Biol. Soc.* **2003,** *4*, 3145–3148.
37. Jezewski, J.; Matonia, A.; Kupka, T.; Roj, D.; Czabanski, R. Determination of Fetal Heart Rate from Abdominal Signals: Evaluation of Beatto-Beat Accuracy in Relation to the Direct Fetal Electrocardiogram. *Biomed. Eng./Biomedizinische Technik* **2012,** *57* (5), 383–394.
38. Jia, W.; Yang, C.; Zhong, G.; Zhou, M.; Wu, S. Fetal ECG Extraction Based on Adaptive Linear Neural Network. *2010 3rd Int. Conf. Biomed. Eng. Info.* **2010,** *2*, 899–902.
39. Kasthuri, A. Challenges to Healthcare in India-the Five a's. *Indian J. Commun. Med.* **2018,** *43* (3), 141.
40. Kotsiantis, S. B.; Zaharakis, I.; Pintelas, P. Supervised Machine Learning a Review of Classification Techniques. *Emerg. Artif. Intell. Appl. Comput. Eng.* **2007,** *160* (1), 3–24.
41. Li, J.; Si, Y.; Xu, T.; Jiang, S. Deep Convolutional Neural Network Based ECG Classification System Using Information Fusion and One-Hot Encoding Techniques. *Math. Prob. Eng.* 2018.
42. Lo, F.-W.; Tsai, P.-Y. Deep Learning for Detection of Fetal ECG from Multi-Channel Abdominal Leads. In *2018 Asia-Pacific Signal and Information Processing Association Annual Summit and Conference (APSIPA ASC)*, 2018; pp 1397–1401.
43. Lopez-Larraz, E.; Sarasola-Sanz, A.; Irastorza-Landa, N.; Birbaumer, N.; Ramos-Murguialday, A. Brain-machine Interfaces for Rehabilitation in Stroke: A Review. *Neuro Rehab.* **2018,** *43* (1), 77–97.
44. Lukoˇseviˇcius, M.; Marozas, V. Noninvasive Fetal QRS Detection Using Echo State Network. In *Computing in Cardiology*, 2013; pp 205–208.
45. Marchon, N.; Naik, G. Electrode Positioning for Monitoring Fetal ECG: A Review. In *2015 International Conference on Information Processing (ICIP)*, 2015; pp 5–10.
46. Marsland, S. *Machine Learning: An Algorithmic Perspective*; CRC Press, 2015.
47. Modi, I. Health Care Delivery System: India. In *The Wiley Blackwell Encyclopedia of Health, Illness, Behavior, and Society*, 2014; pp 816–823.
48. Muduli, P. R.; Gunukula, R. R.; Mukherjee, A. A Deep Learning Approach to Fetal ECG Signal Reconstruction. In *2016 Twenty Second National Conference on Communication (NCC)*, 2016; pp 1–6.
49. Najafabadi, M. M.; Villanustre, F.; Khoshgoftaar, T. M.; Seliya, N.; Wald, R.; Muharemagic, E. Deep Learning Applications and Challenges in Big Data Analytics. *J. Big Data* **2015,** *2* (1), 1.
50. Olivas, E. S.; Guerrero, J. D. M.; Martinez-Sober, M.; Magdalena-Benedito; J. R.; Serrano, L. et al. *Handbook of Research on Machine Learning Applications and Trends: Algorithms, Methods, and Techniques: Algorithms, Methods, and Techniques*; IGI Global, 2009.
51. Pieri, J.; Crowe, J.; Hayes-Gill, B.; Spencer, C.; Bhogal, K.; James, D. Compact long-term Recorder for the Transabdominal Foetal and Maternal Electrocardiogram. *Med. Biol. Eng. Comput.* **2001,** *39* (1), 118–125.
52. Sameni, R.; Clifford, G. D. A Review of Fetal ECG Signal Processing; Issues and Promising Directions. *Open Pacing, Electrophysiol. Ther. J.* **2010,** *3*, 4–20.
53. Selvan, S.; Srinivasan, R. A Novel Adaptive Filtering Technique for the Processing of Abdominal Fetal Electrocardiogram Using Neural Network. In *Proceedings of the*

IEEE 2000 Adaptive Systems for Signal Processing, Communications, and Control Symposium (Cat.No. 00EX373), 2000; pp 289–292.
54. Swarnalatha, R.; Prasad, D. Maternal ECG Cancellation in Abdominal Signal Using Anfis and Wavelets. *JApSc* **2010,** *10* (11), 868–877.
55. Symonds, E. M.; Sahota, D.; Chang, A. *Fetal Electrocardiography*; World Scientific, 2001.
56. Vafeiadis, T.; Diamantaras, K. I.; Sarigiannidis, G.; Chatzisavvas, K. C. A Comparison of Machine Learning Techniques for Customer Churn Prediction. *Simul. Model. Practice Theory* **2015,** *55*, 1–9.
57. Vijila, C. K. S.; Kanagasabapathy, P.; Johnson, S. Adaptive Neuro Fuzzy Inference System for Extraction of FECG. In *2005 Annual IEEE India Conference-Indicon*, 2005; pp 224–227.
58. Vo, K.; Le, T.; Rahmani, A. M.; Dutt, N.; Cao, H. An Efficient and Robust Deep learning Method with 1d Octave Convolution to Extract Fetal Electrocardiogram. *Sensors* **2020,** *20* (13), 37–57.
59. Webster, J. G. *The Physiological Measurement Handbook*; CRC Press, 2014.
60. Willmott, C. J.; Matsuura, K. Advantages of the Mean Absolute Error (MAE) over the Root Mean Square Error (RMSE) in Assessing Average Model Performance. *Clim. Res.* **2005,** *30* (1), 79–82.
61. Yu, C.; Liu, J.; Nemati, S. Reinforcement Learning in Healthcare: A Survey, 2019. arXiv preprint arXiv:1908.08796.
62. Zhang, D. *Advances in Machine Learning Applications in Software Engineering*; IGI, Global, 2006.
63. Zhong, W.; Liao, L.; Guo, X.; Wang, G. A Deep Learning Approach for Fetal QRS Complex Detection. *Physiol. Meas.* **2018,** *39* (4), 045004.
64. Zhong, W.; Liao, L.; Guo, X.; Wang, G. Fetal Electrocardiography Extraction with Residual Convolutional Encoder Decoder Networks. *Aust. Phys. Eng. Sci. Med.* **2019,** *42* (4), 1081–1089.
65. Zhu, X.; Goldberg, A. B. *Introduction to Semi-Supervised Learning (Synthesis Lectures on Artificial Intelligence and Machine Learning)*; Morgan and Claypool Publishers, 2009; p 14.

CHAPTER 11

Blockchain for Wearable Internet of Things in Healthcare

SHUBHANGI KHARCHE[1*] and RIZWANA SHAIKH[2]

[1]*Electronics and Telecommunication Department, SIES Graduate School of Technology, Nerul, Navi Mumbai, India*

[2]*Computer Engineering Department, SIES Graduate School of Technology, Nerul, Navi Mumbai, India, Email: rizwana.shaikh@siesgst.ac.in*

[*]*Corresponding author. E-mail: Shubhangi.kharche@siesgst.ac.in*

ABSTRACT

Background: Security is one of the persistent challenges for the tiny devices embedded in the fabric of Internet of Things (IoT) particularly for the wearables equipped in the healthcare monitoring systems. Blockchain is the state-of-the-art technology to secure and revolutionize the healthcare industries. The choice of communication technologies, operating systems, and the data access technologies is very crucial in wearable IoT (WIoT). Device mobility, reliability, availability, scalability, and the data analytics are persistent challenges for WIoT.

Objectives: To provide fair enough justification for the choice of communication technologies, operating systems, and the data access technologies based on the comparative analysis and to apply Blockchain for the proposed WIoT architecture.

Methodology: Blockchain is applied to provide security and transparency. The immutable nature of Blockchain can be applied to wearable devices to track the mobility and ensure the secure and authorized access. The log

Intelligent Interactive Multimedia Systems for e-Healthcare Applications. Shaveta Malik, PhD
Amit Kumar Tyagi, PhD (Eds.)

record of the traverse can be utilized further to enhance the throughput and reduce the delay at each stage of traverse. Thus, proving the scope for enhancing the energy efficiency of the WIoT devices.

Result: It is observed that the use of Blockchain has ensured security in the challenging environment of WIoT.

Conclusion: With all the challenges, such as device constraints, energy constraints, and mobility, the Blockchain provides improved security and transparency of the traverse of WIoT devices. Using Blockchain, the challenges in WIoT can be solved and hence its overall performance can be improved.

11.1 WEARABLE INTERNET OF THINGS

Wearable Internet of Things (WIoT) networks are an emerging class of wireless sensor and communication networks. The unprecedented growth in the world's population has increased the health problems globally. Precocious detection, monitoring, and medical intervention of health diseases have become a priority. Figure 11.1 shows the block diagram of the wearable IoT healthcare system. Nowadays, different types of wearable devices have been designed and are available for the human body right from head to toe to detect a number of diseases and ailments. Almost more than 140 wearables are available for the human body and the number of worldwide-connected wearables has increased from 325 to 722 million over the past 4 years (2016–2020).[1] The wearable devices have sensing, computing, and communication capabilities. The wearable devices are embedded with different sensors and biosensors for sensing the physical well-being of the person. The sensed information is transferred wirelessly using the communication technologies, such as Bluetooth,[2,3] ZigBee,[4] Wi-Fi,[5] RFID,[6] NFC,[7] and 6LoWPA[8] to the data logging devices, such as the smart phones, laptops, tablet, and personal digital assistant (PDA). The readouts of the sensed information can be viewed using various APPs on the data logging devices. From these devices, the data are uploaded on cloud using wireless broadband standards, such as WiMAX, WiFi, or cellular standards like 3G/4G. The user (patient) information must be authorized and secured before pushing it on cloud for analytics. The physician provides the diagnostics from the patient data and suggests treatment and medication to the end users on their mobile handsets or laptops. In the case of emergency, alerts are generated and the emergency services are contacted.

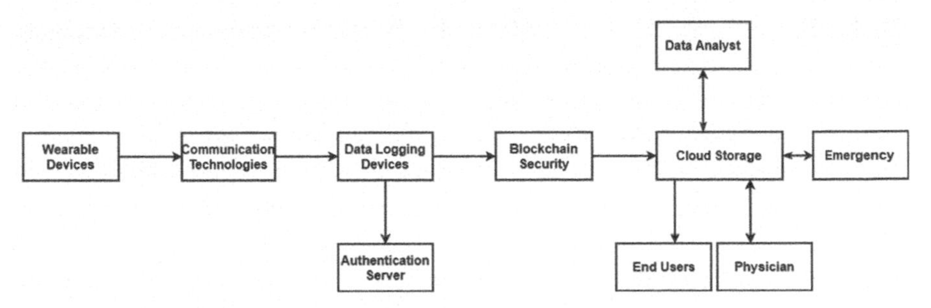

FIGURE 11.1 Wearable internet of things.

11.1.1 WEARABLE DEVICES

The wearable devices consist of a number of sensors to sense the patients' physical health. The sensors can be used for monitoring the physical parameters, to measure and characterize the chemical compounds and to measure and characterize the organic materials. Accordingly, the sensors are classified as the physical sensors, the chemical sensors, and the biosensors. Most of the physical sensors have monitoring as well as the diagnostic abilities. The health monitoring application has a reach from urban areas to rural areas.[9] The wearable sensors monitor the patients' health status by gathering their physiological parameters and the movement data. The sensors are deployed on different locations on the patient body according to the clinical application of interest. For example, a variety of headbands, wristwatches, and smart jewelry are available for monitoring the heart and respiratory rates. For movement data capturing, the sensors are usually available in smart watches and phones. The smart sensors are microelectronics circuits that are capable of performing the functions of signal amplification, processing, and wireless transmission. For example, a wireless electrocardiogram (ECG) sensor has an amplifier circuit, a microcontroller, and a wireless transceiver with antenna subsystem. With the advent of microelectromechanical systems (MEMS), the cost and size of sensors are significantly reduced. The sensors can be embedded in the textile with the help of electrodes to collect and print the ECG and electromyography (EMG) data. To monitor the physical data, sensors can be used in two ways. First, the body worn sensors, and second, the integrated body worn and ambient sensors. The sensor readings are communicated to the data logging devices using the low cost, low power consumption, and with small size transceiver wireless communication technologies. Bluetooth, Bluetooth low energy (BLE), ZigBee standard,

ZigBee pro, ZigBee IP, ultra-wideband (UWB), radiofrequency identification (RFID), near-field communication (NFC), WiFi, and 6LoWPAN are the wireless communication technologies for the short-range communication from the wearable sensors to the data logging devices. Dominantly BLE is used for communication with wearable sensors owing to its prolonged battery life, low cost, reduced power consumption, and bidirectional communication capability.

For the understanding of the wearables, some of the head-mounted devices are discussed hereunder.

- Head-mounted devices or headbands improve the care coordination of the patients. There are numerous such devices available for monitoring a variety of health parameters. A head-mounted device can be used by the medical toxicologist as it allows for video relay of the patient's information and totally hand-free assessment. As per the clinical and research arena centers, there are challenges on the privacy and security of the wearable devices.[10] Another challenge is to employ comparable encryption in head-mounted wearable devices. Connectivity is another barrier to study the effectiveness of head-mounted devices in telehealth.
- The head-mounted devices mostly use Bluetooth communication technology, potentially limiting the bandwidth unlike the smart phones and tablets which use higher bandwidth 3G/4G LTE networks for communication. Moreover, head-mounted devices exhibit greater fluctuations in bandwidth when they are used for video streaming than the smart phones and tablets.
- Another kind of head-mounted device is the bioglass which extracts physiological parameters like Ballistocardiagram (BCG), heart rate, and blood volume pulse (BVP), and respiration rate of the person.[11]
- An EEG band worn on the forehead can be used for headache prevention, sleep management and depression treatment using the collected EEG signals.[12] The band consists of dry electrodes that can be used for rapid and easy monitoring on a regular basis. The band is also used for notifying people with episodic migraines about an imminent migraine headache hours well in time by monitoring forehead EEG dynamics. The band consists of a data acquisition system (DAS), a chargeable battery, and a Bluetooth module to transmit wirelessly the EEG recordings to end users device (computer, laptop, or cell phone) allowing them to move freely. The DAS consists of analog to digital converters, amplifiers, and a microcontroller unit.

- A smart helmet with embedded electrodes can be used to record real-time ECG, respiration, and electroencephalogram (EEG) from face-lead locations.[13] The output signals from electrodes are amplified by a wearable bioamplifier and communicated via a Bluetooth module to a laptop or computer. Such smart helmets are used to avoid traffic accidents particularly when the state of mind of drivers, such as drowsiness, anxiety, stress, and state of their body, such as sickness prevents them from concentrating on while driving.
- Google glass[14]: A wearable computer designed glasses. It allows user interaction through a touchscreen and voice control and it is able to take photos and videos. It weighs 36 g and uses a Li-ion battery[15] which has a capacity of 570 mAh. It has WiFi 802.11b/g, Bluetooth, and micro-USB connectivity modules. It has the following sensors: Bone conduction transducer, accelerometer, gyroscope, magnetometer, ambient light sensor, proximity sensor, touchpad, and camera.

11.1.2 WIRELESS COMMUNICATION STANDARDS IN WEARABLE INTERNET OF THINGS

The communication technologies in Figure 11.1 are the short-range wireless communication standards that must be chosen to minimize latency, minimize adverse effects on the human body and should be secure. Low latency is required so that the emergency case actions can be taken in critical health conditions. Strong security is required to avoid hacking of the patients' sensitive data by the attackers. Thus, the sensed healthcare information should be logged securely in the data logging devices. The information from these devices should be forwarded securely to the cloud storage using block chains to avoid imitation and alteration in patients' sensitive data, via long-range wireless communication standards, such as LoRaWAN,[16] 3G/4G, and WiMax.[17] These wireless standards should provide low latency over long distance communication to avoid detrimental effects on health of patients caused due to high delays. The data should be stored on the cloud for longer durations so that the medical practitioners can benefit from the medical history of the patients. Long time storage on the cloud generates large amounts of data that can help the analyst to achieve optimization in the health estimates using machine learning. Blockchain security is required for long time data storage on the cloud.[18] Table 11.1 depicts the short-range wireless technologies their operating frequency bands, communication range, security and data protection algorithms. Table 11.2 shows the long-range wireless communication standards with their operating frequency bands, communication range, and security protocols.[19]

TABLE 11.1 Short-Range Wireless Communication Standards.

Wireless technology	Frequency band	Range	Security	Data protection
Bluetooth	2.4 GHz	1–100 m	64 and 128 bit AES encryption	16-Bit CRC
Bluetooth low energy	2.4 GHz	>100 m	protocol level security, 128 bit AES	16-Bit CRC
NFC	13.56 MHz	<10 m	Encrypted data is stored in RFID tag, algorithms used are MIFARE, DESFire, etc.	-
Zigbee	868/915 MHz, 2.4 GHz band	10–100 m	128 AES plus application layer security	16-Bit CRC
RFID	860–960 MHz	upto 100 m	Hardware level and protocol level	16-Bit CRC
WiFi	2.4 GHz, 5 GHz	50–100 m	WiFi protected access, WPA and WPA2 encryption	32-Bit CRC

TABLE 11.2 Long-Range Wireless Communication Standards.

	LoRaWAN	WiMax	3G/4G
Security	128-bit AES encryption	Both AES (Advanced Encryption Standard) and the 3DES (Triple Data Encryption Standard) are supported	128-bit AES block encryption algorithms, Transport Layer Security (TLS)
Frequency band	433 MHz, 868 MHz, 915 MHz, 923 MHz	10–66 GHz, 2–11 GHz	1.8–2.5 GHz for 3G, 2–8 GHz for 4G
Range	More than 10 km	50 km (30 miles)	Macrocell coverage: 50–150 km

11.2 OPERATING SYSTEMS FOR WEARABLE INTERNET OF THINGS

Operating systems play a very crucial role in the WIoT in healthcare. An operating system on a device provides an interface to its hardware for the user. Operating systems allow applications for reading the health-related measures on various devices in WIoT. The security of patient health data is of utmost importance in WIoT where operating systems play a major role in embedding security in the devices. Different kinds of operating systems are required due to heterogeneous devices in WIoT. The operating systems used on wearable devices differ from those on the data logging devices/end-user devices and the cloud servers. The varied requirements of operating systems are mainly due to the device memory, energy, cost, and size constraints. Different levels of security are required at the wearable devices that depend on the memory availability at the wearable devices. Some of the wearable operating systems have inbuilt security. The wearable devices are resource constraint, application specific, mobile, and dynamic. So it becomes very challenging to build an operating system that can fulfill the diverse requirements of applications and security. The wearable IoT comprises a wide spectrum of devices based on different architectures. It becomes difficult for only one operating system that can fulfil the diverse requirements and services. Figure 11.2 shows the operating systems required at different devices in wearable IoT (also Tables 11.3–11.5).

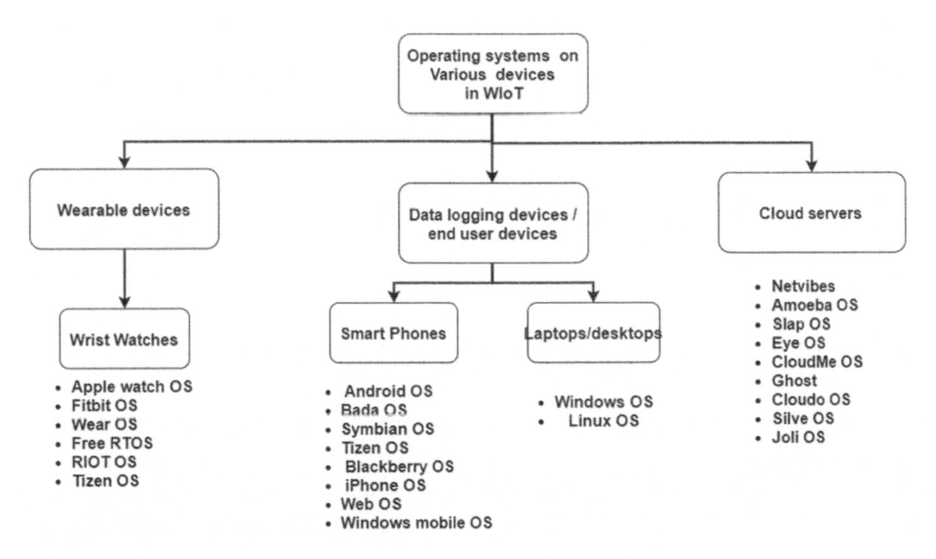

FIGURE 11.2 Operating systems at different devices in wearable IoT.

TABLE 11.3 Wearable Device Open-Source Operating System Comparison for Memory Requirements, Programming Support, Communication Protocols, and Architecture.

Operating system (OS)	RAM	ROM	Programming support	Protocols	OS architecture	Open source
Lite OS[20]	4KB	128 KB	C	BMAC, multi-route	Modular	Yes
RIOT OS[21]	1.5 KB	5 KB	C and C++	IPv6, RPL, COAP, 6LoWPAN, TCPIP, UDP	Microkernel	Yes
Tizen[22]	–	–	HTML, JavaScript, CSS, XML		Layered	Yes

TABLE 11.4 Wearable Device Open-Source Operating System Comparison for Scheduler, Real Time, Multithreading, File System, Energy Efficiency, and Event Driven Support.

Operating system (OS)	Scheduler present	Real time	Multithreading	File system	Energy efficient	Event driven
Lite OS	Yes	No	Yes	Hierarchical	Yes	Yes
RIOT OS	Yes	Yes	Yes	Flash	Yes	-
Tizen	Yes	-	-	EXT4	-	Yes

TABLE 11.5 Wearable Device Open-Source Operating System Comparison for Security.

Operating system (OS)	Contiki OS	RIOT OS	Lite OS	Free RTOS	Tiny OS
Security	Transport layer security (TLS)	Cyber physical ecosystem (CSL)	Embedded chip security	Wolf SSL	Tiny Sec

11.2.1 ROLE OF OPERATING SYSTEMS IN DIGITAL FORENSICS

The digital forensics also known as computer forensics classify the crime here such that the computer is the target of the crime, the computer is a tool used to commit a crime, and it is the repository of information during the crime commission.[23] The operating systems has a crucial role in digital forensics. The data written to the operating system involves the temporary Internet files and the registry entries may be stored within the virtual environment provided by the cloud that can be lost when the user exits the cloud. In the computer forensics, the analyst can be concerned with either the client or the host computer operating system. If the client is involved in the forensic investigation, then the analysis can be performed on the host operating system. The users associated with the cloud, access the data and relevant services through a variety of devices, that is, from smart phones, desktop PC or a laptop. Lack of standardization of operating systems on these devices creates problems to write programs for extracting forensic data.

11.3 CHALLENGES IN WEARABLE INTERNET OF THINGS

In this section, we will discuss several challenges in WIoT in detail.

11.3.1 CHALLENGES FOR WEARABLE DEVICES

A large number of challenges are associated with the use of wearable devices. The challenges for the wearable devices are owing to their small size, low cost, small weight, energy efficiency, and memory constraints.[24] First and foremost challenge is the fabrication of energy efficient, high-performance sensors and memory modules for the wearable devices. The wearable devices have sensing, computing, and communication abilities. The sensors must be low cost and robust. Challenges also lie in the computations performed at the wearable devices. High speed and intelligent computing facilities must be embedded in the small size micro-controllers. Design of compact, lightweight devices that can provide all intelligent functions and can be conveniently worn is a challenge. Challenge is in designing intelligent algorithms utilizing the limited power budgets and memory space.[25] Wearable device operating lifetime is the biggest challenge owing to the continuous use and high-energy consumption. Limited screen space of wearable computers, mobile devices is a challenge. Design of ultra-low-power RF receiver front end for wearable devices is a challenge. Challenge is to operate battery-limited wearable devices. Challenges are also faced in medical use of wearables. Challenges are in utilization and management of multiple wearables for users. Challenges are in predicting AI-based black box operations in smart wearable devices. Challenges for computer architects are to embed the state-of-the-art analytic capabilities in wearable computers, maintaining the energy efficiency and system performance. Challenges are also in developing lightweight operating systems for the wearable devices that can provide various applications to the users.

11.3.2 COMMUNICATION CHALLENGES

Major challenges in IoT healthcare include the device-to-device (D2D) communication, machine to machine (M2M) communication with seamless data transmission involving no data loss. There are also the interoperability challenges among wearable devices as each publishing device has different communication mechanisms and different communication formats. New challenges arise in upgrading the innovative technologies and resources in wearable IoT healthcare systems. Challenges are in efficient and seamless information flow from wireless body area networks (WBANs) to the Internet.

11.3.3 CHALLENGES IN HEALTHCARE INFRASTRUCTURES

There are challenges for integrating wearable devices in hospital infrastructures. Challenges are in determining the physiological correlation between sweat components and the blood. Detecting human emotions from heartbeats alone is a challenge. Smart connected healthcare pose a number of challenges w.r.t. sensing, communication technologies, and AI techniques. Open challenge is in protecting patients safety and privacy in real-time health monitoring. Challenges are in using new technology/communication models to transmit patients' biomedical information directly to a hospital where it is monitored and diagnosed. In ubiquitous healthcare, key challenges are in tracking carefully the wearable device energy drain and packet loss ratios. Data collection is a challenge in mobile health due to inadequate WiFi access points.

Example of Wearable Devices

- **Smart tattoo:** It analyses the chemicals in sweat and provides the health performance
- **Smart wristband:** It is used for monitoring health and taking care of elderly people. For elderly people, it determines their number of sleep hours, sleep time, the wakeup time, how many hours they are active, where they are walking and sitting, daily pattern, and the change in pattern.
- **Smart nicotine patch:** It is used to get rid of smoking. It is attached to body parts like hands and linked to smartphones and also monitors sleep and wake up times.
- **Clip on clothes:** It vibrates, corrects body posture, and sends alerts for sitting straight in the correct position.
- **Pollution sensor:** It determines the quality of air and helps to decide whether to continue walking in the same environment or not. It is

clipped on clothes to sense the pollution and send the notification on smart phones about unsafe or safe environments.

- **Smart contact lenses:** These monitor the eye and correct the eyesight adaptively.
- **Smart dentures:** These are embedded with sensors that determine whether the person has complete teeth, some teeth, or no teeth at all. They also test the saliva to monitor diabetes and communicate the information about it with smartphones.
- **Anti-radiation under garments:** some people have the habits of keeping their cell phones in their pockets. The radiations from cell phones cause harm to the human body. Anti-radiation undergarments provide a shield against harmful radiations.
- **EKG reader:** It provides a lot more information than just providing the heart rate. It also gives different kinds of heart arrhythmias and irregularities.
- **Wearable air purifier:** It rests on the nose and has a transparent body that helps people to identify whether we are smiling or talking. It works even if a person has facial hair or sweat. It helps breath clean in the presence of wildfires and a lot of pollution around.
- **Mental stress headset:** It uses a light-based neurotechnology. Continuously monitors and tracks the mental workload for any activity. The wearable rests on humans' prefrontal cortex and determines oxygenation in brain tissues. It is totally safe to use and helps to translate information about how much mental effort (stress) we are using to complete any current task. The information from the headset is communicated to mobile handsets or desktop with an easy to understand APP. One can easily understand the stress-related information. It sends alert whenever mental break is required. It can be worn alongside our glasses, on headphones or even with our headband.
- **Wearable music creator (WMC):** It is worn on hand. With WMC, one can create beats using any style, sound, or instrument. It is wrapped around the hand and can be turned to use 3-D space accelerating and slowing down the music and when the grip beats are pressed different notes are activated. It can be used as a super mini-piano if it is used on a flat surface. Bracelets like wearable devices connect via Bluetooth to smart devices or computers.
- **Smart vibrating weight loss device:** It keeps the human mindful about the nutrients to put in the body. It reminds for exercise and to eat better. It is paired with an APP which can be fed with the information

of the times when a person wants to eat. It is worn around the neck or connected to the clothes. Also, tracks the steps and calories burnt keeping the person motivated to remain fit.

- **Smart concentration headband:** It is used for improving attention, concentration, and focus on work. It is also used to keep away from cocaine and other toxic habits. It is lightweight, stylish, compact, and low cost.
- **Wearable keyboard and mouse:** The communication for wearable keyboard and mouse is set via bluetooth and can be used with tablet, computer, and laptops.
- **Smart deep sleep headband**: It is paired with an app to improve sleep quality. It consists of 30 adhesive sensors and can be used several times.
- **Electronic cooling and warming scarf:** Its use is two-in-one. It is a neck accessory that heats and cools the human body. It is lightweight, comfortable to wear throughout the day.
- **Smart wristwatch:** It monitors health rate and EKG. The smart wristwatch is very important for patients with irregular heart rate conditions such as artrial fibrillation. The watch takes a step further to provide an advanced sleep tracking system for monitoring sleep apnea. More than 2.2 million Americans have sleep- apnea according to sleepapnea.org.
- **Smart watch for glucose monitoring:** It monitors blood glucose noninvasively. It needs to be calibrated unlike continuous glucose monitors, wherein a tiny sensor is inserted under the skin. The watch does not require any insertion under the skin.
- **Smart relief heating pad:** It can be worn and placed on the back for warning pain relief. It connects to a smart phone through bluetooth and can be turned ON and OFF. It uses a rechargeable battery. The heat level and the time can be adjusted on demand.
- **Smart dialysis stenosis detection device:** It is a portable medical device to detect dialysis stenosis invasively. Stenosis is found in 40% of patients and is very expensive to treat. Stenosis means narrowing the vein. The device works in 10 s and provides intelligent charge to the physician for better decision making. It can predict a completely blocked thrombotic efficiently with 97% accuracy.
- **Smart ring for blood glucose:** It uses nanotechnology-based gas sensors. It is very tiny which can be clipped to the back-pack or worn as a ring. The input to the device is blood glucose readings based on

which customized alerts are sent to the patients through the device itself and the phone APP when blood glucose level is moving outside the range.

- **Smart brain nerve cell treatment device:** This device delivers microcurrent through small clips worn on the earlobe to the brain for stimulating specific groups of nerve-cells treatment with the device within 20 min and can be used at home. The device is safe and effective for relieving pain, anxiety, and depression.
- **Smart patch for chronic wounds:** It continuously monitors chronic wounds. The electronic (smart) wound patch measures and stores wound-bioparameters that are sent to a cloud-based system using a Smartphone APP where doctors remotely monitor wound- healing evolution. The device sends alerts to caregivers and doctors any hours of infections to prevent complications. The smart (electronic) patch technology synthesizes pure grapheme into a noninvasive dressing that stimulates wound healing. The ultra-flexible properties of the bandage means that it can adapt to any shape of wound and can react to slight change in wound parameters with high sensitivity.
- **Smart spirometer:** It is a portable spirometer for asthma, as a spirometer is an apparatus for measuring the volume of air inspired and expired by lungs. The smart spirometer has a smartphone APP and provides real-time data that can be shared with a doctor. The device provides insights on asthma event triggers. The device is meant for kids with asthma.
- **Smart bracelet:** It is worn on the upper-arm during sleep and can provide caregivers with possible epileptic seizures. The bracelet detects toxic seizures, hypermotion seizures, and clustered myoclonic seizures and closely monitors heart rate and motion while users are in rest position. It features an audiovisual alert system which sends alerts on cell phones and call systems. It can be connected to a web portal where users get the heart rate and motion data number.
- **Smart hemoglobin wearable:** It is a hemoglobin sensing system that can sense invisible blood stools from home. The system is designed to identify early signs of disease like colorectal cancer, kidney disease and bladder cancer number.
- **Smart sinus relief wearable:** It is a small handheld device that uses microcurrent wave and low current electrical stimulation that treat adults having allergic rhinitis. The drug-free solution is an over the counter sinus pain treatment that offers no chemical side effects.

- **ECG headband:** It is a wearable EEG headband that accompanies how people naturally learn to increase brain waves that can impact sleep. It makes use of neuro-feedback therapies with real-time displays of brain activity, audio, and visual cues. This help people to identify and modify the behavior.

11.3.4 SENSORS FOR WEARABLE DEVICES

Several types of sensors used in WIoT, which can be discussed here as:

- **Emotion sensor**: Human emotions are determined from the skin conductance test, temperature measurements, and speech and voice measurements with the help of low-cost emotion sensors. The emotion sensors help to avoid the road accidents by detecting the drowsiness and concentration levels of the drivers. The emotion sensors also help to sense the positive and negative moods of the patients.
- **Capacitive sensors:** The respiratory rates can be detected from the contractions and expansions of chest measured using capacitive sensors.
- **Gyroscopes and Accelerometers:** These are the motion sensors that are usually a part of wrist-watches and smart phones. Accelerometers sense vibrations can be used in step counting and also in heart rate monitoring. Gyroscopes sense rotations can be used to monitor the heart rates.
- **Thermal sensors:** These are embedded in the inner and outer parts of the clothes to sense the body temperature.
- **EMG sensors:** The EMG sensors are used for monitoring the human activities, their facial expressions, and muscle strengths. The EMG sensors are embedded in the textile and in the smart phones.
- **Breathing rate sensors:** These are embedded in smart garments. These directly measure the respiratory rate from the gut that occurs during breathing and the deflections of the chest.
- **Step counter sensors:** These are inbuilt in the smart phones and the smart wristwatches to count the number of steps taken by a person since the last reboot of the device. The step counts are communicated to the users via an app on the smart devices.
- **Macro-bending sensor:** This is embedded in the clothes to monitor the cardiac rates and the respiratory movements. These sensors are based on fiber Bragg gratings in polymer optical fiber.

- **Optical heart rate sensors**: These are used for continuous heart rate monitoring and are part of smart wristwatches and smart jewelry.
- **Microphone:** Microphone sensors are used to capture body sounds. These are part of smart wristwatches, E-tattoo, neckwear, smart jewelry, smart headbands, and smart phones.
- **Proximity sensors:** These are inbuilt in smart devices, such as headbands, earbuds, and eyewear. The sensors give an indication whether the device is in use or the user has worn it or not and whether the device is charged or not.
- **Ambient light sensors:** These are part of smart wristwatches, smart eyewear, and smart phones. The sensors sense the ambient light and just the device screen light accordingly.
- **GPS sensors:** Global positioning system (GPS) sensors are used in healthcare to track the location of patients, particularly to note their movements. GPS sensors help to locate the accident spots. The GPS sensors are part of many smart wristwatches, smart insoles, and smart phones. Some wristwatches that do not have any inbuilt GPS module get paired with the smart phone GPS modules to communicate the geographic location.
- **Altimeters:** These are part of smart wristwatches and are used to obtain the patients' exact position, space by removing the drift errors continually.
- **Barometer:** This is part of smart wristwatch that helps in tracking the atmospheric pressure which is often required in healthcare.
- **Magnetometer:** This is also part of a smart wristwatch that is used to sense the magnetic field of the patient, based on which cancer therapies work.
- **Pulse oximeter:** These are part of smart wristwatches that help to monitor the blood oxygen levels of the patient. These are also used to monitor oxygen levels of the people in COVID-19 situations so that low oxygen levels can be taken care early.
- **Warmth sensor:** This is part of a hug shirt and is used to sense the warmth of the wearer.
- **Glucose sensor:** This is part of smart contact glasses and is used to monitor the glucose level of the wearer
- **Ultrasonic sensors:** These are part of smart hats and are used to provide the spatial information of the wearers surrounding area.

11.4 SECURITY CHALLENGES FOR WEARABLE IOT IN HEALTHCARE

Challenges are for healthcare industries in implementing new technologies, security, data privacy, standardization, regulation, and adaptation of the systems for growing information in healthcare practices. Many privacy challenges are posed due to continuous widespread use of wearable devices. Sensitive healthcare data get transferred via various channels leading to many security challenges. These challenges become crucial when they move to cloud computing. Though there are lots of advantages in moving the wearable devices data to cloud, but security challenges are also a concern. Lots of security protocols are proposed and used that provide the security at the cloud level. As the security challenges are emerging, the solution to provide better protection is also enhanced. The Blockchain is such an emerging technology that provides security by the elimination of trusted third parties in the communication. The next section deals with the architecture and security solution by blockchain technology.

11.5 SECURITY IN WEARABLE INTERNET OF THINGS

In the WIoT, the security is needed at the device level, the communication level and the storage level. At device level, security issues are related with the data logging devices, for example, smart phones and the end-user devices, for example, laptops, computers, and the smart phones. As shown in Figure 11.1, the wearable devices are the unique computing devices that provide numerous benefits to the users. At the same time, these devices bring risk of security for the data as well as the wearer. At communication level, the security issues are with the short-range wireless communication standards like Bluetooth which pairs the sensed and processed health data with the smart data logging devices. The security issues are varying with respect to various levels in the wearable device data flow. Figure 11.3 below demonstrates the security requirements at various places in wearable devices.

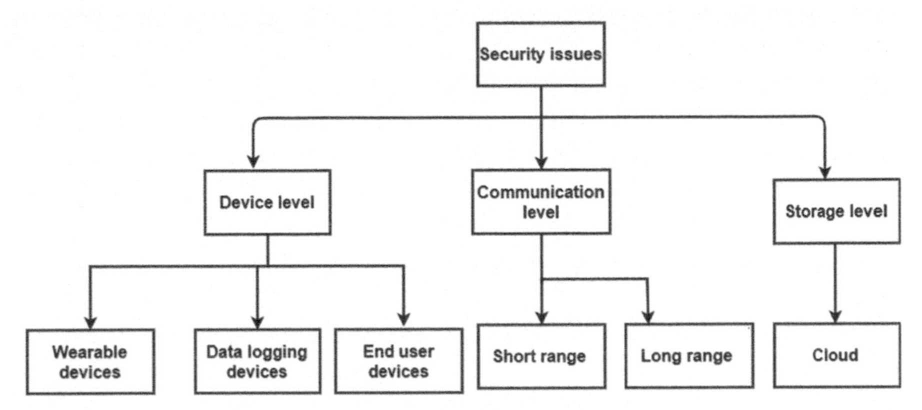

FIGURE 11.3 Security at different levels in wearable IoT.

11.5.1 SECURITY AT DIFFERENT LEVELS IN WEARABLE IOT

Wearable devices come with varying size and computational power. Depending on the usage and applications, the size may vary. Most of the devices like smart watches, etc. use personal data of the users in terms of health and fitness. Many users also perform the transaction using these devices. As the size and computational capacity of wearable devices is less, a strong authentication scheme that needs to protect the sensitive data stored on the devices is impossible to implement. Many times, the authentication password is also not used because of unavailability of the keypad or keyboard.

Also, these devices are connected with the other applications for various reasons. Many users are directed to other applications especially in the healthcare where their data are collected and processed for monitoring their health-related parameters. The open environment for the data transmission and the direct availability of data on these devices leads to various loopholes that are prone to attack. Attackers seek all these issues and can take the advantage of lack of security related to data secrecy and privacy issues. The security of these applications, and so the data privacy and secrecy are the major concerns of wearable devices.[26, 27]

Data collected and received from the sensors of wearable devices are stored in a data logger. The data are transmitted to the data logging devices where they are processed, analyzed, and they display the statistics to the users. Data loggers are the electronic devices that can be used to store data. These devices automatically monitor and record environmental parameters. These devices can include any data acquisition devices, such as plug-in devices or serial communication devices that use computers. A wireless transmitter can convert any smart phone into mobile data logger. A data logger collects the data independently and stores it in nonvolatile memory.[28] Data can be transferred or downloaded from the device to a computer. The stored data can be used for conditional monitoring and analysis. Electronic data logger replaces manual data logging by human observation and recording. It is more effective and accurate to monitor the observations from the wearable devices.[29]

When it is required to discuss the security of acquired data from the wearable devices. It essentially needs the data logger security from which the data can be tempered or viewed. Data logger security arises because:

1. It needs to log sensitive information from the wearable devices that should not be available to malicious users.
2. It needs to maintain the integrity of the data logged to determine on these devices if it was tampered with by an intruder, etc.
3. To capture output at one level for normal operations and at other levels for greater debugging in the event of a failure or an attack.
4. To centralize the control of logging in the system for management purposes.
5. To apply cryptographic mechanisms for ensuring confidentiality and integrity of the logged data.[30]

The preventive measures like cryptographic hash functions and other encryption mechanisms can be used to apply secure logging of data in a timely manner. The authorized access and alteration can be enforced by the implementation of ACL. The logger can be made more secure by applying digital signature while logging the data from the device. As the device memory capacity and computational capacity are substantial, the high-end security features can be applied to these logged data. Multifactor authentication, biometrics, and secure hash calculations like MD5 and SHA can be implemented on these smart devices. The logging devices can be applied with the access control policy and encryption to make the data secure and free from malicious intruders.

There are also end-user devices that process the logger data and apply the statistics. These devices could be smart phones or laptops or servers. The security issues related with these are with respect to the stored data and applications that are required to process these data. Data storage security can be dealt with strong encryption key algorithms and using checksum like MD5 are SHA. Application security can be incorporated by using strong multifactor authentication and role-based access control. The third-party software that is involved in the processing of data and evaluating the statistics are generally controlled at the other end of the wearable devices. These devices perform various operations on data and the statistics will be returned to the user in the form of any visual tool or plain data. The third-party is handling the crucial private user data, it should be trusted and comply with specific security standards to achieve substantial security.

The data transmission and other communications between various entities in the system takes place via short-range and long-range communication devices. Short-range technology includes communication in a smaller region of 1 mm. Various technologies are UWB, Wi-Fi, ZigBee, and Bluetooth. These communication technologies come with their own security mechanisms and application range. As these are used for very short-range communication and the size is also a concern, and limited security is available with these devices. Various encryption schemes are used by different technologies to achieve security. The size of the technology and the distance covered is a compromise between the level of security and the privacy risk associated. A detailed analysis is provided by the author in.[31]

The data are passed to the long-range communication devices if the wearable devices are at distinct locations. Here, the encryption techniques and other session management protocols are used to provide the security. The data once processed and displayed to the user is stored in the cloud for the persistent storage and availability. The cloud service provider has its own security services that are discussed and provided at the time of service-level-agreement (SLA) between the cloud service provider and consumer.

11.6 BLOCKCHAIN TECHNOLOGY

Blockchain is a ledger that is shared and immutable that is used for the process of recording transactions. It can also be used for tracking assets in a business network. An asset can be anything that is tangible or nontangible. A tangible entity could be house, a car, cash, and land, etc. Intangible entities could be intellectual property, patents, copyrights, and branding. All the entities can be tracked and the process can be analyzed and known to all the parties that are involved.

Blockchain derived from Bitcoin is basically a cryptocurrency. Cryptocurrency is an electronic cash or e-cash. It is different from the e-cash used in net banking. The cryptographic hash function is used to generate the token that is used in the form of cryptocurrency.

The current transaction system does carry trust between the entities. Exchange of goods, purchase of items, etc. by credit card, net and mobile banking system by virtually eliminating distance between the buyers and sellers. In spite of these application and features following are the imitations:

1. Duplication of efforts and the requirement for third-party validation.
2. Malicious attacks and mistakes make the business complex.
3. The major problem of the central system is exposing all participants in the network to risk if a central system is compromised.
4. Credit card and other online transaction organizations are incorporating lots of security features that require third-party involvement with the extra cost.
5. Limited transparency and inconsistent information tracking in the movement of goods in the shipping industry lacks the control in business.

As the number of transactions are increasing exponentially, the management and control requires trusted third parties in the communication to provide sufficient security measures.[32]

11.6.1 BITCOIN

It is a digital money that evolved in the year 2009. It addresses the various security issues. It is a decentralized digital currency that is invented by Satoshi Nakamoto.[33] It is managed by a huge distributed network of computers. Joining and participating in bitcoin is permission less. Anyone in the network can perform sending and receiving of the funds. The bitcoin fund is accessed by the individual by e-purse and wallet. A digital wallet (or e-wallet) is an application or software that securely stores user credentials for various payment methods and websites. It makes the user to complete purchases easily and quickly using any communications technology. Users can create stronger passwords that can be accessed through the wallet without remembering.[34] Following features of digital wallet makes it more suitable for bitcoin and blockchain applications:

1. Digital wallets are financial accounts that are used to store funds or digital cash. They can also be used to make transactions and track payment logs.
2. The digital wallets are the software that may be included in a bank's mobile app. They can also act as a payments platform like PayPal.
3. Digital wallets are the most important interface for using cryptocurrencies or digital tokens such as Bitcoin.

Digital wallet is accessed by the hexadecimal digits without using any username. This leads to anonymity in the system. This concept of wallet is used in bitcoin to perform the transaction and communication.

Bitcoin has no central control and is enabled by peer-to-peer network of computers. It has following features:

1. Cost-effective: As there is direct communication between peers, there is no cost required for intermediaries.
2. Efficient: Availability of transactions on public ledger makes it accessible by all peers in the network. Transaction information is recorded once and efficiently accessed through the distributed network.
3. Safe and secure: Once the data are stored on the ledger, they are stored permanently using one-way hash. This tamper proof data cannot be changed. It can just be visible by all the peers making. This makes the system safe and secure.

11.6.2 BLOCKCHAIN

Bitcoin invented the foundation of blockchain, which acts as a Bitcoin-shared ledger. Blockchain (Fig. 11.4) provides the means for recording the Bitcoin transactions. This blockchain is acquired, available, and accessed by all the peers in the bitcoin network.

Blockchain is a chain of blocks that are stored in a public database. Blocks carry digital data and include three parts:

1. Blocks store information about transactions like the transaction hash and other accounting details.
2. Blocks also store the identity information about the participating entities in transactions. Digital signatures are used to include sender and receiver information.

3. Blocks store the digital hash value that is irreversible and distinguish them from other blocks. Each block stores a unique code called a hash that is different for individual records. Hashes are cryptographic codes that are one-way function. Once the hash is generated for a data, it is impossible to get the original data back using the hash.

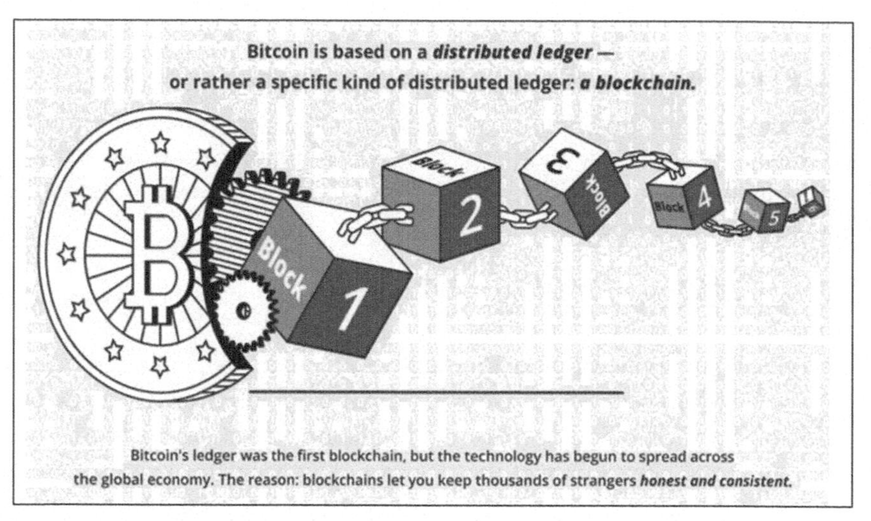

FIGURE 11.4 Blockchain technology.

Source: Image courtesy of Investopedia

11.6.3 HOW BLOCKCHAIN WORKS

Blockchain is a collection of blocks that are linked together using the hash value. Each block consists of a set of transactions. Once the transactions are stored, they have been validated and processed and a block is generated. The validation of transactions is done by the predetermined nodes or computers in the blockchain network. These nodes are called miners. Miners are special computers with high computing facilities that perform the process of mining. In mining operation, miners check and validate the transactions of blocks by calculating a cryptographic puzzle or algorithms that are decided in the beginning of the blockchain creation. Multiple miners will compete to perform the operation on mining, but only one miner solves the puzzle correctly wins the mining. The winner of the miner inserts the processed block at the end of the earlier chain that is already validated. The final blockchain is updated and is broadcast to all the nodes in the network maintaining the consistency and immutability.

11.7 WEARABLE DEVICES SECURITY USING BLOCKCHAIN

The above discussion of the blockchain indicates a secure and transparent data sharing without trusted third party. The wearable devices data security can be enhanced by incorporating blockchain in the process of data access and transmission.

Figure 11.5 below indicates the flow of data from wearable devices to the blockchain.

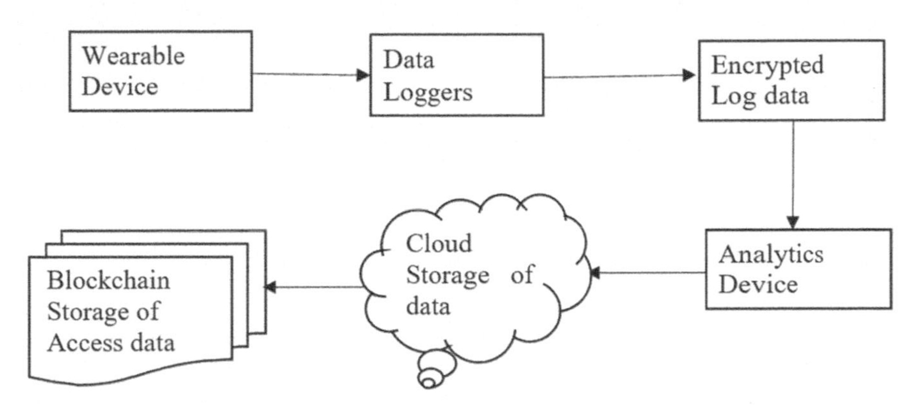

FIGURE 11.5 Data flow from the wearable devices to the blockchain.

The data from wearable devices are passed to the communication devices. Here, the encryption techniques and other session management protocols are used to provide the security. The data once processed and displayed to the user are stored in the devices or node that processes these data. The node data can be in the encrypted form or plaintext depending on the type of user. The analysis and the statistical processing are done on the device. The processed data along with the log records are transmitted to the cloud. Cloud can be used for the persistent storage and availability. The data once are moved toward the cloud are pronc to various security threats of the cloud computing environment. The agreement between the service provider and the cloud is used to provide the security of data and access control. These access control and data travel from the cloud service is managed and controlled by the blockchain technology. The blockchain technology can be combined with the cloud for stored logged data of the wearable devices. Authorization and access control is managed by the blockchain and smart contract. Access control by the authorities involved in the process is used to control secrecy and privacy of the personal data and user health log records. The logged user data security can be enhanced by incorporating the concept of blockchain for the storage and access control.

11.8 CONCLUSION

Wearable devices are used at large by many countries. Security aspects are growing as the application and the usage is increasing. This security becomes more critical when it comes to the cloud environment. Here, the various aspects of security issues are discussed at various levels in the wearable device architecture as well as cloud. Blockchain technology that has captured a large market in the field of security is presented. A solution for achieving security in the cloud can be incorporated by the application of blockchain in the cloud computing services.

KEYWORDS

- **wearable IoT**
- **blockchain**
- **healthcare**
- **communication technologies**
- **operating systems**
- **data access technologies**

REFERENCES

1. Tankovska, H. Connected wearable devices worldwide 2016–2022, Sept. 23, 2020. https://www.statista.com/statistics/487291/global-connected-wearable-devices/
2. IEEE Standards Association. IEEE 802.151-2002-IEEE Standard for Telecommunications and Information Exchange between Systems-LAN. *MAN-Specific Requirements-Part* 15.
3. Chang, K. H. Bluetooth: A Viable Solution for IoT? [Industry Perspectives]. *IEEE Wireless Commun.* **2014**, *21* (6), 6–7.
4. Alliance, ZigBee. *Zigbee Smart Energy Profile Specification.* ZigBee Standards Organization, 2008.
5. WiFi Standards. Sept 30, 2020. https://www.ieee802.org/11/

6. RFID. Regulatory Status for Using RFID in the EPC Gen 2 Band (860 to 960 MHz) of the UHF Spectrum (PDF). GS1.org. 2014-10-31, retrieved 2020-09-30
7. NFC Definition, NFC Record Type. NFC Forum Technical Specification, 2006.
8. Kushalnagar, N.; Montenegro, G.; Schumacher. C. Rfc 4919: Ipv6 over Low-Power Wireless Personal Area Networks (6lowpans): Overview. *Assump. Problem Statement Goals* **2007**, *31*, 45–75.
9. Patel, S.; Park, H.; Bonato, P.; Chan, L.; Rodgers, M. A Review of Wearable Sensors and Systems with Application in Rehabilitation. *J. Neuroeng. Rehabil.* **2012**, *9* (1), 1–17.
10. Chai, P. R.; Wu, R. Y.; Ranney, M. L.; Porter, P. S.; Babu, K. M.; Boyer, E. W. The Virtual Toxicology Service: Wearable Head-Mounted Devices for Medical Toxicology. *J. Med. Toxicol.* **2014**, *10* (4), 382–387.
11. Hernandez, J.; Li, Y.; Rehg, J. M.; Picard, R. W. Bioglass: Physiological Parameter Estimation Using a Head-Mounted Wearable Device.In *2014 4th International Conference on Wireless Mobile Communication and Healthcare-Transforming Healthcare through Innovations in Mobile and Wireless Technologies (MOBIHEALTH)*; IEEE, 2014; pp 55–58.
12. Lin, C. T.; Chuang, C. H.; Cao, Z.; Singh, A. K.; Hung, C. S.; Yu, Y. H.; Nascimben, M. et al. Forehead EEG in Support of Future Feasible Personal Healthcare Solutions: Sleep Management, Headache Prevention, and Depression Treatment. *IEEE Access* **2017**, *5*, 10612–10621.
13. Von Rosenberg, W.; Chanwimalueang, T.; Goverdovsky, V.; Looney, D.; Sharp, D.; Mandic, D. P. Smart Helmet: Wearable Multichannel ECG and EEG. *IEEE J. Transl. Eng. Health Med.* **2016**, *4*.
14. Varsha R., Nair S.M., Tyagi A.K., Aswathy S.U., RadhaKrishnan R. (2021) The Future with Advanced Analytics: A Sequential Analysis of the Disruptive Technology's Scope. In: Abraham A., Hanne T., Castillo O., Gandhi N., Nogueira Rios T., Hong TP. (eds) Hybrid Intelligent Systems. HIS 2020. Advances in Intelligent Systems and Computing, vol 1375. Springer, Cham. https://doi.org/10.1007/978-3-030-73050-5_56.
15. Nitta, N. et al. Li-ion Battery Materials: Present and Future. *Mater. Today* **2015**, *18* (5), 252–264.
16. Farrell, S. Low-Power Wide Area Network (LPWAN) Overview. IEEE, 2018. https://buildbot.tools.ietf.org/html/rfc8376
17. Nuaymi, L. *WiMAX: Technology for Broadband Wireless Access*; John Wiley & Sons, 2007.
18. Tyagi, A. K.; Rekha, G.; Sreenath, N. Beyond the Hype: Internet of Things Concepts, Security and Privacy Concerns. In *International Conference on E-Business and Telecommunications*; Springer: Cham, 2019.
19. Sreenath, N.; Muthuraj, K.; Vinoth Kuzhandaivelu. G. Threats and Vulnerabilities on TCP/OBS Networks. In *2012 International Conference on Computer Communication and Informatics*; IEEE, 2012.
20. Cao, Q. et al. The Liteos Operating System: Towards Unix-like Abstractions for Wireless Sensor Networks. In *2008 International Conference on Information Processing in Sensor Networks (ipsn 2008)*; IEEE, 2008.
21. Baccelli, E. et al. RIOT OS: Towards an OS for the Internet of Things. In *2013 IEEE Conference on Computer Communications Workshops (INFOCOM WKSHPS)*; IEEE, 2013.

22. Vashisht, G.; Vashisht, R. A Study on the Tizen Operating System. *Int. J. Comput. Trends Technol. (IJCTT)* **2014**, *12*.
23. Huebner, E.; Henskens, F. The Role of Operating Systems in Computer Forensics. *ACM SIGOPS Operat. Syst. Rev.* **2008**, *42* (3), 1–3.
24. Kharche, S.; Pawar, S. Node Level Energy Consumption Analysis in 6LoWPAN Network Using Real and Emulated Zolertia Z1 Motes.In *2016 IEEE International Conference on Advanced Networks and Telecommunications Systems (ANTS)*; IEEE, 2016.
25. Kharche, S.; Pawar, S. Optimizing Network Lifetime and QoS in 6LoWPANs Using Deep Neural Networks. *Comput. Electr. Eng.* **2020**, *87*, 106775.
26. Wearable Technology Devices Security and Privacy Vulnerability Analysis. *Int. J. Netw. Security Appl.* May **2016**, *8* (3), 19–30. doi: 10.5121/ijnsa.2016.8302.
27. Wearable Devices: Security Risks. Expert Offers Mitigation Advice for Healthcare Organizations Marianne Kolbasuk McGee (HealthInfoSec), June 24, 2015. 10 Minutes
28. Omega Engineering, Omega Data Loggers, downloaded on July 2020, from https://in.omega.com/prodinfo/dataloggers.html#:~:text=Introduction%20do%20Data%20Logging%20Devices,real%20time%20data%20recording%20system
29. Tiny Tage. Gemini data loggers, downloaded on July 2020, from https://www.geminidataloggers.com/info/what_is_a_data_logger
30. Secure logger, downloaded on July 2020, from https://distrinet.cs.kuleuven.be/software/securitypatterns/catalog-html/Secure%20Logger.html
31. Yua, Y.; Zheng, L.; Zhu, J.; Cao, Y.; Hu, B. Technology of Short-distance Wireless Communication and Its Application Based on Equipment Support. In *Advances in Materials, Machinery, Electronics II AIP Conference Proceedings*; 1955. 040135-1–040135-5; https://doi.org/10.1063/1.5033799 Published by AIP Publishing. 978-0-7354-1654-3/$30.00
32. IBM Blockchain, downloaded in July 2020, from https://www.ibm.com/downloads/cas/OK5M0E49
33. Bitcoin, downloaded in July 2020, from https://www.bitcoin.com/get-started/a-quick-introduction-to-bitcoin/
34. Digital Wallet. https://www.investopedia.com/terms/d/digital-wallet.asp

CHAPTER 12

AI-Based Robotics in E-Healthcare Applications

P. PRAVEEN KUMAR, T. ANANTH KUMAR*, R. RAJMOHAN, and M. PAVITHRA

Department of Computer Science and Engineering, IFET College of Engineering, Tamil Nadu, India

**Corresponding author. E-mail: ananth.eec@gmail.com*

ABSTRACT

The rise of IoT systems for centralized and integrated medical devices and sensors in the medical industry has changed the scenario. Several IoT applications based on the healthcare platform were recently developed. One of the goals of IoT systems is to assist medical professionals with patient care activities. Like many IoT-based solutions built for healthcare, healthcare advancements are becoming more scalable and cost-effective. To assist elderly patients, we have proposed a new deep learning-based medical robot framework called DeepBoT. This model uses live monitoring sensors for measuring patient health condition and deep learning algorithm for analyzing recorded patient's health information. This chapter also provides an overview of current robot devices and artificial intelligence (AI) technologies and principles in the medical industry. Secondly, robotic, AI and machine learning (ML) technologies will be checked, which will allow for new medical science technology assistance applications. In the course of the procedures, robots are used to assist doctors with the use of instruments in a multimodal communication system. The induction of the concept of AI using

Intelligent Interactive Multimedia Systems for e-Healthcare Applications. Shaveta Malik, PhD Amit Kumar Tyagi, PhD (Eds.)

intelligent devices and the adoption of networking techniques for high-speed data transmission in the healthcare unit has set a new standard for healthcare ideologies.

12.1 INTRODUCTION

Developments and implementation of emerging technology and enhancement of the quality of life of individuals in healthcare units help the people to a safer way of living. AI-based embedded devices like intelligent wearable equipment with highly integrated active sensors that help track, recollect, and diagnose disease from sensory data symptoms. Robotic nurses that track patients' well-being and record patient's well-being in the absence of physicians and help the users to know the status of their health irrespective of their place.[3] The integration of AI techniques to identify and predict disease reliably makes a model wise. The intelligent design increases the capability of the healthcare organization across these strategies so that everyone can benefit. This chapter discusses the existing shortcomings and potential capacities of AI in relation to the numerous and complicated criteria of those who are trying to maximize the beneficial effects of cognitive education. One approach is to create new theoretical models that scientists can describe roughly how learning occurs in the human brain. These models loosely use a neural framework. In an exponentially more complex and "meaningful" multilevel network of neuromorphic chips, this paradigm has processed data, allocating a logical compilation of information at every point. The neuromorphic chip stores and processes information using the same processor or node at the location before sending it to the next node.[16] Instead of the traditional design of a regular computer that puts storage and processing into different separate systems, we switched to a system that minimizes the amount of work and maximizes the processing and storage in one single coherent system. In conjunction with each other, the machine drives itself into a more active state in which sharp learning can take place. Applying deep learning (DL) to medicine promises to be fruitful in the present and future. Scientists also accept that for various assignments, AI will compete with trained pathologists. The use of machine learning (ML) for computer-based diagnostics (CAD) is an example of this. One year before, the human technicians and doctors' official diagnoses were identified by the use of CAD mammograms of women who acquired CANCERS for breast cancer.

This chapter introduces a new IoT aware, intelligent system for automated patient and biomedical monitoring and tracking within hospitals and health organizations. Keeping to the IoT vision, we offer a Smart Hospital System (SHS), which uses various but complementary technologies, especially RFID, WSN, and smart mobile, interoperable across network infrastructures from CoAP/6LoWPAN/REST. This chapter demonstrates the human robot architecture as an innovative smart medical service system that overcomes the limitations in medical applications through conventional multisensor fusion methods. Ultimately, the healthcare model 5.0 summarizes the leading technologies such as AI, IoT, and 5G connectivity.

12.2 ROBOTS FOR TRACKING/MONITORING PATIENTS

Robot and automation technologies expand daily in the health and allied sectors. For the next couple of years, the International Robotics Federation (IFR) expects an ever-expanding market pattern in medical robots with an estimated USD 9.1 billion in 2022.[18] Robots enable not only doctors and healthcare professionals to conduct complicated and reliable. The application depends on the requirements of the physical properties of the medical robot. In several tasks from surgical, restoration and service robots, standard as well as concurrent robots are used. In the hospital, the majority of the service robots are versions of high-payload, but limited (DOF) mobile robots. However, multi-DOF surgical robots are flexible and accurate systems that show decent efficiency and effectiveness like a well-trained human surgeon, typically with a minimum estimated range within mm thickness.

Medical robots use sophisticated technologies to perform different tasks necessary to tidy up, sterilize, transportation, administer, repair, and work.[20] Evolutionary versatile integrated interfaces for managing and handling such dynamic and versatile robotics are commonly used. Figure 12.1 is the multifunctional device that operates in a resurgence of healthcare. Two settings, the patient's home and the medical office are included with this situation. Both ecosystems are interconnected by a multirobot network, with one or more operators certain in each ecosystem. The framework is a conduit between the patient and the doctor. It is worth noting that the inner layer of this formulation represents the smarter part of a basic teleoperative structure. The doctor uses his console to enter the world of the patient and retrieves information on the state of the patient.[5]

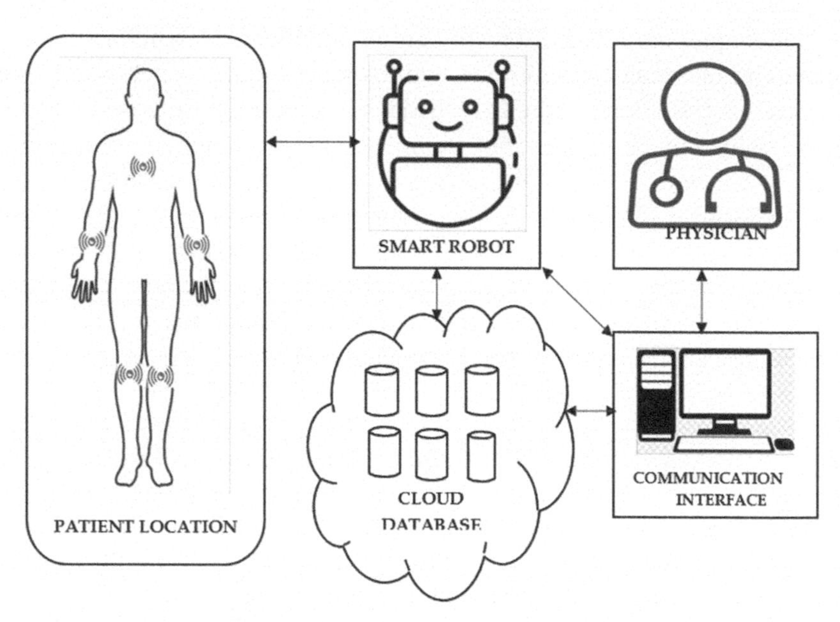

FIGURE 12.1 Health monitoring using intelligent smart robot.

The patient approaches the robot directly and the other way around too. It is responsible for the processing of environmental information. To archive the recorded data on the cloud server, it is compiled and protected. The data manager organizes and changes the records to be aligned with the method with the current interpreted findings for the therapeutic function of medical professionals.

12.2.1 AI-BASED PATIENT MONITORING SYSTEM

Smart robot networks are network structures that are decentralized. Organization may be transferred to other devices which host a device connectivity node or group of nodes. The smart robot framework can be deployed via Internet and remote locations.[26] The new method utilizes the web for exchanging information, data, and warnings between the physician's office and the location of the patient. The interface of the physician is linked by the smart robot to the multifunctional device. It imposes all the objectives and research questions to shape the network of agents and collects information over the internet from the patient's area.

Patient data are collected for medical purposes at places available to the doctor from a computer. For confidentiality purposes, data are often processed on cloud computers. The system's capabilities are coordinated such that the whole system is constantly synced, tested, and exchanged, and future disputes in terms of the existence of values are eliminated. In reality, values will evolve in the course of contact with the world or the user, ensuring that the proper alignment of information between all processes is assured. The Robotic Assistant is installed as a device program in the physician's home computer. It functions as an extension of a touch screen which provides the doctor with access to the atmosphere of the patient.[9] The Robotic Adjutant allows doctors to work with the robot in the location of the patient remotely. Operations that have been approved encourage doctors to control the device, such as the tracking of device activity, interrupt, or restart if appropriate. There are however, most autonomous mobile bots exist, which allow the doctor to work autonomously throughout patient contact into the patient's atmosphere. In this way, the robot is self-sufficient and will recommend such behavior for the doctor. In our framework, doctors are willing and teleoperate and change prior procedures or to introduce new treatments to improve the health of the patient. The robotic caretaker is a good option for the supervision and assistance of patients in lieu of the physicians and should not be called a professional's robotic avatar.

The strategist agent shall understand the robot's intellect. It includes data syncs, changes, and stored in the theology for the Information Manager to pick the right solution that suits the current scenario.[26] To accomplish the system purpose, the Strategist Agent finds the correct answer using a rational formula, which, once the context has been defined, allows the system to progress toward the goal. Logical calculations sometimes called schedules in multiagent structures that are usually transferred to a Design Database (Design Database) repository. Robots are embedded with strategies that use sight and sensing agent expectations and knowledge. The robot in the atmosphere of the patient is indeed a robotic nurse. The design of multiagents offers modules for information control, planning, orientation, perception, and robot feel. Every agent uses its logical model in order to achieve the given objectives which reflect the agent's desires. The robot technology utilizes algorithms to stop and pass obstacles. The agent can use an artist's rendition of the state-of-the-art robot SLAM (Simultaneous Localization And Mapping) algorithm[26] to maneuver in unspecified indoor sites. The object reconnaissance model utilizes the DL models YOLO. YOLO is a state-of-the-art object identification and recognizing implementation framework.

Object recognition is useful for moving tasks, such as stopping or improving the positioning algorithm or using such artifacts as targets or fixed points.

The perception agent may also sense the patient's state, such as knowing that the patient is seated, resting, or walking. The vision agent still cares about other chores. Other sensors are used to detect the activities and improve the consistency of the experience on the multirobot network. An entity who is assigned to handle sensed data constantly detects the area with a view to maintaining up-to-date telemetries on the system. The collected knowledge is stored and exchanged by all agents. In addition, the robot distributes multiple vision tasks and strong programming operations through other design nodes to manage the complexity of the robot. Credence exchanged by participants in the network is used to choose the right strategy for the system. Programs are considered as individual actions or agent procedures. Sensors are used to gain experiences and an extract input from the world, and of experience is converted into the appropriate belief in a series of convictions treated as defined by the data Manager.[6]

12.2.2 DEEP LEARNING-BASED ROBOT FOR PATIENT MONITORING

DL-based Intelligent Robotic (DeepBoT) Systems in Figure 12.2 are generally designed to enable elderly, disabled persons, and people with medical illness to live safely at home. AAL systems also allow for collaborative support of a health-controlled environment.

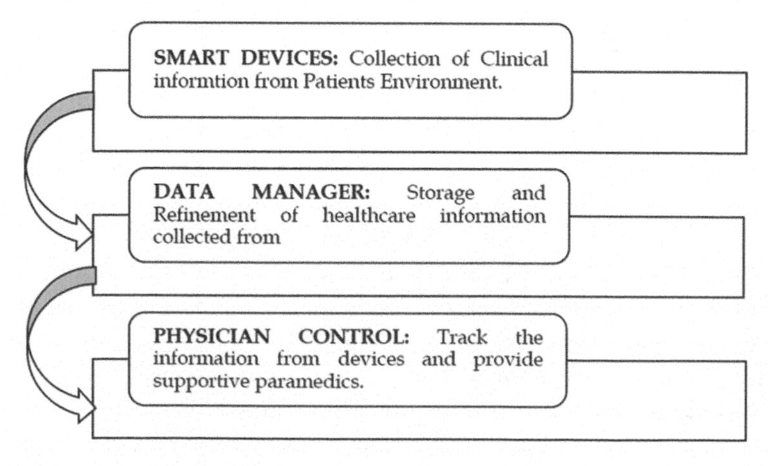

FIGURE 12.2 DeepBoT framework.

The platforms consist of a range of portable sensor, smart phones with wired embedded with sensors, networking equipment, and software platforms, which are linked in an atmospheric-aided ecosystem to interchange information and provide services. The wearable devices are attached to the DeepBoT and patient portals for the transmission of health information to the Physician.[15] In the following section, the architecture of the proposed DeepBoT model is discussed in depth. In order to monitor the elderly people in real time, DeepBoT uses intelligent robots and embedded devices to improve the performance of the medical care by providing comprehensive and comfortable care platforms that support home-based surveillance. DeepBoT has the primary task to regulate and retain compliance by an officially approved healthcare provider and physicians to neurobiological data gathered from the portable sensors. The intelligence gathering is then generated in the data-center governed by the physician. The framework comprises of three main strata which accomplish the real-time objective.

Smart devices: The patient will be hooked up with a wearable device for the collection of clinical information. These sensors measure vital signs, such as the attenuation of the oxygen consumption, the body temperature, the respiratory rate, and a wide range of health sensors. It is very crucial to observe these health issues in patients' body as any anomalous data might eventually wind up with an illness. Senses are transferred via Smart Robot and eventually to a cloud database to the physician interface. In addition, devices work on a regular basis without the interest of patients to optimize anything.

Data Manager: The Database includes the location of the device data and stores the refined data. The platform collects health information from its mobile via the network and the information is filtered for physician examination. Furthermore, the data collection and preparation across all condition, identification of patients, and cloud storage were carried out so that irregular variations in the health information were classified according to the patient condition and illnesses. Any patient and/or physician's website or doctor's office, or both may be submitted with all details/details collected from the patient status.[25] This can be seen in simpler medications and direct clinical information notifications. Database thereby facilitates the technology to connect and exchange expertise, and encourages physicians to store records, insights, and analysis on patients, enabling those of a common preference to automatically retrieve the content. This can be seen in simpler medications and direct healthcare information notifications.

Physician's Control: It is a forum for physicians to track the information and rhetorical devices data of patients. The physicians will review the data from the database generated by the system and take measures. Data integration occurs in real-time at this site by taking all information to the database until it is able to be used to hold physicians informed, and to support paramedics.[12] In the emergency situations, take an immediate step to avoid inpatient care until the situation gets worse.

12.2.3 COMMUNICATION MODULE IN DEEPBOT

DeepBoT is suggested to track human conditions through an IoT public health system. The healthcare agency called Nursing Faith of its residents in care facilities, based on statistics for disease-affected patients. The framework in Figure 12.3 is based on the variability and low-disruption resources.

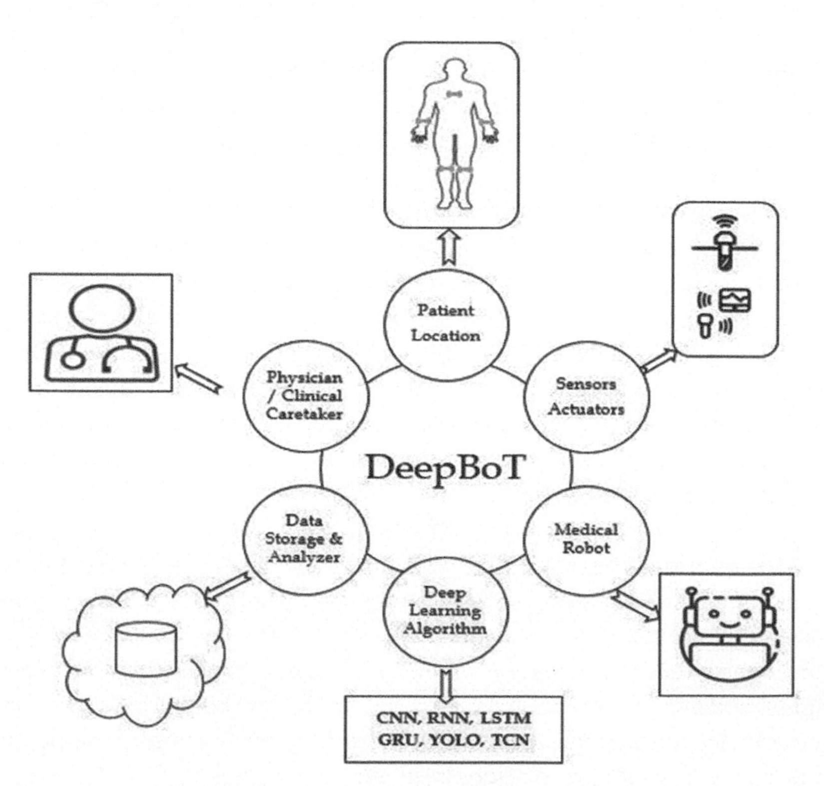

FIGURE 12.3 Communication framework in DeepBoT.

Patients Environment: This reflects the manifested component, for which patients are linked and worked over IoT's (i.e., being capable of sending and receiving information through this system) all the clinical devices and wearables.[11] A specific identity and information exchange capacity across the platform to communicate with the portal is given for any transmitter/portable. Biotechnological devices and/or environmental (circumcision) data may be the information provided in this sheet. The information collected in this level is transferred via network routing protocols to the Intelligent Robot.

Smart Robot: This component is to communicate with the detectors used by patients to detect signs of the condition and conduct main data analysis. This section consists in a review of the circumstances of patients referred to the clinicians. Furthermore, the robot is often capable of responding to signs of deformity, for example, whenever a requisition for assistance (e.g., an apprentice care manager request) or distress request (e.g., a paramedics call) is observed.

Storage Manager: This section has been implemented on the smart robot info. The key collection and identification task of diseases is then conducted. The information collected in the database is processed by DL computer processing and research tasks to classify the illnesses and anomalies.

Physician Monitoring: This provides both a forum for patients and a forum for clinicians. Interface also offers regulations of the captured sensed data for statistical modeling, tracking, and accessibility. For example, the patient may have a primary user permission, while the healthcare professional has limited access as per the approved access permission, across various levels.

12.3 BIOMEDICAL DEVICES IN HOSPITALS

From the traditional hardware-based medical systems to include, or be, biological material has become the new way of healthcare. In more recent years, biology and IT have converged with chemistry, engineering, and nano-technology to create biomedical devices that are becoming more biology than hardware. Biomedical systems consist of equipment, machinery, implants, in vitro reagents, software, materials, and other associated devices for the prevention, diagnosis, treatment, and recovery of human diseases in a safe and efficient manner. These devices can accomplish many purposes, such as diagnosis of illness or injury, monitoring of treatment, intensive care, sampling, replacement of body parts, etc. In the biomedical device industry, biomedical output has come into being and has played an important role.[10]

With the implementation of various production technologies for medical care and biomedical device manufacturing, protection, consistency, quantity, cost, efficiency, and speed may be improved.

Prevention, diagnosis, treatment, and control of physiological disorders are the primary purpose of biomedical devices. As most of these devices touch the human body, it must be a non-error process to manufacture them, and the biomedical device guidelines are very stringent. Failure to comply with these recommendations can lead to severe problems in patients, such as metal hip implants, which can release toxic metal particles during wear and cause significant health effects at and probably in the implant site.[10] Biomedical instruments are developed primarily for the diagnosis or treatment of humans. The human body is a dynamic structure. The average human being has around 37.2 trillion different organism-forming cells. In fact, all these organ systems may be affected by diseases and injuries that require diagnostic and care, and require medical instruments to allow proper operation and recovery. Examples of devices which have been used for a long time are dental implants, sutures and bone casts. Without external gas supplies, the TV-100 has an internal compressor which supplies air. Oxygen can also be supplied at O_2 concentrations from a 50 psi gas supply, distributed from 21% to 100% via an internal mixer.[6]

Sonendo's GentleWave system is a new option in endodontic technology which features a unique fluid delivery and control system for root canal therapy. The GentleWave system has SoundFlow from the company, providing more efficient and effective delivery of fluid. In combination with the SoundBAR of the company, the system combines macro and microscopic cleaning. The HeartLogic diagnosis of heart failure by Boston Scientific is a technology bundled into implantable CRT-Ds and ICDs. Potential weekly heart failure can be predicted and early signs of worsening heart failure can be identified. To monitor physiological patterns, diagnostic heart failure features multiple sensors. Sensors for heart sounds display high filling pressure and weakened ventricular contraction. Sensors for thoracic impedance test fluid and pulmonary edema.[6] Respiration sensors detect short patterns of breathing associated with short breathing. Heart rate monitors are recommended to detect cardiac problem and arrhythmia.

Edwards Lifesciences Inspiris Resilia aortic valve is designed for patients with heart valve disease and a valve replacement surgery is recommended. The valve is made of bovine heart tissue preserved by Edwards technology to serve as a storage system to reduce the build-up of calcium in the heart. While at the same time serves to prevent the valve from becoming brittle.

The technology blocks calcium from tissue deposition, and according to the company, prevents further exposure to glutaraldehyde. Unlike mechanical valves, the Inspiris aortic valve typically does not require the heart valve replacement patient to take long-term blood thinners.

Glaukos' iStent, which is called a glaukos' iStent, is designed for safely and effectively reducing the intraocular pressure in patients diagnosed with primary open-angle glaucoma, pseudoexfoliative glaucoma, and pigment glaucoma. In some cataract operations or as a standalone procedure, the injection is implanted through a single shot only. It creates a bypass through the primary blockage site to improve the company's physiological outflow.[6] Glaukos' iStent Inject system features two preloaded stents in a sterile single-use inserter. With a central inlet and four outlets, iStent Inject optimizes channel access to flow and collector. The preloaded inserted cam drive is used to implant two stents and the insertion tube features a window to optimize visualization during implantation. A trocar tip guides implant easily through trabecular meshwork.[1]

The Magtrace and a Sentimag Magnetic Location System for biopsies is a magnetic system used for guiding lymph node biopsies in breast cancer surgery. The system consists of a magnetic tracer, a magnetic sensor, and base device to help with sentinel lymph node biopsies, according to the FDA. Magtrace is a solution containing iron nanoparticles and carboxydextran. The Magtraceis injected into the tissue and the lymph nodes absorb the magnetic particles. Surgeons can then use a Sentimag probe on their skin near the tumor to detect the sentinel lymph node magnetic particles. The lymph node can be removed for testing of cancer cells.[7]

Medtronic deep brain stimulation is a therapy intended for people with six or more seizures per month on average over a period of 3 months. It is intended for use in patients of 18 years and older who have been diagnosed with a partially occurring seizure and are refractory to up to three and more antiepileptic medicines. The stimulation showed that the frequency of seizures in adults with medically refractory epilepsy was decreased. The most severe seizures, complex partial seizures, and the incidence of epilepsy injury have been reduced. People who used deep brain stimulation in Medtronic have also reported to have long-term improved quality of life. Medtronic deep brain stimulation works to stimulate epilepsy by targeting the anterior nucleus of the thalamus, part of the brain involved in the seizure process.

The Biotronik PK Papyrus covered coronary stent system is a single stent system with 58% more flexibility and a low crossing profile that enables surgeons to more efficiently dress drillings according to the company. The stent system of Biotronik has a diameter of 1.25 mm, which decreases by 23% from traditional stents of sandwich design. It is used in acute coronary artery perforation.[23]

Ear Therapeutics reSET-O is an 84-day Prescription Digital Therapeutic for opioid use disorder. In addition to ambulatory treatment including transmucosal buprenorphine and contingency management, it is designed to increase the retention of patients in outpacious treatment by providing cognitive behavioral therapy. It is appointed for patients of 18 years of age or older and is supervised by a clinician. The dashboard for the smartphone app, accessible to clinicians, contains information about completed lessons, use of patient-reported substances, patient-reported cravings and triggers, use of patient-reported medication, compliance rewards, and input to clinical data such as urine screens.[5]

12.4 DEEPBOT IN REHABILITATION ROBOTS

Rehabilitation is dedicated to the restoration of natural shape and efficiency following accident or disease, and rehabilitation engineering is dedicated to the provision of disabled and elderly assistance equipment. Rehabilitation is increasingly required internationally, although costs, the number of healthcare employees have been predicted to be un-benefited in the opposite direction. In the field of rehabilitation, robotics was then widely adopted. Mechanical reconstruction equipment covers visually impaired assistive gadgets, prosthetics and orthotics, recovery systems, and other support mechanisms such as savvy wheelchairs. Many study groups around the world have implemented a number of rehabilitation robots, accompanied by a few noteworthy examples chosen to be discussed.[19] The "smart wheelchair," an intelligent wheelchair that incorporates a variety of sensors and human-interface systems to support people with disabilities, was unveiled at the University of Pennsylvania. Their novel structure has the shared control model. The human intelligence user can reason about the environment and monitor the wheelchair, but the human augmentation program is used to perform the navigation-based low-level tasks showing three-dimensional data, planning safe routes, continuous control, and execution. This helps the human user to think in "robot-centered" coordinates, that is, in terms of sensory data and thus the planet's model is known to the augmentation program.[24]

DeepBoT can be used for therapeutic purposes in hospitals. Because of the provision of off-schedule monitoring sensors and DL approach for patient records, DeepBoT was created to carry out such tasks. The DeepBoT navigation system utilizes the exact structure of an emergency clinic corridor situation and relies on late results on validated sensor-based motion arranging calculations that specifically resolve the route issue in an ambiguous and unstructured situation. Algorithms for managing unmodeled factors, such as sensor noise and sensor inaccuracy are also incorporated into the framework. An improved light direction and range (LiDAR) scanner can be embedded into DeepBoT by researchers. Inside the "eyes" of the robot, LIDAR can be a system that detects light, measures direction, and decides the range of objects in its way. This is always a strong improvement over previous technology, which used sonar to identify shapes. New sensing systems and ways of mixing data from various sensors have also been sponsored by researchers improving navigation capabilities.[21]

12.5 CASE STUDIES (MEDICAL ROBOTS IN HEALTHCARE)

This section will discuss several case studies with respect to medical robots in healthcare in detail.

12.5.1 SURGICAL ASSISTANT

In the recent years, robots are revolutionizing medical industry. Previously, robots are involved in cleaning floors and supplying foods to patients, etc. But nowadays, Robots are involved in clinical research practices for identifying blood specimens and supplying drugs to the patients. In addition, robots are also involved in performing surgery (surgical robots) to the humans either by their own or through telesurgery. Robotic surgery is the process in which the robots will able to follow the surgical procedure controlled by the computer program.[17] Surgical assistant planning generally follows three steps:

1. patient imaging
2. creating 3D model of the patient imaging
3. planning the surgery

A. Telesurgery: It is one part of robotic surgery in which robots are guided by human surgeon remotely. The surgeon guides the robots from a remote location. The surgeon does not have direct interaction with the patients, but through robotic hand, the surgeon will able to interact with the patients.[14]

B. Types of robotic surgery: Robots are involved in different types of surgery which include robotic gynecologic surgery, robotic kidney surgery, robotic prostate surgery, robotic eyes surgery, etc. Robotic kidney surgery has been successfully carried out by the researchers. The robot will be able to remove one kidney or part of a kidney at short time span. Compared with the human surgeon, the pre-trained robots will able to perform surgery at different organs of the human body.

12.5.2 ROBOTIC NURSE

These robots are intended to help specialists in the emergency clinic in a similar way as that of human attendants. Humanoid nursing robots are providing services 24/7 with minimum cost. Friendly humanoid nurse robots, such as Paro (Japan), Pepper (France), and Dinsow (Japan) serve the patient in their unique style. Robot Paro can participate in straightforward discussions, direct gathering interactions, and comfort shy patients with affection.[14] Robot Pepper can work in a more clinical manner, serving to emergency needs, plan arrangements, give understanding training, and decipher designs in vitals/ labs in appropriate clinical setting. Robot Dinsow is assisting desolate and idle elderly patients with disposition, ambulation, ROM works out, provide suggestions to take medication, and inclusion in self-care.

12.5.3 REHABILITATION ROBOTS

Rehabilitation robots are utilized as a therapy aid to reduce the work of therapists by using robots instead of human intervention. These robots are ideal tools involved in the treatment and recovery of patients suffering from stroke, brain injuries, etc. Many rehabilitation robots have been released, and they have shown positive results when they are clinically tested. Generally, two types of rehabilitation robots are available, one is End-Effector (EE) and the other one is Exoskeleton (Exo) categorized according to the mechanical structures. One type is said to be a therapy robot, it makes the patient to perform the body movements. Other type are based on the telemanipulation which follows the spacecraft robot controlling technique to guide the patients.[27]

12.5.4 SANITIZING AND DISINFECTING ROBOTS

To control the spreading of virus, sanitizing and cleaning the hospital environment is important. Robots are utilized to clean the hospitals during the pandemic situation and they also reduce the risk of disease spreading to the humans.[28] UV disinfection robots could deliver disinfection by killing microorganisms and provide a clean hospital environment and reduce the risk of spreading virus to humans.

12.5.5 BOTS IN DISEASE DETECTION AND TREATMENT

In order to detect the disease and identify the treatment for the disease, mini-bots are utilized. Capsule endoscopy can be used by the doctor to examine the disease related to gastrointestinal tract, intestine, etc. Small pill-sized video camera is swollen by the patient and it records the picture of the organ through which it passes using its own light.[2] The recorded images are transmitted to the device worn by the patient. Later, the images are viewed by the physicians to detect the diseases.

12.5.6 CHATBOTS

It is one of the emerging technologies in healthcare in which patients will be able to interact with bots by sharing the symptoms and problems related to them and Chatbots will prescribe the medicines and diet plan to them. The Chatbots are trained for certain types of symptoms previously based on the training, they will react to the patients request.

12.5.7 BOTS FOR STEM CELL GROWTH

Robots have effectively been utilized to mechanize the cycle of development of smaller than usual organs, or organoids, from undeveloped cells. Until now, the methodology was to develop them from grown-up immature micro-organisms. However, bringing robots and computerization into the image has permitted scientists to create these organoids from pluripotent foundational microorganisms, a profoundly adaptable cell that is fit for changing into any organ. Moreover, these robots are likewise modified to examine the organoids they have made, and in this way, they spare specialists a ton of time and increment precision, as well.

12.5.8 BRAIN BIOPSY ROBOT

Brain and Spinal Cord Cancer is the tenth leading cause of death in both genders. Neurosurgeons need to assess the best treatment for a brain tumor, both in terms of tumor type and tumor grade. Brain biopsy cannulas are used in the biopsy of the brain. This histological confirmation test for brain tumors shows a high diagnosis rate and low complication This has been illustrated by several academic studies. A biopsy cannula is inserted into the target area during the process by syringe pressure. Then, the extracted brain tissue is attached to the cannula and then is withdrawn during the endoscopy process. When the sample is surgically drawn into a cannula, the yield is accomplished by slicing. There are two main procedural difficulties to remember when placing the cannula, which is inserting it into the target area. Another problem is the safe collection of the biopsy specimen. The manual biopsy procedure can have many challenges, such as insertion depth control issues, tremor that causes errors, and aspiration pressure that does not stay under control. The USPTO's main results are composed of the biopsy module, which consists of the injection unit and the pressure control unit. To allow the operator to attach the inserter and the robot, the connector that is mounted on the EE of the robot is used. To use a lockable connector, the connector must be inserted through the pressure control system to reenergize it. A linear actuator, which can be operated by the insert, consists of a stepper and a stepper motor screw lead that are operated by the inserter. A Stepper M1 motor helps to lock the outer shell of the cannula using a connecting mechanism while the cannula is in the target position.

12.5.9 RADIO SURGERY FOR TUMOR

Present stereotactic instruments are commonly divided into three fundamental platforms:

A device that has a radiation source of cobalt-60 and a rigid skeletal stabilization (Gamma Knife). Like ever-evolving 3D printer systems, Gantry-based devices are used in the manufacturing of these devices. A linear accelerator (LINAC) is a machine that is used to treat cancer with x-rays. It is a machine that is controlled by computers. The Gamma Knife system is a radiation therapy that is used for the treatment of tumors.. This is a machine that is used to do a type of surgery using x-rays. Cyber Knife System is also a type of robotic radiotherapy device involved in robotic surgery. It is a machine that uses x-rays. A recently published report indicated that the mechanical precision was approximately of about 0.2 mm and the most effective total device precision was approximately of 1.71 mm for an MRI-based care preparation system equipped with an automated positioning system. The Cyber Knife system's accuracy is statistically similar to that of the previous total software accuracy of 0.61 mm of the skeletal monitoring system on anthropomorphic phantoms of 2.3 mm and a CT-based care preparation of 2.7 mm, leading to a targeting error of approximately 0.49 mm. The Cyber Knife uses an image-guided targeting device (targeting device) and a robotic device (delivery system) that is attached to a medical instrument (treatment delivery) to conduct treatment (more or less) faster and safer. The X-ray imaging system consists of coupled x-ray sources and flat panel detectors, while these sources and detectors are mounted on a lightweight, 6 MV X-band LINAC, utilizing a robotic manipulator for delivery within the treatment. Neodymium core-shell integrated circuitry, placed on either side of the patient inside of a diffraction grating picture tube, will quickly provide the physician with two high-resolution, optical real-time images. These images are digitally replicated radiographs recorded with imbedded fiducials or bony anatomy. CT data are collected by aligning embedded fiducials or bony anatomy.[8]

The precision of the Cyber Knife system is comparable to the results of recent studies performed. The overall error of the skeletal tracking system was found to be 0.6 mm because of a bone segmentation within the Phantom and CT-Dependent Intruder Aligned Care and Targeting preparation, and the error 0.5 mm was due to erroneous targeting. The Cyber Knife system is an image-guided aiming platform factory that is joined with a robotic treatment process. The biocompatible, lightweight, compact, and flexible RBS is designed to provide vision-guided targeting system consisting of coupled x-ray signal sources and flat panel detectors, while the treatment delivery system consists of a lightweight, portable 6 MV LINAC mounted on a robotic manipulator. By having a pair of high-resolution, real-time optical images, they can analyze the brain as well as take images of the either side of the patient's head. In this construction, two correct images are aligned based on an inserted reference marker, or fiducial, to completely replicate the original 3D geometry onto the new one. Utilizing transfer of energy to a pair of orthogonal rotational axes and three orthogonal translational axes, deviations in the x, y, and z-axes are calculated and transmitted to the patient care system.

12.5.10 SPINE ASSIST ROBOT

Spine illness, a prevalent disease in modern life, has caused patients to suffer tremendous physical and psychological pain. Owing to the professional expertise of the specialist, spine surgery was extremely uncertain. In addition, it was to show the anatomy of the spine to the observation doctor. The three-dimensional robot-assisted navigation scheme of spine surgery was based on bilateral and three-dimensional medical images. Optical monitoring devices have been used in general to identify the relative position of the surgical instrument and the patient. With the technology of three-dimensional visualization, virtual space has been developed. The surgical course was planned by the doctor and the robot worked with the surgical instruments to achieve high-precision surgical positioning. Three-dimensional spine surgery navigation can identify the lesion's goal more clearly and reliably and can increase the success rate of spine surgery relative to conventional two-dimensional spine surgery navigation. The three-dimensional reconstruction module's key functions were as follows: First, the CT device was used for sectional analysis of the spine model and its importation into the public machine. The CT image definition section was read by VTK. Second, the three-dimensional ray-projection reconstruction was completed. Thereafter, the contour extraction was done. The Marching Cube Surface Rendering Algorithm was used for 2D CT and 3D reconstruction. The ray-projection algorithm was applied to the surface, rendering model structure observation.[4]

12.5.11 SURGICAL ROBOTS

Surgery robots are often categorized into two classes, (1) the robots that perform surgery, and (2) the robot that assist the surgeon within the surgery, however, here we discuss only the second group. Many physicians and also patients are still doubtful about the security in utilizing such machine within the very sophisticated and skillful work.[27] Truly, the robot has several abilities superior to human physicians, such as the precision, repeatability, controls, and data-expeditions. However, the experienced physical doctors could decide better decisions over the robot AI system in several circumstances. This leads many researchers to mix both advantages together by letting physicians to figure in collaborating with special-built robots for surgical applications. The following are some samples of surgical robots developed and prepared to use within the real surgery. At Johns Hopkins University, robot-integrated robotics has been investigated and developed over the past years, Taylor introduced the “Steady Hand” robot to extend the physicians’ performances in microsurgery.

CISST Lab at Hopkins also developed and implemented a surgical robot in the real total hip replacement surgery. Better results have been found in terms of consistency and precision over the manual surgery. They show the environment during the operation using a surgical robot system. Finally, a few commercial surgical robotic systems are in the market. Examples include RoboDoc system, and da Vinci, Etc. They reviewed a variety of robots currently utilized in the medical aid, while the aim of scripting this work was to introduce the sector of robots within the medical applications. Medical robotics can be a promising tool for transcending human healthcare personnel) restriction and also can enhance the continuity and efficiency of healthcare facilities. Therefore, the medical robotics research should be increasingly supported and investigated from now. Moreover, the robotics requires a multidisciplinary background in engineering and a radical knowledge of the task that the robot is meant to perform. Therefore, the medical robotics require collaboration among engineers and physicians.[22]

12.6 CONCLUSION

In this chapter, the DeepBoT multisensor fusion approach for human–robot interfaces has been established as an innovative smart medical service framework to address the limitations of traditional sensor fusion approaches in medical applications. This model uses live monitoring sensors to measure patient well-being and DL algorithm to interpret recorded patient health information. This chapter also explains evolving robot systems and artificial intelligence (AI) technologies and principles. During procedures, robotics is used to help doctors to use devices in multimodal communication. AI model induction using intelligent machines and introduction of high-speed data transmission networking technologies in healthcare unit set a new norm for healthcare philosophies. This chapter defined a basic modeling framework 5.0 which consists of robot nurses, smart IoT computers, and 5 G networks.

KEYWORDS

- **artificial intelligence**
- **robotics**
- **healthcare**
- **machine learning**
- **IoT**

REFERENCES

1. Adimoolam, M.; John, A.; Balamurugan, N. M.; Kumar, T. A. Green ICT Communication, Networking and Data Processing. In *Green Computing in Smart Cities: Simulation and Techniques*. Springer: Cham, 2020; pp 95–124.
2. Alistair, A. V.; Alnajjar, F.; Gochoo, M.; Khalid, S." Robots, AI, and Cognitive Training in an Era of Mass Age-Related Cognitive Decline: A Systematic Review". *IEEE Access* Jan **2020**.
3. Ashish, G.; Chakraborty, D.; Law, A." Artificial Intelligence in Internet of Things". *ET Res. J.* Sept **2018,** 1–11.
4. Bhagyashree, M.; Patnaik, S." Healthcare 5.0: A Paradigm Shift in Digital Healthcare System Using Artificial Intelligence, IOT and 5G Communication". In *2019 International Conference on Applied Machine Learning (ICAML)*, 2019.
5. Kandzari, D. E.; Birkemeyer, R." PK Papyrus Covered Stent: Device Description and Early Experience for the Treatment of Coronary Artery Perforations". October 1, **2019**, *94* (4), 564–568.
6. Jun Kim, H.; Lim, S. H." Clinical Outcomes of Trabecular Microbypass Stent (iStent) Implantation in Medically Controlled Open-Angle Glaucoma in the Korean Population". *Medicine* Aug. 14, **2020**, *99* (33), e21729.
7. Hernando, J.; Aguirre, P.; Aguilar-Salvatierra, A.; Leizaola-Cardesa, I. O.; Bidaguren, A.; Gómez-Moreno, G. Magnetic Detection of Sentinel Nodes in Oral Squamous Cell Carcinoma by Means of Superparamagnetic Iron Oxide Contrast. *J. Surg. Oncol.* **2020**, *121* (2), 244–248.
8. John, A.; Kumar, T. A.; Adimoolam, M.; Blessy, A. Energy Management and Monitoring Using IoT with Cup Carbon Platform. In *Green Computing in Smart Cities: Simulation and Techniques*. Springer: Cham, 2020; pp. 189–206.
9. Jiang, G.; Yin, L.; Jin, S.; Tian, C.; Ma, X.; Ou, Y. A Simultaneous Localization and Mapping (SLAM) Framework for 2.5 D Map Building Based on Low-Cost LiDAR and Vision Fusion. *Appl. Sci.* **2019**, *9* (10), 2105.
10. Vashishth, K.; Jain, A.; Garg, V." Status of Biomedical Devices Industry: Current Scenario, Way Forward in India a Wider Perspective", 2018.
11. Kumar, T. A.; John, A.; Ramesh Kumar, C. "2. IoT Technology and Applications". In *Internet of Things*. De Gruyter: Berlin, Boston, 2020. doi: https://doi.org/10.1515/9783110677737-002

12. Luo, L.; Zhang, X.; Yang, X.; Yang, W. "Deepbot: A Deep Neural Network Based Approach for Detecting Twitter Bots". In *IOP Conference Series: Materials Science and Engineering*; 2020; p 719.
13. Markan, S.; Verma, Y. Indian Medical Device Sector: Insights from Patent Filing Trends. *BMJ Innov.* Jul 1, **2017**, *3* (3), 167–175.
14. Ye, M.; Li, W.; Tat Ming Chan, D.; Wai Yan Chiu, P.; Li, Z. A Semi- Autonomous Stereotactic Brain Biopsy Robot with Enhanced Safety. *IEEE Robot. Automat. Lett.* April **2020**, *5* (2).
15. Kirtas, K.; Passalis, T. N." Deepbots: A Webots-Based Deep Reinforcement Learning Framework for Robotics". In *IFIP International Conference on Artificial Intelligence Applications and Innovations AIAI 2020: Artificial Intelligence Applications and Innovations*, May 29, 2020; pp 64–75.
16. Mania, N.; Singh, A.; Nimmagadda, S. L." An IoT Guided Healthcare Monitoring System for Managing RealTime Notifications by Fog Computing Services". *Int. Conf. Comput. Intell. Data Sci. (ICCIDS 2019), Procedia Comput. Sci.* **2020,** *167*, 850–859.
17. Taylor, R. H. A Perspective on Medical Robotics. *Proc. IEEE* Sept. **2006**, *94* (9).
18. Samuel, T. A.; Pavithra, M.; Mohan, R. R. LIFI-Based Radiation-Free Monitoring and Transmission Device for Hospitals/Public Places. In *Multimedia and Sensory Input for Augmented, Mixed, and Virtual Reality*. IGI Global, 2021; pp 195–205.
19. Hyeyoung Lee, S.; Park, G.; Youn Cho, D. "Comparisons between End-Effector and Exoskeleton Rehabilitation Robots Regarding Upper Extremity Function among Chronic Stroke Patients with Moderate-to-severe Upper Limb Impairment". *Sci. Robot.* **2020**, Article number 1806.
20. Srimathi, B.; Ananthkumar, T. Li-Fi Based Automated Patient Healthcare Monitoring System. *Indian J. Publ. Health Res. Develop.* **2020**, *11* (2), 387–392.
21. Staubli, P.; Nef, T.; Klamroth-Marganska, V.; Riener, R. Effects of Intensive Arm Training with the Rehabilitation Robot ARMin II in Chronic Stroke Patients: Four Single-Cases. *J. NeuroEngineering Rehab.* **2009**, *6*, 46. https://doi.org/10.1186/1743-0003-6-46b.
22. Taggart, W.; Turkle, S.; Kidd, C. In *An Interactive Robot in a Nursing Home: Preliminary Remarks*. CogSci Android Science Workshop, 2005.
23. Barbero, U.; Cerrato, E.; Secco, G. G.; Tedesch, D." PK Papyrus Coronary Stent System: The Ultrathin Struts Polyurethane-covered Stent". *Future Med.* May 12, **2020**.
24. Veerbeek, J. M.; Langbroek-Amersfoort, A. C.; van Wegen, E. E.; Meskers, C. G.; Kwakkel, G. Effects of Robot-Assisted Therapy for the Upper Limb after Stroke: A Systematic Review and Meta-analysis. *Neurorehab. Neural Repair* **2017**, *31*, 107–121.
25. Shi, W. C.; Sun, H. M." DeepBot: A Time-Based Botnet Detection with Deep Learning." *Soft Comput.* **2020**.
26. Wu, Z.; Chen, X.; Gao, Y.; Li, Y. Rapid Target Detection in High Resolution Remote Sensing Images Using Yolo Model. *ISPAr* **2018**, *42*, 1915–1920.
27. Yang, C.; Zhang, J.; Chen, Y.; Dong, Y.; Zhang, Y. A Review of Exoskeleton-Type Systems and Their Key Technologies. *Proc. Inst. Mech. Eng. C J. Mech. Eng. Sci.* **2008,** *222* (8), 1599–1612.
28. Lu, Y.; Yeung, C.; Radmanesh, A.; Wiemann, R.; Black, P. M.; Golby, A. J. "Comparative Effectiveness of Frame-Based, Frameless, and Intraoperative Magnetic Resonance Imaging Guided Brain Biopsy Techniques." *World Neurosurg.* Mar. **2015**, *83* (3), 261–268.

CHAPTER 13

Automated Health Monitoring System Using the Internet of Things for Improving Healthcare

VERGIN RAJA SAROBIN M.[1,*], SHERLY ALPHONSE[2], and JANI ANBARASI L.[1]

[1]*School of Computer Science and Engineering, VIT Chennai, India*

[2]*Assistant Professor, Senior Grade 2 School of Computer Science and Engineering, VIT University, Chennai, India*

[*]*Corresponding author. E-mail: verginraja.m@vit.ac.in*

ABSTRACT

The aim of this chapter is to investigate the applicability of the four pillars of Industry 4.0 namely, Internet of Things (IoT), Big data, Cloud Computing, and Artificial Intelligence, in the healthcare domain with suitable authors' findings. The industry 4.0 architecture for healthcare domain includes health sensors that gather data from the patient. Thus collected data is relayed to a microcontroller and then sent to the cloud wirelessly. At present almost 70% of the production data in the healthcare domain are unused. Industry 4.0 paves the way for maximum utilization of collected data for on-time diagnosis using machine learning algorithms. The experimental results of authors' findings will be outlined by a comparative analysis of various machine learning algorithms such as Naive Bayes, Random Forest Tree, Support Vector Machine (SVM), and Classification and Regression Trees (CART). Apart from remote health monitoring, it is possible to connect medical devices to help clinicians to make data driven decisions using IoT.

Intelligent Interactive Multimedia Systems for e-Healthcare Applications. Shaveta Malik, PhD
Amit Kumar Tyagi, PhD (Eds.)

A case study of integrating an IoT-based pill dispensing unit to use the compared techniques and protocols is detailed.

13.1 IOT IN THE FIELD OF HEALTHCARE

In recent times, Industry 4.0 is gaining momentum and Internet of Things (IoT) is an integral part of it. Especially healthcare domain can be greatly benefited by the use of IoT. Healthcare systems see an exponential increase in the rate of health issues like cardiac failure, lung failures, and other heart-related diseases. IoT in such circumstances can greatly ease the process of time-to-time health monitoring, which is very essential. The IoT can connect to any device in real-time and convert it into a smarter one. It is very similar to the wireless sensor network (WSN) in which the end devices such as sensor nodes capture the data from the environment and transfer it to the remote location for further processing.[1,2] IoT is the dynamic system behind various applications such as smart home, advanced smart city automation, Industry 4.0, e-health care, and smart grid.[3]

The IoT has created a revolution in the technological prospect of healthcare. Recently, there has been a lot of emphasis on the significance of IoT for healthcare, industrial automation systems, etc. Current venture of IoT, Big data, Cloud Computing, and Artificial Intelligence (AI) which are the core components or the pillars of Industry 4.0 has paved the way for effective and continuous health monitoring remotely and wirelessly. This new business model, Industry 4.0 for healthcare domain, will be studied and explored in detail in this chapter.

An advanced health monitoring system is described in this chapter that includes health sensors that gather data from the patient. Thus collected data are relayed to a microcontroller and then sent to the cloud wirelessly. At present, almost 70% of the production data in the healthcare domain are unused. The fourth revolution in healthcare industry, Healthcare 4.0 paves the way for maximum utilization of collected data for on-time diagnosis using machine learning algorithms. The experimental results of the authors' findings will be outlined by a comparative analysis of various machine learning algorithms such as classification and regression trees (CART), random forest tree, and artificial neural network (ANN). Apart from remote health monitoring, it is possible to connect medical devices to help clinicians to make data-driven decisions using IoT. A case study of integrating an IoT-based pill dispensing unit to use the compared techniques and protocols is detailed here.

13.1.1 CHALLENGES IN HEALTHCARE DOMAIN OVERCOME BY IOT

With the increase of sophisticated services in the modern world, the number of diseases affecting humans is also on the rise due to lack of exercise and proper diet. For older people, a routine medical checkup is very necessary which consumes a lot of money and time. The usage of IoT in healthcare helps to access the medical services from home itself and to diagnose the diseases as earlier as possible. IoT, when applied in the domain of healthcare, helps in parallel monitoring and reporting. This application also helps to transfer the medical reports to a physician directly for immediate medical treatment. This real-time monitoring can send the heart rate information, pressure details, body temperature, Electrocardiogram (ECG) details, blood sugar levels, etc., to the concerned persons like a physician or an insurance firm for further actions in spite of the time or place. The cutting down of hospital visits and the mobility of healthcare facilities creates a next-generation healthcare system. The deployment of IoT can generate a revolution in the healthcare domain.

In the case of manual data analysis, it requires an enormous amount of data to be handled which is very difficult. In IoT applications, the cloud storage can be used for storage and analysis, where only the final reports are presented as graphs to the physicians that are easy to access. The use of artificial intelligence techniques makes the IoT applications less prone to errors. The decision-making techniques help in diagnosing the diseases.

13.1.2 APPLICATIONS OF IOT IN HEALTHCARE

The applications of IoT in the healthcare domain[4] are

- Medical nursing system
- Smart rehabilitation system
- IoT-based kidney abnormality detection system
- The posture detection system for patients
- Physiological condition monitoring
- The medical monitoring system for patients with neurological diseases
- An IoT-based medical healthcare system for autism patients
- ECG monitoring for remote patients
- The secured and fully automated healthcare system
- Smart health band
- Monitoring obstructive sleep apnea (OSA)

- Mobile healthcare system
- Inexpensive cardiac arrhythmia management (ICarMa)
- Medical Bot
- Ubiquitous medical healthcare monitor system
- Continuous monitoring of patients with chronic diseases.

There are several advantages of IoT in healthcare like

- Avoidance of frequent visits to the hospital
- Real-time monitoring and tracking of patient's health
- Availability of quality treatment to all the patients irrespective of the location
- Easy supervision of medical drugs
- Better quality of treatment due to the easy availability of medical devices
- Cutback the flooding of patients to a hospital
- Easy availability of medical services to the areas hit by the storm, flood, tsunami, etc.
- Better interpretation and diagnosis of the medical problem
- Prediction of the arrival of patients to an intensive care unit (ICU)
- Improvement in the quality of living
- Cost savings
- Optimized operations of a hospital

The usage of wearable devices has aided in better monitoring of patients. There are also ingestible sensors that help in tracking the irregularities within a human body due to diabetes. Computer vision techniques help in better visual perception, and the application of AI techniques helps in better decision making. The combination of all these techniques with IoT improves the precision of service. Thus, these wearable IoT devices can monitor temperature, pressure, heart-beat, breathing patterns, etc., and serve the patients through the feedbacks obtained from physicians in emergencies. The preparation time of patient's reports has significantly reduced.

13.2 BACKGROUND OF IOT IN THE FIELD OF HEALTHCARE

In a smart environment, all the devices around us will be connected to one network or another. Sensors and radio frequency identification (RFID) embed the communication systems with the environment. So this generates a massive amount of data. 4G-Long Term Evolution (LTE) wireless and Wi-Fi access help the vision of the IoT to accomplish quickly, and smart portables are evolving widely in various sectors of this modern world. Among the different sectors using IoT, the healthcare sector is the most vital one. A survey by Islam et al.[5] has reviewed different architectures, platforms, and state-of-the-art techniques in IoT-based healthcare systems. This work has analyzed the various security threats and privacy features in an IoT-based healthcare system. The article discusses that the other techniques like big data, artificial intelligence, and wearable devices that can also be incorporated into an IoT-based healthcare system. This work also highlights the efforts taken by the industries to embrace IoT and the technologies that are needed to reshape IoT. It also discusses the visions of different technology firms like Cisco, Microsoft, Google, Samsung, Apple, IBM, etc. This work gives details on recent research activities, policies, and regulations of the IoT-based healthcare system.

According to Sun et al.,[6] the amount of data handled by the cloud storage in IoT healthcare applications increases more exponentially. The security issues are of great concern in IoT healthcare applications. There are three layers in IoT healthcare applications namely perception layer, network layer, and application layer as given in Figure 13.1. The perception layer does the process of collecting healthcare data. The network layer processes the input and transmits it to the application layer. The application layer integrates the data with medical devices and other services. The security requirements that are taken into account while designing a healthcare application are data integrity, data usability, data auditing, and patient information privacy. The solutions for security problems are data encryption, access control, trusted third-party annealing, data search, and data anonymization.

Kang et al.[7] have addressed the different platforms in the IoT-based healthcare systems. This work gives details about the various platforms of healthcare systems that use IoT and collects patient health data for the opinion of doctors. This paper summarizes information about recent devices and developments. They have also discussed the combination of cloud computing with medical services. The combination of IoT with real-time healthcare services helps in improving the on-demand services. The usage of smart devices in the healthcare industry is developing at a faster pace with the Food and Drug Administration (FDA) endorsement. The usage of the cloud is also accelerated in healthcare platforms. A secure open-source like Watson Health Cloud has better performance by sharing and translating patient data. The application programming interface (API) introduced by healthcare, provides more accessible storage and access of data. This platform enables usage of machine learning algorithms and data analytics algorithm. Flex is a manufacturer of medical devices to which Google has signed a deal for cloud computing.

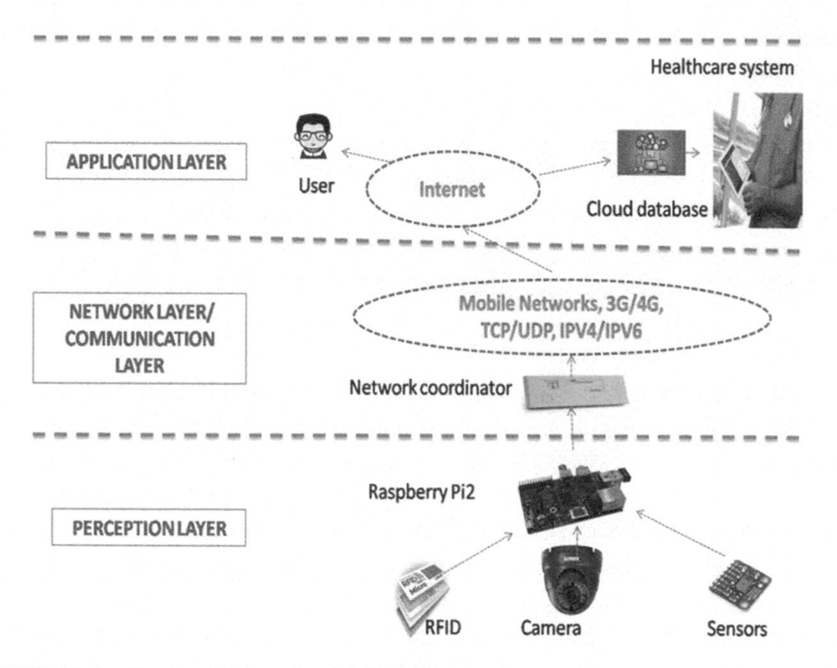

FIGURE 13.1 Systematic three layer IoT healthcare.

Abdelgawad et al.[8] proposed architecture for a wearable IoT system. This system facilitates the living of senior citizens. This architecture comprises a low-power system that can be worn the whole day and can also turbo-charge whenever necessary. This system is very light in weight and is also very comfortable. It is also very affordable as it is made up of low-cost components. A pulse oximeter sensor was utilized to compute the quantity of oxygen in blood and ECG sensor was used to monitor the cardiac status. The ECG sensor also monitors the damages in heart muscles. A nasal sensor was used to measure the breathing rate, and a temperature sensor was used to measure the temperature. The temperature sensor can be moved to any body part in which the temperature needs to be measured. In addition to the medical sensors-mentioned above, fall detection sensors and light sensors can also be used. The fall detection sensor functions like an accelerometer that detects whether the user had a great fall or a lighter one and helped to provide all the services. The light sensor helps the patients to adjust the room light. Bluetooth low-energy (BLE) iBeacon reads the ID of the user by transmitting a Bluetooth signal to track the user's location. The analog signals are converted to digital form by the sensor interface circuits and given as input to the micro-controller. The micro-controller communicates these data to the cloud storage. The system has IEEE 802.11 Wi-Fi and BLE wireless transceivers attached to the Raspberry Pi 2 microcontroller. The data collected from the sensors are then passed to the cloud storage that can be accessed anywhere in the world through the internet. The cloud server has proper API through which the users can access the data quickly. The cloud storage has a more extensive database for storing vast amounts of data and is also connected to API like Google sheets and data analysis algorithms for tracking the user's health data easily. In this implementation, both MySQL and Apache run on the virtual machine (VM) that runs Ubuntu 14.04. This VM is one among the multiple VMs which together constitute a huge VSphere. They have implemented and evaluated a prototype of an IoT-based healthcare system to prove the performance of their architecture. This system was capable of monitoring the falling and lighting conditions easily. The results demonstrate the better working of the system in spite of being a low-cost one.

Guillén et al.[9] have analyzed the IoT protocols and architecture for the transmission of medical data. The patient's data are organized, and the general models are also improved in this work. The main parameters in telemedicine are simple architecture, security protocols, data transmission, and real-time monitoring. Sensors are the hardware or software applications validated by a physician. The results are optimized in this work. In an IoT-based healthcare system, the quality of service should be analyzed. The arrival of new technologies has changed the wired test. The patient can perform their daily activities because of this new architecture. This paper examined the IoT-based system and also proposes a secured architecture. Data transmission has been experimented between the cloud server and various IoT controllers to finalize the main parameters. The conditions were monitored during testing, and the features were selected. A tolerance error was used to measure the clinical variables using different devices. Also, Analog to Digital adaptation is also done based on the media channel, data storage, servers, and other memory devices. The technologies in IoT-based healthcare system are sensor technology, data processing technology, and network technology.[10] The IoT-based healthcare system has five features like stability which involves continuous and trust-worthy monitoring, the continuity that has constant connectivity between users and internet, confidentiality which includes great and confidential storage for patient data, reliability that converts patient data into helpful information, efficiency in diagnosis and medical care. All IoT applications for healthcare include these five features. The steps in data collection are accepting new sensor, dropping a sensor, canceling the utility of a sensor, resetting of a sensor, enabling a sensor, disabling a sensor, putting data, sending the signal, checking the signal, and ignoring the signal. Some of the hospitals, which are using IoT services in the hospital, are Colchester General Hospital of The United Kingdom, Aventura Hospital of the USA, and North Shore University Hospital of the USA, and Apollo Hospitals of India.

The frequency of hospitalization and the effects due to the medical fees are minimized by IoT-based healthcare system. The medical services are more of home-centric type than hospital-centric type because of an e-healthcare system. This application involves embedding sensors and actuators with medical services. De Morais Barroca Filho and de Aquino Junior[11] have presented the architecture as seven layers. The first layer is the requirements layer that is dependable for features to make the system more secure. This layer is composed of mechanisms that guarantee scalability, ubiquity, reliability, interoperability, portability, performance, robustness, privacy, availability, authentication, integrity, and security. The actors like doctors, nurses, and families are responsible for the user's layer. The communication layer is composed of protocols like 6LoWPAN, Zigbee, IEEE 802.15.4, RFID, Bluetooth, Ethernet, WI-FI, IEEE 802.15.6, GPRS, NFC, 3G/4G, and IrDA. In the monitoring layer, data are acquired from heart rate, pulse oximeter, galvanic skin response, muscle activity, transpiration, oxygen saturation, blood pressure, airflow, body movement, body temperature, breathing rate, blood glucose, and ECG sensors. For the environment monitoring, the data regarding temperature, humidity, light, body position, SPO_2, location, motion, pressure, and CO_2 sensors. Figure 13.2 gives a glimpse of IoT sensing and communication module. The middleware layer receives the data from the monitoring layer and passing on them to the systems and services layer. The systems and services layer does the services and formatting. It also uses formatting protocols like XML, EHR, HL7, CSV, HER, and PHR. The application layer protocols are REST, YOAPY, CoAP, XML-RPC, HTTP, and web services. The patient's layer is composed of patients with respiratory diseases, rehabilitation, arterial hypertension, obesity, elderly, and diabetes.

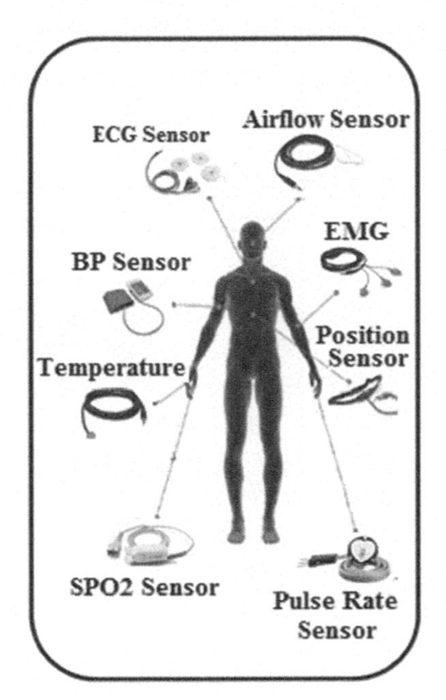

FIGURE 13.2 Sensing and communication module in IoT healthcare.

Hou and Yeh[12] have addressed IoT as a ubiquitous network comprised of interconnected objects. In this paper, a sensor-tag based architecture is used for healthcare applications. A novel authentication scheme based on a single secured sign was proposed, and coexistence proof protocols for robust IoT-based healthcare applications were proposed. This model guarantees a robust IoT-based healthcare system. This paper suggested two secure protocols for communication in an IoT-based healthcare system. A Single Sign-On (SSO)-based scheme and a coexistence proof mechanism were proposed. These proposed techniques can be used in an IoT-based healthcare system consisting of sensors, tagged items, and sensor tags. These techniques helped to provide a secured communication and robust authentication service. Also, a coexistence proof protocol was proposed for assuring multiple sensor-tagged objects existing in the same place. These techniques ensure patient safety and better treatment of medical complications. WSN is used in healthcare systems for receiving medical advice. Security attacks are a hard problem in WSN. In a paper by Sharma and Bhatt[13], hashing and multipath routing techniques were used to overcome the challenges related to security in WSN. After splitting the healthcare data into multiple components, some hashing technique was applied to it. If there was a change in the obtained hash value, then it was detected that there is a change in it. Then the hash value for each component was sent to the servers. These techniques like the secret split and multipath routing helped to preserve the security and privacy of WSN. In this type of healthcare systems, the sensor nodes that monitor the patient's health can communicate among themselves. There are some limitations of WSN like distributed denial of service (DDOS) attack in which the intruder tries to attack the source by sending multiple requests.

Sahi et al.[14] have reviewed different techniques for privacy preservations in IoT-based healthcare devices. In most of the methods used for e-healthcare, security and confidentiality are very important. Two critical aspects in e-healthcare are the quality of the medical services and trust. Privacy is one crucial issue while obtaining a patient's trust. Privacy needs access control, non-repudiation, authentication, and accountability. Privacy requires a wide range of works as IoT-based healthcare devices include IoT, WSN, data storage, and access. If the patient's data are corrupted either intentionally or unintentionally, it affects both the patient and the hospital. Most of the researches focus on only the system parts that have failed. In recent investigations, these have shifted from the organization level to the patient level. Thus the patients are given more authority in maintaining the privacy of their records. The techniques are divided as anonymization and access control techniques for data privacy. This paper explored the solutions offered by research for patient privacy.

While using cloud computing techniques in IoT-based healthcare systems, data privacy becomes a more critical issue. According to Masood et al.[15] because of the usage of cloud computing in IoT, limitations like scalability, storage, and computing are overcome. This combination of cloud technology with WSN helps in the early diagnosis of diseases. But the distributed processing gives way for threats. This paper reviewed the different techniques related to data privacy. The methods were classified according to their areas of application as pair-wise key establishment, dynamic probability packet marking, multi-biometric key generation, hash function, chaotic maps, attribute-based encryption, number theory research unit, hybrid encryption, tri-mode algorithm, and priority-based data forwarding techniques. Their advantages and limitations were also listed. According to this study, IoT-based healthcare has a six-step framework as (1) choosing the preliminaries, (2) selecting the entities of the system, (3) choosing the technique, (4) connecting to point-to-point protocols (PPPs), (5) checking the security, and (6) calculating performance. Gubbi et al.[16] have presented a user-centric model for interacting with the public and private clouds. Scalable cloud is used to satisfy the competing needs of different application areas. These works have separated networking, computing, visualization, and storage in a shared setting for better access. This does not affect the standardization and cloud storage. The scheduling algorithm used to access the cloud computing facilities should have multi-objective optimization, good response time, low cost for usage, availability of more number of resources at low cost, penalties in case of degradation of service, fault tolerance with task duplication, and replication of critical tasks on different resources to avoid error due to failure of one resource.

IoT-based healthcare system increases the productivity, reliability, and accuracy of the medical devices. The resources and the medical services are interconnected in IoT-based healthcare system. The benefits of IoT-based healthcare system are making life more cheap and convenient, improving the outcome of patient, reducing the cost, improving the medical care for patients, creating lives healthier and longer, saving lives, managing maximum number of diseases, real-time management of diseases, prevention of diseases, automatic alerts in significant changes of health in heart patients, saving lives and time, resources simplicity and affordability, easy to use, managing records easily, efficient usage of resources and money and off time services from doctors.

The challenges of IoT-based healthcare system are managing device diversity scale, flexibility and development of applications, data isolation, confidentiality, availability of medical expertise, CPU power, system memory, performance, constraint over network bandwidth, data exchange, resources availability, data volume, hardware implementation, design optimization issues, security challenges, interoperability, technical issues, modeling of the relationship between the sensed measurement and diagnosed diseases, implementation of software for medical analytics, intelligence in decision making, real-time processing of the system, predictability, low power consumption, data integration, unstructured growth of diverse data at an exponential rate, participatory sensing, geographic information systems (GIS)-based visualization, data analytics, cloud computing, architectural challenges, security, energy efficiency, quality of service and protocols. Also, The Plug n' Play objects used should unify with other objects in the other environment. The primary challenge is the standardization of protocols and frequency bands. The primary concern is the security attack which disables the network, pushes the false data, and accesses the personal information. The WSN, RFID, and cloud are highly prone to security attacks. Cryptography is a defense against security attack. RFID is the most vulnerable component for attack as it leads to person identity detection. So, more research on cryptographic algorithms that can be used to encrypt patient data is needed. The updating of current sensor algorithms is also required. Cloud security is another area where there is more scope for research in the IoT-based healthcare system.

13.3 FOURTH REVOLUTION IN HEALTHCARE TECHNOLOGY: HEALTHCARE 4.0

The early industrial growth from (1) water and steam power which is the mechanization era (Industry 1.0), (2) rise of electricity which is the electricity era (Industry 2.0), and (3) semiconductor devices and improvement in computer technology which is the automation era (Industry 3.0) has paved the way for modern industrial revolution, the cyber-physical systems era (Industry 4.0). Industry 4.0 has largely revolutionized the manufacturing industry by disrupting several existing systems. The tremendous and exponential growth of the cyber-physical systems era had a vast influence on society and nearly all the sectors of the economy to increase the global income and the quality of life. Of course, IoT, digital infrastructure, and communication technology are the prime need for Industry 4.0. This new revolution not only impacted the manufacturing sector, but also revolutionizing healthcare technologies (Healthcare 4.0) for servicing the human population. Figure 13.3 gives a relationship of industrial and healthcare technologies growth and transformation.

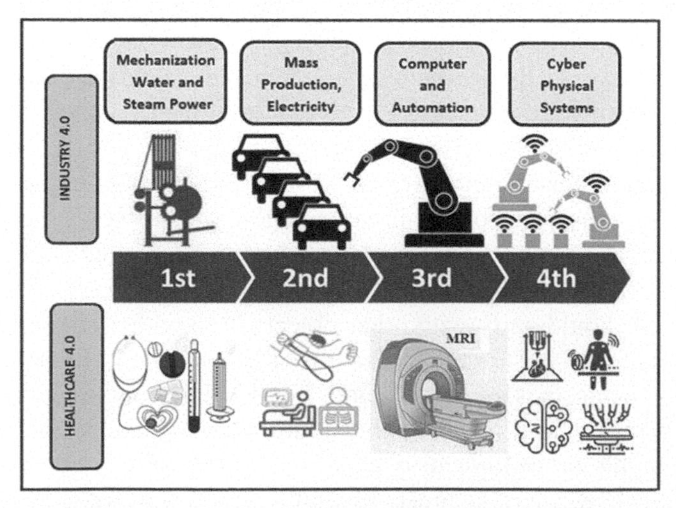

FIGURE 13.3 Technology revolution in manufacturing and healthcare industry.

In the fourth healthcare technology revolution, enormous cyber and physical systems (CPS) are intently merged by Internet of Healthcare Things (IoHT), automated medical production, intelligent sensing, healthcare big data, artificial intelligence, healthcare robotics, and cloud computing to not only provide digitalized healthcare services and to produce digitalized healthcare products. As an outcome of Healthcare 4.0, the whole healthcare sectors have accomplished considerable growth in the direction of 8-P Healthcare vision: patient centered, personalized, predictive, preventive, precision, participatory, pervasive, and pre-emptive healthcare. An illustration of contemporary Healthcare 4.0 is given in Figure 13.4. Following are the major enabling technologies of Healthcare 4.0.

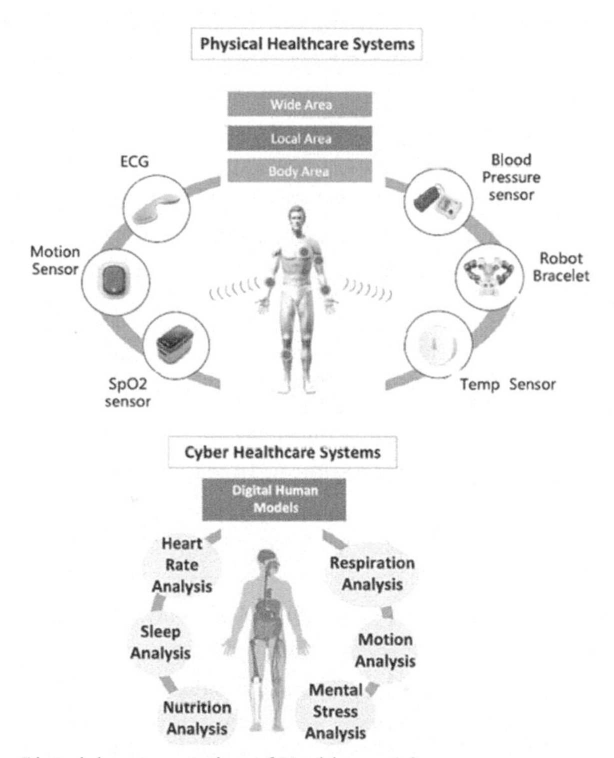

FIGURE 13.4 Pictorial representation of Healthcare 4.0.

13.3.1 CYBER PHYSICAL SYSTEMS

According to Wang,[17] CPS is bringing together both computation and physical processes through Information and Communication Technology (ICT).

In a CPS, the events are observed continuously, controlled, coordinated, and incorporated through computing and communication strategy. Application of CPS comprises manufacturing, military systems, automotive systems, medical devices, traffic control, power generation and distribution, HVAC (heating, ventilation, and air conditioning), water management systems, aircraft, and distributed robotics.

13.3.2 ADVANCED AUTOMATED IN HEALTHCARE

The new venture of Healthcare 4.0 will largely impact the medical equipment manufacture and the production of medicine. Medical care resource efficiency, accuracy in decision making, effective productivity, product and process reliability, quality and service, and cost-effective procedures will be achieved by incorporating advance automation, IoT, and robotics technologies. As an example, an automated human pain assessment system has been developed to help the ICU patients who may fail to self-report.[18] Using a wearable device as a facial mask, the patient's facial expression will be captured using a surface electromyogram (sEMG). The signals are then analyzed to predict the pain intensity automatically from the remote location.

13.3.3 INTERNET OF HEALTH THINGS

Recently, the healthcare industry for its ambient assisted living is seeking the attention of IoT through digital technology progress. For the medical care area, things such as wearable devices, artificial organs, smartphones, biosensors, and smart pharmaceuticals are of sure and such devices are connected through internet. This concept is primarily known as IoHT.[19] Major functionality of medical care things are gather data from the patients; converts the raw information into simple, processed data and transfer the processed data to the application developed for a customer through smartphones, computers, smartwatches, etc. These electronic gadgets are connected to the gateway through communication protocols like IEEE 802.15.4 standard, 6LoWPAN (IPv6 over Low Power Wireless Personal, Zigbee, or Bluetooth). The gateway if further connected to a cloud platform or remote clinical server for data storage and processing.

13.3.4 INTELLIGENT SENSING

The precise physiological and psychological health examination is possible with the development of advanced smart devices and biomedical sensing systems. As of technology perspective, the development of these sensing systems is to be highlighted as ease-of-use, comfortable, long-lasting, miniaturized, and networked to meet the necessity of unobtrusive and accurate sensing. Typically, the sensing and smart devices are integrated in the garments or in the living area. Such the devices are tethered with the human body for a long period of time, consequently more effort is rendered to reduce the size of the device and to improve user experience. On the other hand, there are many challenges in these unobtrusive sensing systems like multisensory data fusion, device power consumption, security and privacy of gathered data, material's biocompatibility system's wearability, and smart sensor miniaturization. These issues have to be resolved to enhance efficiency, reliability, accuracy, interoperability, and security of the unobtrusive sensing systems. As an example of intelligent sensing, a novel design strategy of noncontact wearable ECG system that is proposed by Majumder et al.[20], is advancement in ambulatory ECG systems.

13.3.5 HEALTHCARE BIG DATA

Effective and maximum usage of the enormous amount of health data drawn from the patient is essential to enhance healthcare services in a meaningful way and to render ambient-assisted living. The healthcare big data is emerging from the tremendous volume of information collected from wearable devices.[21] Since the data are gathered on a daily basis from the bio-medical and wearable devices, it becomes very high volume of patient details. Data acquisition, processing of enormous structured and unstructured data from various sources is a big challenge in data analytics. This is well handled with big data analytics in diagnosing patient conditions, prior detection of diseases, and matching treatments with outcomes.

13.3.6 HEALTHCARE ARTIFICIAL INTELLIGENCE

AI signifies the machine finding solution for complex problems through knowledge and reasoning, more or like human intelligence solving real-time problems. This promising technology at present has been used in different applications like industry, military, environment, and also in healthcare systems, to solve various problems for patient monitoring, device handling, and the healthcare industry automation. AI systems are employed to analyze complex health data for the advanced and critical treatment techniques, like disease diagnosis phase, medicine development phase, treatment procedure, and patient monitoring and care. Furthermore, AI-enabled applications and tools are emerging and appearing in healthcare area which could be embedded on wearable and connected smart devices.[22]

13.3.7 HEALTHCARE ROBOTICS

Through technological advancements, healthcare robotic is becoming one among the quickest emerging domain of robotics transforming the traditional medical treatment procedures with the highly evolving robot technology, comprising catheter robotics, companion robots, surgical robots, and medication management.[23] Humans are bound to errors due to lack of attention, memory problem, and at times due to cognitive errors. Currently, surgical robots are becoming perfectly precise today. The human surgeon with the help of surgical robots can carry out lesser invasive surgery and can handle more precise surgery procedures than ever with minor slits and quick healing. In case of emergency, with the help of catheter-automated robots, the surgeons are able to accomplish medical procedures inside a patient's blood vessels without open surgery.

13.3.8 CLOUD COMPUTING

There is no doubt that cloud computing technology is offering its customers the required computing capacity and capabilities with limited investments in infrastructure. The economic perspective of cloud computing can be substantial that provides pay-for-use basis data storage and other resources which progress the resource utilization and performance capabilities. Also, the operational benefits of cloud services provide scalability and security for health-related data.[24] Typically, the cloud service provider data centers are more protected against various attacks and threats using high-level security measures, comprising data encryption. E-health systems developed with the support of cloud services can offer web access to data anytime and from anywhere.

13.4 AUTOMATED HEALTH MONITORING SYSTEM USING IOT FOR IMPROVING HEALTHCARE

The tremendous population growth and their expectation of longer lifetime have driven the need of infant and old-age health monitoring, organ transplantation, artificial organ embedment, and drug dispensing. World Health Organization (WHO) reveals that there is a significant increase in global life expectancy from 66.4 years in 2000 to 71.4 years in 2015.[25] As the elderly population is increasing, the possibility of life-threatening diseases like heart diseases, diabetes, neurological disorder, and strokes also raises. Hence, it is clear that there is an urgent need of an advanced healthcare system for detecting, curing, and preventing such diseases. The major concern here is the preferred system should be cost-effective, robust, reliable, reachable, and harmless. Combining IoT with healthcare engineering can bring up wireless patient monitoring and capture medical data for further analysis. In addition, the network connectivity helps the clinical expert to monitor the patient's health condition remotely. By 2025, 41% of IoT usage will be focused on the healthcare domain.[26] In IoT-driven healthcare systems, the traditional hospital environment can be replaced by smart homes and human body-centered huge monitoring systems will be replaced with miniaturized implanted bio-sensors.

The Automated Health Monitoring System (AHMS) architecture discussed here is a fog computing-based automated health monitoring system. The fog computing-based patient health monitoring system is a new concept in this era. Deploying a fog server reduces the bandwidth requirement and increases the efficiency of the network by providing real-time information to the mobile users nearer to the edge of the network. It is basically three-layer architecture: end device layer, fog layer, and cloud layer as given in Figure 13.5.

13.4.1 END DEVICE LAYER

Sensors and actuators comprise the major functional elements of the end device layer. The most important function of the end device layer is to retrieve the information from the surrounding especially the patient's healthcare-related direct and indirect data with the employed end devices. The most common healthcare data are blood pressure, temperature, glucose level, ECG, Electromyogram (EMG), Electroencephalogram (EEG), pulse oximetry, cholesterol level, and toxins in the body that can improve prior detection and prevention of critical chronic diseases. In this AHMS model, data such as patient's physiological parameters, environment-related data and behavioral-related data are gathered from several wireless hardware smart devices, deployed at various spots at the smart home, and from the body area network of the patient as given in Table 13.1. In a robust healthcare monitoring system, wireless smart sensors can be integrated inside or outside the patient's body as wearable. These hardware devices work on wireless sensing phenomenon and have the capability of sensing and transmitting data in real-time. Each sensor node is integrated with biosensors and other medical sensors. In due course, the gathered information will be relayed to the fog layer.

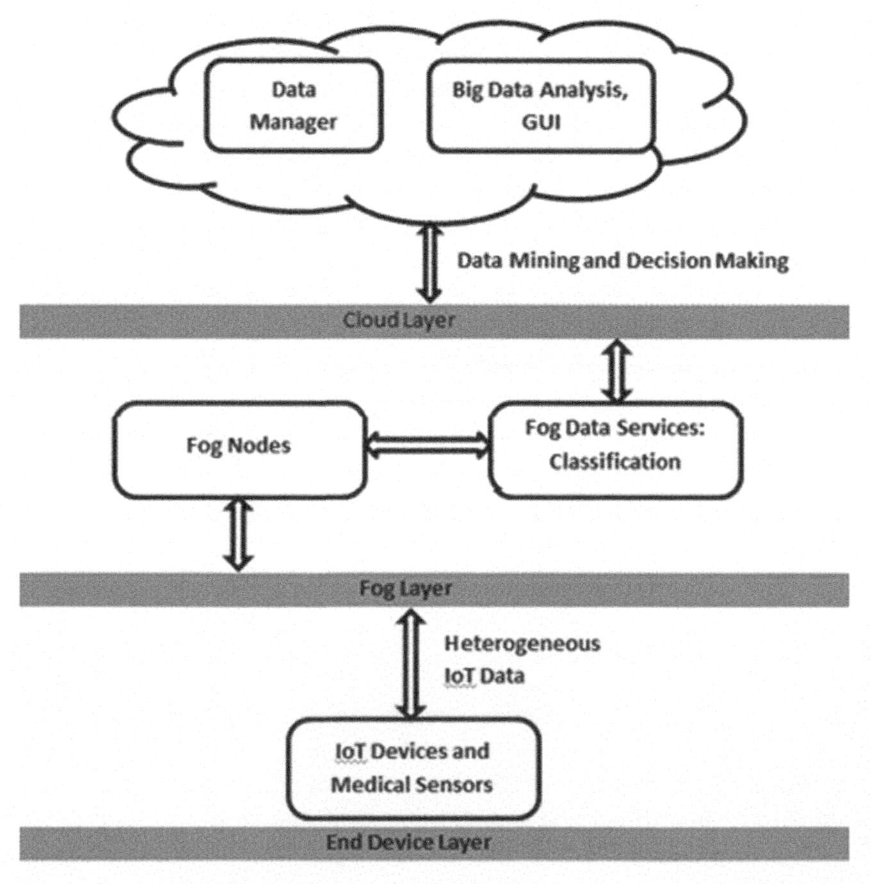

FIGURE 13.5 AHMS architecture.

TABLE 13.1 AHMS Model's Dataset.

AHMS dataset	Health factors	IoT end devices
Health sensitive data	Blood pressure, heart rate, body temperature, glucose level, ECG, and respiration rate	Smart wearables, heart rate sensors, temperature sensor, ECG monitor, and respiratory sensor
Environment-related data	Noise level, environment temperature, toxic waste, and air quality	Temperature sensor, noise sensor, chemical detector, and humidity sensor
Behavioral-related data	Anxiety and stress.	Q sensor, bio-sensors, and smart wearable.

13.4.2 FOG LAYER

The fog layer is an intermediate layer between the end device layer and cloud layer. The fog tier performs processing, storage, and services at the edge devices such as access point, gateway, and routers. Through the fog computing method, the heterogeneous data are pre-processed to generate the required knowledge. In this AHMS model, a smart IoT Gateway is acting as the fog layer to handle data preprocessing, communication, and event processing. The data gathered in the end device layer are of both structured and unstructured data which are converted into the required format by the fog node before sending it to the cloud layer. The fog node performs a data handling process in which the real-time range of the measured data from the edge devices is compared with the guideline range of data prescribed by authorized practitioners. The datasets then will fall in either normal event group or abnormal event group. The sensitive events like blood pressure, glucose level, and temperature if fall under abnormal range lead to patient health to the worst state. Bayesian Belief Network (BBN) classifier based on Naïve Bayes classification method is used in this study to classify normal and abnormal event data with the help of degree of impact (DOI) score. Algorithm 13.1 depicts the procedure to determine the patient state as either safe or unsafe based on patient data acquired at the fog node.

The IoT gateway when encounters the patient data to be abnormal, it sends the notification immediately to the caretaker of the elderly person with the least latency and response time through message queue telemetry transport (MQTT) broker. MQTT is considered in this work because it is a secure and lightweight protocol specially designed for IoT applications. The implementation is based on the Publish-Subscribe IoT communication model to send and receive messages based on topics. MQTT offers end-to-end secured reliable communication based on secure socket layer (SSL).

Algorithm 13.1 Patient State Determination at Fog Layer.

The patient data are gathered from end devices to fog nodes

For each patient
Compare the health attributes with the threshold range
Calculate the Degree of Impact (DOI)
If DOI is in normal range
Then patient_state = safe
Return patient_state
Else patient_state = unsafe
Create an alert signal to the patient
Patient data (ECG and EEG) send to cloud for further analysis.
End for
Output: Current state of the patient.

13.4.3 CLOUD LAYER

Information mining is a primary function in the cloud layer that focuses on transforming the diversified data from the fog nodes to a common analyzable format. Since the patient health attributes are time-sensitive, temporal health index (THI) which is based on the attributes selected from various datasets of the patient is determined in the cloud layer for critical alert generation. Algorithm 13.2 shows fog computing- and cloud computing-based decision-making processes through THI value.

Algorithm 13.2 Alert Generation at Cloud Layer using THI Value.

Event set is brought to cloud layer (Health+Environmental+Behavioral)
For each patient

Step 1: Determine the current state of patient using Algorithm 1
Step 2: If patient_state=unsafe
Step 2.1: Create a checkpoint
Step 2.2: Compute the patient's THI value
Step 2.3: If THI >permissible limit
Patient shifted for emergency medical handling
Early warning signal send to family members
Else continue step 2.2 after some time interval

Step 3: Else continue step 2 after some time interval
End for

Further, the cloud layer patient data are subjected to a big data analyzer with machine learning algorithms. The common analyzable data in the cloud layer are then normalized and divided as training and test data using a random division of 80:20 proportions. Different classification algorithms like Random forest, CART, and ANN have been implemented to investigate the accuracy level. The inference in the following Table 13.2 shows that ANN with 20 Multilayer Perceptron units based on ReLU activation function gives better stability than the other models.

TABLE 13.2 Score Table.

Weights assigned	**5**	**2**	**3**	
Out of 10	**Stability**	**Training performance**	**Validation performance**	**Total**
ANN	9	8	8	90
CART	8	9	9	85
Random forest	7.5	9	9	80

13.5 IOT-BASED PILL DISPENSING SYSTEM

To demonstrate the working of the topics discussed from Sections 13.1–13.4, a case study of a pill dispensing system is considered in this section. The architecture and implementation details for an IoT-based pill dispensing system are explained in detail here. A pill dispensing system is an automated medication dispensing system for users who take medications regularly. They may not be closely supervised by a professional on a day-to-day basis. Statistics have shown there is a probability of mistakenly missing doses or taking overdoses due to the lack of supervision. The device proposed here will avoid these medication consuming errors due to the lack of supervision. The device proposed here takes into account the schedule of the user's medication timing and dosage. It enables easier for the user to adhere to the medication timings. A website is interfaced to connect to the proposed device. Using the web portal, the professional practitioner can prescribe the medicines and the users can see the medicines prescribed. To provide an error-free environment, a barcode facility is used for identifying the transactions. Each and every prescription is associated with a barcode. The barcode will be scanned to identify the prescribed medicine and the dosage accordingly will be given to the users from the pill dispenser.

13.5.1 METHODOLOGY AND IMPLEMENTATION

The proposed system utilizes the BeagleBoard is used for the processing purpose. Servo motors are used for making the mechanical rotation to drop the medicines as required. The hardware components and their interaction in the proposed system are shown in Figure 13.6. The overall workflow is represented in Figure 13.7.

BeagleBone is used in the proposed system because it is one of the processors with low power requirements. Another advantage of using BeagleBone is that the software is open source. The servo motor in the proposed system will aid in the counting of the number of pills. It is a closed-loop motor mostly used in robotics and automated manufacturing. To aid the dispensing of medicines to the dispensing tray, PVC pies are used. Firebase is the platform used for providing the web app. Automatic data synchronization, file storage, messaging, and analytics of the proposed work are done using the firebase.

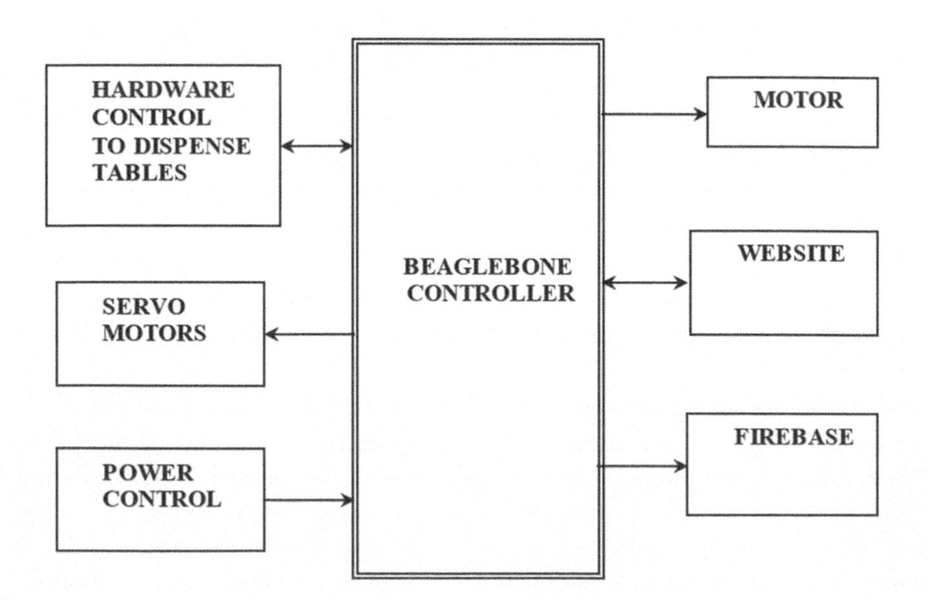

FIGURE 13.6 Module of pill dispenser system.

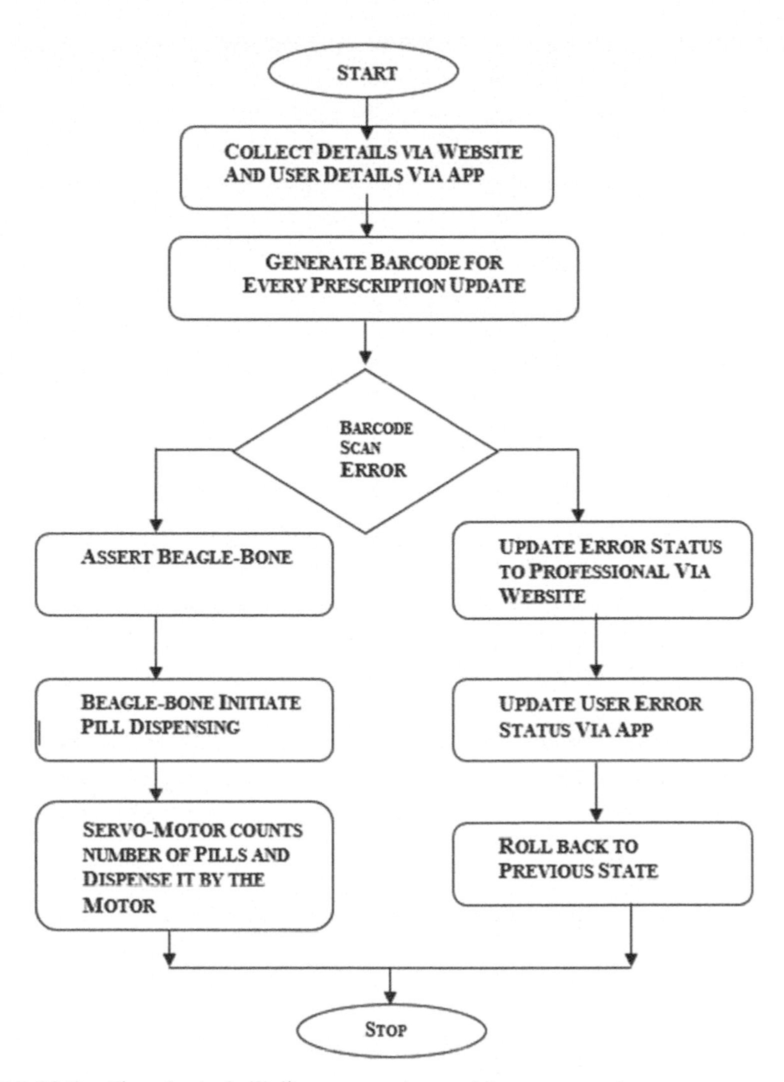

FIGURE 13.7 Flowchart of pill dispenser system and its components.

13.5.2 SYSTEM ANALYSIS

The proposed system is analyzed for scalability, error-free performance, and privacy. The system became scalable because the interfacing is done through web services. A website was created for the prescriptions to be updated by the professional practitioner. Whenever there was a change in the prescription, the details are updated to the user via a web app. This can be scaled easily for any number of customers. We have synthetically included test cases and there was no downtime visible to the end-user due to scalability. With respect to security and privacy, each and every customer detail is encrypted and stored in the cloud storage. Without the username and password, it will not be possible to get the information about the user. The system was providing error-free dispensing of pills. This was made possible mainly due to the use of a barcode to represent an update in the prescription. Only the doctors are allowed to modify the prescriptions in the webpage provided for the purpose. The corresponding encrypted credentials will be provided to the individual professionals.

13.6 CONCLUSION

The modern ICTs have revolutionized the healthcare industry prominently. As a proof of concept for the above idea, a remote mobile health monitoring system with mobile phone and cloud service capabilities is discussed in this chapter. It provides an end-to-end solution; specifically, physiologic parameters, including respiration rate and heart rate, are measured by wearable sensors. The readings are recorded by a mobile phone which presents the graphical interface for the user to observe their health status more easily. Also, it provides doctors and family members with necessary data through a web interface and enables authorized personnel to monitor the patient's condition and to facilitate remote diagnosis.

KEYWORDS

- **Healthcare 4.0**
- **Internet of Things (IoT)**
- **fog computing**
- **cloud computing**
- **data analytics**
- **machine learning**

REFERENCES

1. Vergin Raja Sarobin, M. Optimized Node Deployment in Wireless Sensor Network for Smart Grid Application. *Wireless Pers. Commun.* **2020,** *111*, 1431–1451.

2. Sarobin, M. V. R.; Ganesan, R. Swarm Intelligence in Wireless Sensor Networks: A Survey. *Int. J. Pure Appl. Math.* **2015**, *101* (5), 773–807.
3. Sarobin, M. V. R.; Ganesan, R. Deterministic Node Deployment for Connected Target Coverage Problem in Heterogeneous Wireless Sensor Networks for Monitoring Wind Farm. In *Advances in Smart Grid and Renewable Energy*; Springer: Singapore, 2018; pp 683–694.
4. Joyia, G. J.; Liaqat, R. M.; Farooq, A.; Rehman, S. Internet of Medical Things (IOMT): Applications, Benefits and Future Challenges in Healthcare Domain. *J. Commun.* **2017,** *12* (4), 240–247.
5. Islam, S. R.; Kwak, D.; Kabir, M. H.; Hossain, M.; Kwak, K. S. The Internet of Things for Health Care: A Comprehensive Survey. *IEEE Access* **2015,** *3*, 678–708.
6. Sun, W.; Cai, Z.; Li, Y.; Liu, F.; Fang, S.; Wang, G. Security and Privacy in the Medical Internet of Things: A Review. *Secur. Commun. Netw.* **2018,** *2018*.
7. Kang, M.; Park, E.; Cho, B. H.; Lee, K. S. Recent Patient Health Monitoring Platforms Incorporating Internet of Things-Enabled Smart Devices. *Int. Neurol. J.* July **2018**, *22* (Suppl 2), S76.
8. Abdelgawad, A.; Yelamarthi, K.; Khattab, A. IoT-Based Health Monitoring System for Active and Assisted Living. In *International Conference on Smart Objects and Technologies for Social Good*; Springer: Cham, Nov 30, 2016; pp 11–20.
9. Guillén, E.; Sánchez, J.; Ramírez, L. IoT Protocol Model on Healthcare Monitoring. In *VII Latin American Congress on Biomedical Engineering CLAIB 2016*; Torres I., Bustamante J., Sierra D., Eds.; Bucaramanga, Santander, Colombia, October 26th–28th, 2016. IFMBE Proceedings, Vol. 60; Springer: Singapore, 2017.
10. Gondalia, A.; Dixit, D.; Parashar, S.; Raghava, V.; Sengupta, A.; Sarobin, V. R. IoT-Based Healthcare Monitoring System for War Soldiers Using Machine Learning. *Procedia Comput. Sci.* **2018,** *133*, 1005–1013.
11. de Morais Barroca Filho, I.; de Aquino Junior, G. S. Proposing an IoT-Based Healthcare Platform to Integrate Patients, Physicians and Ambulance Services. In *International Conference on Computational Science and Its Applications*; Springer: Cham, July 3 2017; pp 188–202.
12. Hou, J. L.; Yeh, K. H. Novel Authentication Schemes for IoT Based Healthcare Systems. *Int. J. Distrib. Sensor Netw.* Nov 1, **2015,** *11* (11), 183659.
13. Sharma, N.; Bhatt, R. Privacy Preservation in WSN for Healthcare Application. *Procedia Comput. Sci.* Dec 31, **2018,** *132*, 1243–1252.
14. Sahi, M. A.; Abbas, H.; Saleem, K.; Yang, X.; Derhab, A.; Orgun, M. A.; Iqbal, W.; Rashid, I.; Yaseen, A. Privacy Preservation in e-Healthcare Environments: State of the Art and Future Directions. *IEEE Access* **2018,** *6*, 464–478.
15. Masood, I.; Wang, Y.; Daud, A.; Aljohani, N. R.; Dawood, H. Towards Smart Healthcare: Patient Data Privacy and Security in Sensor-Cloud Infrastructure. *Wireless Commun. Mobile Comput.* **2018,** *2018*.
16. Gubbi, J.; Buyya, R.; Marusic, S.; Palaniswami, M. Internet of Things (IoT): A Vision, Architectural Elements, and Future Directions. *Future Gener. Comput. Syst.* Sept 1, **2013**, *29* (7), 1645–1660.
17. Wang, Y. Trust Quantification for Networked Cyber-Physical Systems. *IEEE Internet Things J.* **2018,** *5* (3), 2055–2070.

18. Yang, G.; Jiang, M.; Ouyang, W.; Ji, G.; Xie, H.; Rahmani, A. M.; Liljeberg, P.; Tenhunen, H. IoT-Based Remote Pain Monitoring System: From Device to Cloud Platform. *IEEE J. Biomed. Health Inform.* **2018,** *22* (6), 1711–1719.
19. Rodrigues, J. J.; Segundo, D. B. D. R.; Junqueira, H. A.; Sabino, M. H.; Prince, R. M.; Al-Muhtadi, J.; De Albuquerque, V. H. C. Enabling Technologies for the Internet of Health Things. *IEEE Access* **2018,** *6*, 13129–13141.
20. Majumder, S.; Chen, L.; Marinov, O.; Chen, C. H.; Monday, T.; Deen, M. J. Non-Contact Wearable Wireless ECG Systems for Long Term Monitoring. *IEEE Rev. Biomed. Eng.* **2018**.
21. Ravi, D.; Wong, C.; Lo, B.; Yang, G. Z. A Deep Learning Approach to On-Node Sensor Data Analytics for Mobile or Wearable Devices. *IEEE J. Biomed. Health Inform.* **2017,** *21* (1), 56–64.
22. Jordanski, M.; Radovic, M.; Milosevic, Z.; Filipovic, N.; Obradovic, Z. Machine Learning Approach for Predicting Wall Shear Distribution for Abdominal Aortic Aneurysm and Carotid Bifurcation Models. *IEEE J. Biomed. Health Inform.* **2018,** *22* (2), 537–544.
23. Enayati, N.; De Momi, E.; Ferrigno, G. Haptics in Robot-Assisted Surgery: Challenges and Benefits. *IEEE Rev. Biomed. Eng.* **2016,** *9*, 49–65.
24. Chen, M.; Ma, Y.; Li, Y.; Wu, D.; Zhang, Y.; Youn, C. H. Wearable 2.0: Enabling Human-Cloud Integration in Next Generation Healthcare Systems. *IEEE Commun. Mag.* **2017,** *55* (1), 54–61.
25. M. M. Nair, A. K. Tyagi and N. Sreenath, "The Future with Industry 4.0 at the Core of Society 5.0: Open Issues, Future Opportunities and Challenges," 2021 International Conference on Computer Communication and Informatics (ICCCI), 2021, pp. 1-7, doi: 10.1109/ICCCI50826.2021.9402498.
26. Zhang, Y.; Yu, R.; Xie, S.; Yao, W.; Xiao, Y.; Guizani, M. Home M2M Networks: Architectures, Standards, and QoS Improvement. *IEEE Commun. Mag.* **2011,** *49* (4).

CHAPTER 14

Biomedical Data Analysis: Current Status and Future Trends

AMIT KUMAR TYAGI[1,2*], S. U. ASWATHY[3], and SHAVETA MALIK[4]

[1]*School of Computer Science and Engineering, Vellore Institute of Technology, Chennai Campus, Chennai 600127, Tamil Nadu, India*

[2]*Centre for Advanced Data Science, Vellore Institute of Technology, Chennai 600127, Tamil Nadu, India*

[3]*Department of Computer Science and Engineering, Jyothi Engineering College, Thrissur, Kerala*

[4]*Department of Computer Engineering, Terna Engineering College, Navi Mumbai, India*

[*]*Corresponding author. E-mail: amitkrtyagi025@gmail.com*

ABSTRACT

Nowadays, data analysis is the trend and is used in many applications for different purposes. When data analysis is used in e-healthcare, defence sectors, and the like, it becomes crucial and has more expectations, that is, to provide more accurate results to users/people. Modern biological information has many special characteristics in contrast with data from general application areas, which has received the attention of many research communities to work/analysis for finding efficient (useful) solutions for critical diseases. Data analysis is used for health and clinical informatics, that is, in biomedical applications. Biomedical informatics/biomedical imaging data need to be analyzed properly and effectively with efficient tools, algorithms for

Intelligent Interactive Multimedia Systems for e-Healthcare Applications. Shaveta Malik, PhD
Amit Kumar Tyagi, PhD (Eds.)

making (generating) useful decisions. In biomedical informatics, we mainly focus on bioinformatics, imaging, and genetics (BIG). Note that medicine and biomedical sciences are two different fields (for data-context). Usually, biological data are characterized as large quantities, complex structures, high dimensionality, evolving biological concepts, and inadequate data modeling. In fact, there is a thing in common in medicine and biomedical sciences, that is, sophisticated (relevant) data analysis and data mining/learning methods are needed for each field. Biomedical informatics gives a consistent interdisciplinary platform for integrating data and knowledge to provide practical decision-making support in hospitals, health care, and translational science in the analysis of available information. This work provides/presents a detailed overview of bioinformatics aspects of data accumulation and data-driven approaches in medical information science, data and knowledge integration, methodological concerns for data mining model evaluation, transitional bioinformatics, and genetic epidemiology. In summary, this article addresses various useful components such as data characteristics, opportunities, and challenges for statisticians, and recent advances and active areas of bioinformatics, biomedical imaging, and statistical genetics research. These components or topics play an essential role in understanding the concept of biomedical data analysis.

14.1 INTRODUCTION—BIOMEDICAL DATA ANALYSIS

In the last few years, bioinformatics has become an all-encompassing term for something connected with both computer science and biology, that is, bioinformatics. Big Data research has been a hot topic in the world of research, technology, and business. Healthcare and biomedical sciences are becoming increasingly data-intensive as researchers produce and use broad, complex, high-dimensional, and diverse domain-specific datasets. The healthcare sector, from electronic health records, testing tools, support systems for clinical decision making, etc., has always been a major source of biomedical data. Biomedical developments share the fields of human genomics[1]—genome project, medical imaging[2] growth in m-Health, telehealth, and telemedicine—trillions of data points have been developed as a result of recent advances in biotechnology and the advent of new computing sources. This vast volume of data is used by users to create some potential association rules and extract some useful information. Data analysts and data scientists are prepared for such roles and activities. Big Data technology refers to the vast volumes of data obtained, processed, combined, and analyzed from different distributed sets of data. Today, Big Data science has been increasingly recognized as an evolving field and discipline and can be one of the most critical resources, not only in life sciences such as medicine and healthcare, but also in other fields such as education quality, government prospects, social sciences, and the financial sector and business opportunities—reduced rates, improved patient outcomes (less medical errors and readmissions), improved outcomes, protection, efficiency and innovation of healthcare facilities, and related systems through Big Data Science and in the healthcare and medical sectors.

It is important to develop Big Data science infrastructure, smart analytical tools, and advanced computer systems that can use the grounded theory framework to conceptualize, theorize, and model Big Data due to the unique characteristics of knowledge discovery potential for Big Data and make it actionable and operational for better life sciences solutions. Big questions have been raised here now, such as:

a) How we can get creativity and imagination with Big Data science in some other sensitive apps, such as bio-medical/medicare applications.
b) How real-time characteristics affect researchers' use of Big Data during the processes of collection, storage, analysis, and visualization.

Both targets are very difficult to accomplish, so study groups/community must discuss them. Along with this, there are several special features of Big Data or conventional medical science-based sampling inference:

a) As the sample/data size becomes larger, research becomes better;
b) The longer the (real) follow-up time, the closer the clinical outcome is, and the greater the clinical importance of the relevance and benefit. The statistical significance is not understood to show the clinical

14.1.1 BIOMEDICAL DATA SCIENCE OR BIOMEDICAL INFORMATICS

It is an interdisciplinary field in which the productive use of biomedical information, expertise, and practise for scientific study, problem-solving, and decision-making are investigated and discussed, motivated by attempts to enhance public health.

14.1.2 BIOMEDICAL IMAGING

The dynamic chain of collecting, processing, and visualizing conceptual or functional images of living organisms or objects, including the collection and processing of information related to images, comprises biomedical imaging.[5] From the early, the basic applications of X-rays for the diagnosis of injuries and the detection of foreign bodies, biomedical imaging has become a compendium of useful methods, not only for health care but also for the study of biological structure and function, and to address fundamental biomedical questions. A variety of non-invasive research techniques have been developed with technological advances such as radiography, X-ray computed tomography (CT), nuclear (including positron emission tomography (PET)) and ultrasound, and optical and magnetic resonance imaging (MRI). New phenomena have been used by many modern microscopes, such as nonlinear photon interactions and the sensing of atomic forces on structures. Imaging may provide accurate, useful knowledge of the tissue structure, anatomy, and function, as well as detailed explanations of many basic biological processes. In recent years, biomedical imaging technology has grown into a separate and coherent set of ideas and concepts and has assumed a central role in most medical research. Continued advances in biomedical and other sciences, such as molecular biology and nanotechnology, have expanded imaging technologies to other fields, such as studies of gene expression or brain functional organization. This book mentions a number of significant recent advances in biomedical imaging technology, and the relevance of biomedical science and imaging specialists in clinical medicine is well illustrated. In particular, there are many examples of how visualization progresses from representations of abstract visual anatomy into research, leading to a variety of biomedical processes being quantitatively calculated.

Biomedical imagery is an effective way of visualizing the inner organs of the body and its diseases. The imaging technologies of today provide pioneering insights into biological processes. Biomedical imaging enables biological processes to be imaged in vivo, including improvements in receptor kinetics, molecular and cell signalling, and membrane molecular interactions and function. Biomedical imaging, often non-invasive, allows reliable detection of metabolites that can be used as biomarkers for the diagnosis, occurrence, and response of diseases. For instance, in order to complement X-rays and medical ultra-sonography, CT scanning is an effective method of medical imaging. More recently, it has been used for preventive medicine or disease screening. Head CT scanning is typically used for infarction, tumors, calcification, hemorrhage, and diagnosis of bone injury. Not only do CT scans demonstrate the presence of a tumor, but also its size, spatial position, and duration. Dynamic structure and velocity images are given by fast, lightweight, and inexpensive techniques such as ultrasonic imaging, while magnetic resonance imaging is the most flexible modality that gives programmable structure and function. On the other hand, in clinical neurology, MRI is useful, including segmentation and explanation of brain tissue; calculating volumes of brain structure; analyzing diseases such as multiple sclerosis, neuro-degeneracy, stroke, and tumors; and describing the function of cardiac pathology and cardiology. Remember that morphometric or densitometric measures are the basis for quantitative analysis of biomedical imaging outcomes.

14.1.3 *BIOMEDICAL IMAGE PROCESSING*

In nature, biomedical image processing is analogous to multidimensional biomedical signal processing. It requires processing, upgrading, and viewing x-ray, ultrasound, MRI, nuclear medicine, and optical imaging images.

14.1.3.1 INTRODUCTION TO NEXT GENERATION SEQUENCING DATA

- Next-Gen Sequencing (NGS)[6]: streamline next-gen data statistical analysis through a range of personalized simulation approaches and workflows.
- Link Mapping[7]: establish optimal linkage maps for experimental populations, view marker maps, and perform comprehensive quantitative trail loci (QTL) analyses.
- Predictive Modeling[8]: classify biomarkers from data sets with large dimensions, use cross-validated predictive modeling reviews with robust predictor screening capabilities, and create and compare multiple models for different traits.
- Pharmacogenomics[9]: examine integrated DNA, RNA, metabolite, and protein expression genomic patterns to establish biological functions in disease response and drug response.

14.1.3.2 GENOME AND GENOMICS

We should identify the genome within a cell as the complete set of genes. The study of the genetic structure of animals, in simple terms, is genomics. In an organism, the analysis of genes is genomics. It gives a summary of the whole range of DNA genetic instructions, whereas transcriptomics looks at patterns of gene expression. It is modern science that deals with the discovery and notation of a single organism as a whole with all sequences in the genome. Determination of genomic sequences is just the beginning of genomics. If this is done, the genomic sequence is used to analyze the role of different genes (functional genomics), to compare one organism's genes with those of another organism (comparative genomics), or to determine the 3-D structure of one or more proteins in each protein family, providing evidence of their role (structural genomics). At http:/rgp.dna.affrc.go.jp/IGRSP, we can access the International Rice Genome Sequencing Project. Note that in humans, a copy of the entire genome consists of more than 3 billion DNA base pairs, that is, is present in all cells that have a nucleus.

14.1.3.3 PROTEOMICS AND METABOLOMICS

Proteomics[10] studies and interacts with complex protein components, while metabolomics also offers an intermediate stage in the understanding of species metabolism as a whole. Proteomics is the study of all of a cell's proteins. Studies of proteomics are more useful because proteins are the functional molecules in cells, and they represent real conditions. It can also be used to construct a map of a network of proteins in which protein associations can be formed for a particular living system. Also, it can be used to monitor protein variations to determine the difference between a wild type and a genetically modified organism. It is also used to research protein–protein relationships involving plant defence responses. At Iowa State University, the USA, for example, proteomics research includes:

- Examination of protein changes that are a major concern for young maize seedlings in the maize proteome during low temperatures;
- Analysis of the variations in genome expression in the growth of soybeans stressed by high temperature; and
- Protein recognition expressed in response to diseases such as soya cyst nematode.

Metabolomics[11] also represents an intermediate stage in the comprehension of an organism's entire metabolism. It's one of the newest "omics" sciences. Here, metabolome refers to the whole set of compounds in a sample with low molecular weights. These compounds are substrates of enzymatic reactions that are by-products that have a direct effect on the phenotype of the cell. Metabolomics attempts to determine the profile of a sample of these compounds at any given time under unique environmental conditions (Fig. 14.2).

All the aforementioned terminologies can be understood via an example. Genomics and proteomics have provided extensive genotype information in agriculture but convey limited information on phenotypes. The closest relation to phenotype is between low molecular weight compounds. Metabolomics can be used to determine the differences between the volumes of a healthy and diseased plant among thousands of molecules. The technology can also be used to identify the difference between organic and genetically modified crops in nutrition and to identify metabolites for plant protection.

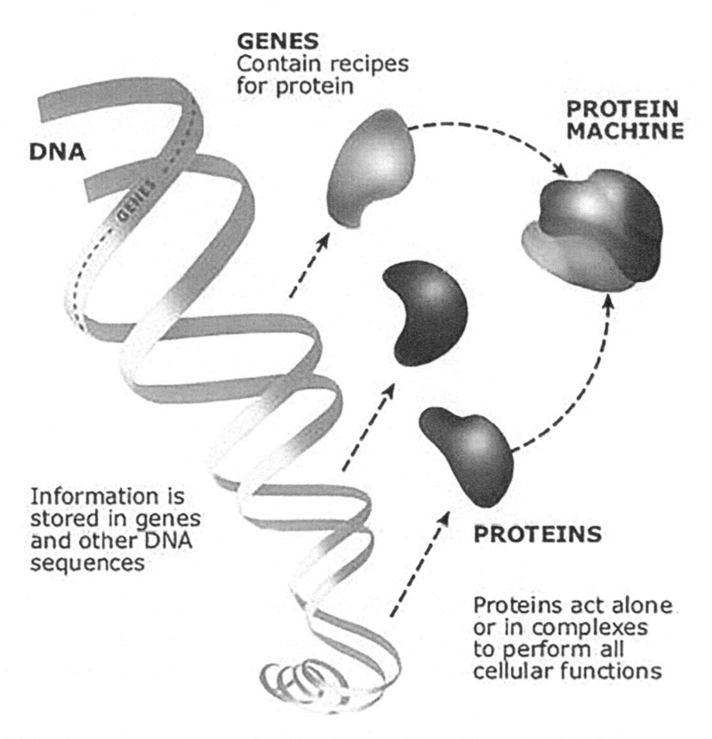

FIGURE 14.1 Genes, proteins, and molecular machines (*Source*: U.S. Department of Energy Genomes to Life).

In a cell, the fundamental stream of genetic information is as follows (Fig. 14.1). The whole sequence of RNA (also known as its transcriptome) undergoes some editing (cutting and pasting) to become a messenger-RNA, which carries information to the cell ribosome, a protein factory, which then converts the message into a protein (Fig. 14.1). The DNA is either transcribed or copied into an RNA type.

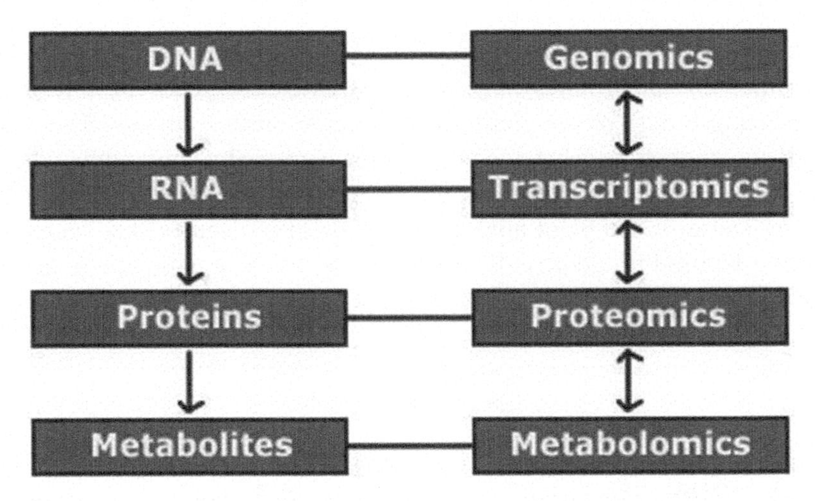

FIGURE 14.2 Example of a metabolic network model for *E. coli* (Source:http://biotech.nature.com).

Hence, now the remaining part of this article is organized as:

- Section 14.2 discusses work related to genomics and biomedical data analysis research.
- Section 14.3 discusses the motivation behind writing this article (or this information).
- Section 14.4 discusses the scope or importance of biomedical data analysis for biomedical imaging applications (in the past, present, and future).
- Section 14.5 explains the role of brain networks, that is, analysis of neuroimaging data with genetics and statistical analysis.
- Section 14.6 discusses the concepts of molecular genetics and population genetics in detail.
- Section 14.7 discusses genomic association studies from a business's perspective.

- Section 14.8 discusses various tools and methods of existing/available for analyzing biomedical data.
- Section 14.9 discusses various open research issues, challenges faced in biomedical imaging, and opportunities for the future as including research directions.
- Finally, Section 14.10 concludes this work with various research gaps and future enhancements.

Hence, in this work, our main goal is to fill many identified research gaps in the current era by providing effective literature on genomic research and medicine.

14.2 BACKGROUND WORK

Biomarkers discovery study involves systematic data analysis across several domains, including clinical data, data on pathology, gene expression, and data on epigenetics. Proper research may help to explain the function of biology and better interpret the effect of the markers on the disease. Realizing the importance of the data in biomedical science and translational biomedicine, we built a pipeline of data analysis with a collection of computational functions and an integrated approach that can serve as a framework for the discovery study of many biomarkers. The pipeline for the data analysis was built using data obtained to classify biomarkers associated with the obesity-related disease.

The research and invention behind the materials, instrumentation, and applications used to achieve biomedical imaging has progressed steadily since the X-ray was first invented in 1895. It only needs milliseconds of exposure time for modern X-rays utilizing solid-state electronics, significantly decreasing the dosage of X-rays initially needed for film cassette study. With increased clarity and contrast detail, the quality of the image has also improved, allowing more efficient and reliable diagnosis. By using a contrast medium to help visualize organs and blood vessels, the disadvantages of what X-rays could show were partially overcome. Comparison agents, which have grown over the years, were first introduced in 1906. Today, digital X-rays allow the sharing and comparing of pictures. In the previous decade, some recent developments in imaging techniques have been constructed (to advance) in other biological sciences, such as molecular biology/discoveries, that is, nanotechnology. These have driven some of the recent trends, including

(i) Using imaging to study function or other properties rather than merely anatomy, such as mapping of brain function or evaluating metabolism,
(ii) The development of advanced animal imaging equipment, for example, genetically engineered mice or awake, non-human primates,
(iii) Development of molecular imaging probes, such as novel positron emission tomography (PET) or optical ligands, to evaluate cellular processes;
(iv) The movement to use quantitative imaging measures as objective tissue indices, such as the mapping in absolute units of regional blood flows: and progressively,
(v) The creation of image databases and the use of computerised data mining techniques to determine genetic risk factors for psychiatric disorders, and the interrogation of electronic health image information and other documents such as genomic data, such as the association of genetic polymorphisms with anatomical and functional differences in the brain. Unlike more conventional radiology scientists trained in, for example, physics or engineering, today's imaging science trainees must become experts in conceiving and linking image-based biomarkers to fundamental biology, metabolism, genotyping, or similar issues. Note that we find (from our previous work) that measuring distinct biological domains remains a considerable challenge in the face of complex mechanisms of biochemical control.

This section addresses background work relating to the study of biomedical data, biomedical data science, etc. Now, the next section, will address the inspiration behind this job, our plan for write an elated section to this field of merging.

14.3 MOTIVATION

Human genomics is an important development of precision medicine research and a perfect example of Big Data analysis for medical applications. More precisely, over time, thousands of genes to identify a handful of genes that reacted to the drug that could be potential drug targets could turn into a temporal fashion computational issue related to the "curse of dimensionality" question. Various statistical learning and data processing methods or approaches to statistical research need to be considered and applied with effective ingenuity and innovation to solve certain problems. Today, large genome data generation tool/technology translational research scheme/process, analytical pipeline, methods, and techniques for acquiring and transforming 300 billion disease data points into diagnostics, therapeutics, and new insights (knowledge) into public health and disease care (e.g., Disease Alzheimer's disease and bi-polar disorder). In the previous decade, we were unable to handle or analyze this much large amount of data.

In contrast, biomedical imaging may aid in the early stages of diagnosing fatal (serious) diseases such as cancer, which could lead to more successful therapies. Some medical imaging techniques include optical imaging (OI), magnetic resonance imaging (MRI), computed tomography with positron emission,[12] computed tomography (CT), and conventional X-ray clinically applied.[13] The accuracy of the imaging in diagnosis and treatment improves significantly with the advent of molecular imagery. Molecular imaging involves the visualization, characterization, and measurement of biological parameters in human tissues at the molecular and cellular levels. Using such innovations in biomedical imaging, that is, data analysis in biomedical applications we can save millions of lives with identifying diseases at early stages.

Hence, this section includes points toward our motivation behind writing this work related to medicare/bio-medical imaging applications. Now, in continuation of the motivation section, the next section will discuss the importance of the scope of biomedical imaging today and tomorrow.

14.4 SCOPE OF BIOMEDICAL IMAGING TODAY AND TOMORROW

From the 20th century to the 21st century, imaging technology has evolved exponentially in depth and value. The advent of faster, more efficient computers has led to sophisticated image recognition and processing algorithms that can be used to extract useful, quantitative information from images, such as measurements of the dimensions of specific brain structures. The use of informatics techniques for mining large image databases[14] now illustrates how they apply to multidimensional data such as genotype and structure. Bio-medical imaging typically focuses on diagnostic and therapeutic image processing. Using sophisticated sensors and computer technologies, snapshots of physiology and physiological processes in vivo can be obtained.

Biomedical imaging techniques use either X-rays (CT scans), vibration (ultrasound), magnetism (MRI), or radioactive pharmaceuticals (nuclear medicine: SPECT, PET) to determine the current state of an organ or tissue and can observe the patient over time for evaluation of diagnosis and treatment. Digital technology has developed a CT scanner that allows doctors to view X-rays on a scan. Current research is no longer limited to fundamental anatomical imaging, but focuses on what can be gleaned from practical imagery. Biomedical engineers use CT and MRI in biomedical imaging to assess tissue profusion in the blood, which is particularly useful after a heart attack or suspected heart attack. To examine different types of brain activity following strokes and traumatic head injuries, researchers may frequently use functional magnetic resonance imaging (fMRI). There are few (parallel) technological advances in cellular and molecular biology that are common innovations in imaging science. The capacity of imaging to identify, diagnose, and control pathological, physiological, and molecular changes is of fundamental significance for disease treatment, personalized therapies, and basic biological research. As new advances in imaging science are being made, we are heading into the discipline of clinical radiology. Many medical centers have developed research-dedicated animal and human imaging facilities that combine different modalities. To create new drugs, pharmaceutical companies regularly use imagery, because imagery provides unique information on non-invasive reactions to compounds.

Imaging science needs three advances, such as, first, imaging science aims to improve and enhance imaging technology to provide new or improved knowledge (e.g., new ways of detecting and mapping neural activity in the brain, such as functional MRI): imaging scientists are continuously developing new instruments or changing techniques to obtain, for example, higher spatial resolution images. Second, imaging science offers an understanding of the quality of image metrics information and the factors that influence them (physical and biological) (e.g., what determines the importance of the signals recorded in MRI experiments, and how are they linked to the underlying neural activity): the perception of image information, the robustness of image inferences, and the ability to infer. Third, imaging science is developing technologies that use image-derived knowledge for both scientific and clinical uses.

Hence, this section discusses the importance of imaging science in various applications like clinical radiology, cellular and molecular biology, etc., for diagnostic and treatment evaluation. Now, the next section will discuss the role of brain networks, how it works, and how neuro-imaging help us to trace brain activity to conclude with implementing imaging Genetics and statistical analysis on neuro-imaging data.

14.5 ROLE OF BRAIN NETWORKS/NEUROIMAGING DATA IN IMAGING GENETICS AND STATISTICAL ANALYSIS OF NEUROIMAGING DATA

Generally, data mining and machine learning are ruling the predicting world. Data mining techniques[15] was popular in the previous decade when data was available with few entries, but as entries grow, the requirement of modern tools/technique also grows. So, machine learning comes into the picture, together with this, for extracting information from the image, deep learning comes into pictures. Data mining typically refers to the application of statistical and analytical methods to assist with the extraction of potentially useful and synthetic information from large quantities of raw data. Potential applications in the biomedical field include the automated detection of candidate biomarkers for a disease or treatment response and the creation of diagnostic or prognostic rules that leverage such biomarkers. Being standardized, these approaches can be extended to different data sources and biological/ biomedical issues.

14.5.1 NEUROIMAGING DATA

MRI image data are stored as either 8- or 16-bit integers in a binary data format. In addition to the raw image data, we also used the subject metadata, image form, image parameters, as well as image dimensions. Note that the dataset contains an anatomical (T1 weighted) scan and fMRI as well as patient/subject behavioral data for each participant. The dataset is shared via OpenNeuro[16] and is formatted according to the standard brain imaging data structure (BIDS). Often, publicly sharing data pre-processed with fMRIprep and quality management reports. For that, we collected fMRI data (from GitHub repository https://github.com/rotemb9/NARPS_scientific_data) to quantify the variability of neuroimaging findings through analytical teams, and to assess peer-related outcomes by running predictive markets. This dataset has been collected for checking replicability and addressing contradictions of previous findings. In addition, two conditions used provide a rare opportunity to compare the interactions between the particular matrix of gains/losses used and the behavior elicited and the neural fMRI operation.

Again, two processes like MRI pre-processing and anatomical data pre-processing were used to analyze neuroimaging data.

14.5.2 BRAIN ACTIVITY PROCESS

FMRI[17] is a functional neuroimaging technique, which tests the function of the brain by detecting changes in blood flow associated with neuronal activity. The assumption is that neurons need more oxygen when they are healthy. In contrast to methods such as electroencephalography (EEG) or magneto encephalography (MEG), FMRI has a trade-off between spatial and temporal resolution, but is a relatively slow neuro-imaging technique. However, the essential quality of magnetic resonance imaging is its excellent spatial resolution. We may recreate all the skull shapes and cortical layers of our respondents with magnetic imaging. The respondents usually have to lie motionless in a magnetic center when emitting radio frequency blasts. The fMRI tests the shift in magnetization between blood rich in oxygen and blood low in oxygen, suggesting a relative difference between various regions of the brain. The fMRI is one technique used to reintegrate anatomy into action, allowing for this connexion. The fMRI readings are sadly not fine. The spatial resolution has only recently progressed to the millimeter stage and the measurements are sadly not in real-time. Between brain activity and related changes in blood flow and oxygenation, there is a delay of about a second, which can be detected by the fMRI. Researchers must, however, be able to quantify activity on an actual, millisecond-by-millisecond basis and a much smaller spatial scale.

On the other hand, machine learning is a term that allows the machine to learn without being specifically programmed from examples and experience (previously collected data). Reinforcement learning is the best example of machine learning, that is, doing reward-based learning in case of learning by itself. Notice that the user feeds data to the generic algorithm in machine learning (a complete procedure) and the algorithm/computer constructs the logic (makes decisions) based on the data provided. Extensions to high-dimensional datasets are few features for machine learning techniques to discover and manipulate structures in the data, allowing productive use of the available data. Now, steps involved in analyzing neuro-imaging data for measuring the performance of a brain, that is, based on its activity, are as follows (Table 14.1):

TABLE 14.1 Procedure Need To Be Taken, for Analyzing Neuroimaging Data.

Steps used	Data preparation	Procedure/action taken	Remarks
1	Data extraction	• from files: tab-separated, comma-separated (CSV), Excel files • by querying a database	
2	Data alignment	• merging various sources of data (joining on patient identifiers) • merging of cohorts (joining on variable names and variable conventions)	
3	Data cleaning	• coding of variables • correction of clearly wrong inputs • creation of additional variables (age from date of birth, BMI from height and weight)	
4	Data exploration	• plotting of data • detection of outliers or other problems in the data	
5	Imputation	• completing the dataset by filling in missing data when possible • based on educated guesses, simple assumptions, or models	
6	Data transformation	• correction of skewness • standardization • rank transformation	

Notice that some of those steps are performed simultaneously and iteratively. Hence, this section discusses brain activity, how it works, and how neuro-imaging help us in tracing footprints of any disease via brain activity. For that, we see that performing imaging genetics and statistical analysis on neuro-imaging data. Now, the next section addresses with detail the fundamental principles of molecular genetics and population genetics.

14.6 BASIC CONCEPTS OF MOLECULAR GENETICS AND POPULATION GENETICS

Molecular genetics[18] is a genetic sub-field used as an "investigative method" for evaluating the structure and/or role of genes in the genome of an organism using genetic screens. It investigates the gene's chemical and physical existence, and the mechanisms by which genes regulate production, growth, and physiology. In contrast, it's the analysis of DNA's molecular structure, its cellular activities (including its replication), and its role in deciding an organism's overall composition. Remember that genetic engineering relies heavily on molecular genetics. Genetic data plays an essential role in molecular genetics. In general, data contain measurements on one or even more variables for biomedical imaging applications. Data are collected from a survey of people that constitute an interesting population.

14.6.1 TYPES OF BIO-MEDICAL DATA

Data may consent if patients agree to join a particular study, or otherwise disagree: use of unconsented data has additional restrictions and privacy concerns.

- Information of patient : date of birth, gender, and date of entry/exit to study
- Medical routine data: height, weight, blood pressure, cholesterol levels, and medicines used
- Laboratory specialty data: proteins, lipids, metabolites, glycans, and imaging
- Genetic information: genotype or sequencing
- Expressions of gene and information of epigenetic (methylation of DNA)

14.6.2 DATA IS SLY

Data is dirty, noisy, and messy, and this is especially true of biomedical data:

- Manual intervention: GPs or nurses' data entry.
- Data sources are varied: data obtained from cancer registries, death reports, research populations, and leading to discrepancies between sources or echoes (slightly different dates of several copies of the same record).
- Sources can vary over time: different doctors/measurements of nurses performance, laboratory changes performing tests, policy updates, and connecting errors.

- Different criteria (even variations between Scotland and England) across continents, such as index of deprivation and date formats.
- Incomplete information: responses to specific questions (i.e., eating patterns, smoking, and alcohol use), or questionnaires that have been poorly constructed.
- Data on medicines: prescribed medications vs. cashed-in prescriptions vs adherence, dosages, and erasure of medications.
- Data extraction or transmission issues, anomalies in file format (management of commas in a CSV file), and so on.
- Data manipulation or lack of accuracy when adding data.
- Data missing: data not reported, data lost, patients traveling, and new variables reported after study commencement.

In clinical practise and policy recommendations, a quantitative variable can be summarized into categories that have been used. Notice that binary and categorical outcomes lead to problems with classification,[19] while other outcomes (like continuous survival) lead to problems with regression. Note that for analyzing data, choice, and quality of data matter a lot, that is, is too important. A study's result is any specified illness, health status, or health-related occurrence, and it is often one of the most useful parts of a dataset. Could be accessed from different sources:

- Medical history (routine examinations, hospital records, etc.)
- Organic steps (blood samples, etc.)
- Procedures for diagnosis (ultrasound, etc.)
- Interviews or questionnaires

Hence, this section discusses the basic concepts of molecular genetics and population genetics, also discusses the type of biomedical data, etc. Now, the next section will discuss genomic association studies from a business perspective in detail.

14.7 GENETIC/GENOMIC ASSOCIATION STUDIES: FROM BUSINESS'S PERSPECTIVE

The result variable usually depends on the research question we want to address. That's the issue of phenotype description preference. In an epidemiological or clinical research, the choice of the phenotype is one of the most important. Changes in the concept of a phenotype can give rise to differences to outcomes (also large). The quality of the assessed phenotype has direct implications for the quality of the findings, particularly as regards their reproducibility and relevance to clinical practise. Over the past ten years, genome-wide association studies (GWAS) have developed into a powerful method for studying the genetic history of human disease. In this thesis, the main concepts underlying GWAS, including common disease architecture, common human genetic variation structure, and genetic information capture technologies, designs for the sample, and the statistical methods used for data analysis.

Studies of genome-wide associations have had a significant effect on the human genetics field. For many modern human diseases, they have identified new genetic risk factors and prompted the genetic community to think on a genome-wide scale. There is entire-genome sequencing on the horizon. Over the next few years, we will see the introduction of cheap sequencing technologies that will substitute one million SNPs with the entire three billion nucleotides genomic sequence. Data storage and manipulation, quality control, and data processing problems will be more complex, challenging the technology and skills of computer science and bioinformatics. The combination of sequencing data with that from other high-throughput transcriptomes, proteome, climate, and phenotypes such as large quantities of neuroimaging data would only complicate our goal of understanding the genotype–phenotype relationship to enhance healthcare. The future of human genetics is the synthesis of these several layers of complex biomedical data together with their combination with experimental systems.

Hence, this section discusses genetic association, and its related areas, applications, where it is mostly used in this current era. Now, the next section will discuss many existing tools, methods, or algorithms to refining or analyzing biomedical data.

14.8 TOOLS/METHODS AVAILABLE FOR ANALYZING BIOMEDICAL DATA

Over the past era, major advances in "omic" techniques such as genomics, proteomics, and metabolomics have allowed a number of molecular and organismic processes to be monitored for high-throughput. Such methods have been widely used to classify biological variants (e.g., biomarkers), to characterize complex biochemical structures, and to investigate patho-physiological processes. The integration of omic-domain data remains difficult, while comprehensive genes (genomics), mRNAs (transcriptions), proteins (proteomics), and metabolites (metabolomics) are the targets of many omic platforms (Figs. 14.1 and 14.2). Epigenetics and mRNA, or post-translational modification, in the face of complicated biochemical regulation, such as organism versus tissue versus cellular processes (Figs. 14.1 and 14.2). The synthesis of experimental outcomes from various "omic" platforms today is an evolving method aimed at helping to recognize latent biological connexions that can only be visible through systematic research that integrates measurements through multiple biochemical domains. This section focuses on selected methods and techniques for incorporating metabolomics with genomic and proteomic findings.

14.8.1 BIG DATA ANALYTIC APPROACHES VS. PREVIOUS APPROACHES

Basic analytics and descriptive analytics (e.g., in the simplest style that allows us to break down Big Data into smaller, more concrete pieces of information) are the function of summarizing "what happened" and relying on the knowledge gained from historical data to provide trend-setting insights into passes by comparing basic level analytics with advanced level analytics in Big Data science.

While the advanced level analytical tools listed earlier in Big Data science concentrate on predictive analytics aimed at analyzing patterns and predicting future outcomes and trends, and answering "what might happen" and "what should we do?" Prescriptive analytics involves functions as the decision support of Eva by quantifying the implications of future decisions to inform about potential results.

14.8.2 BIOLOGICAL NETWORK-BASED INTEGRATION

Another group of ground-breaking approaches used to study a variety of pathways for the species and cells is network-based studies. Biological networks reflect complex interactions, including genes, proteins, and metabolites, between various types of cellular components. Such networks can be used to integrate or map different "omic" experimental results and help to distinguish altered graph neighbourhoods that are not based on any predefined biochemical pathways. For instance, for the computation, visualization, and functional enrichment analysis of biological networks, SAMNet Web and pwOmics support transcriptomic, proteomic, and interactomics data integration. The estimation, understanding, and visualization of the metabolite-to-gene networks are enabled by Metscape, a plug-in for the widely used Cytoscape network analysis software. Except in cases where knowledge of the biochemical domain or molecular annotations is uncertain, another tool such as MetaMapR leverages the KEGG and PubChem databases to provide methods for integrating and visualizing complex metabolomic findings. For instance, MetaMapR was used to combine both biochemical reaction data of molecular structure and mass spectral similarities to identify pathway-independent relationships, including between molecules of unknown structure or biological function. In cases of insufficient domain knowledge of gene, protein, and metabolite interactions, biological-network-based methods alone can, however, provide limited insight and are often extended by incorporating empirical associations or correlations between measured species. Now, some key features of anomic tool collection for data analysis and integration are described in Table 14.2.

Hence, this section discusses several tools for biomedical data analysis or algorithms or methods used in biomedical data science (and biomedical bioinformatics). Now, the next section will deal with open issues and challenges in the required technology. In addition, it will provide several useful directions (for future research) for several researchers/research groups.

14.9 OPEN ISSUES, CHALLENGES, AND FUTURE RESEARCH DIRECTIONS

Today, with the development of technology, Big Data science enables the computational or mathematical model and associated assumptions to be continuously refined with the continuous arrival of new data for a more accurate outcome and better-informed decision-making, due to its evolving and dynamic real-time role. More importantly, it makes it possible not only to consider what has happened and what is happening at the moment, but also to predict what will happen in the future by using predictive analytics. Because of the lack of data, the main problems facing Big Data researchers today are the ability of researchers to find, analyze, integrate, and communicate with both real-time data and associated applications. Today, for an efficient and creative solution, we need scalable databases, intuitive user interfaces, interactive visualizations, machine learning tools, and scientific APIs. Therefore, the practical and actionable approach is to transform billions of data points into solutions, fundamental and advanced levels are required for deep learning and data analysis.[20, 21] At the basic stage, research involves:

(a) Simple real-time online questions, pipeline, flow, and analytical tools;
(b) Pre-processing broad data reduction: detection of missing data, anomalies, outliers; collection, transformation, pre-processing loading of data items, automated filtering of unusable data, accuracy, and correlations;
(c) Statistical methods for summarizing qualitative and quantitative data, reporting trends and patterns;
(d) Automation of data and metadata generations, for example, computer-automated blog post analysis;
(e) Quick and easy to use model visualization tools: understanding and making sense of the data.

Now, some problem we faced concerning data may be mentioned as follows:

- Data issues should be detected and solved prior to review.
- As data set sizes expand and become more complicated, it becomes harder and more onerous to find problems in the data as well.
- However, with larger sample sizes rather than with small ones, some form of outlier detection is more effective.

Big Data research also needs to overcome some hurdles in the evidence-based medical system (practise or medicine) and should be tackled with a better plan to data sharing, transparency, and integrity from a scientific and clinical perspective. To effectively integrate the vast amounts of biochemical information produced from current and next-generation OMI platforms, it is important to develop methods that can deal with both huge, complex, high-dimensional data, and sparse knowledge of the fields of biotechnology for future work (as future research directions). Potential tool developers need to take into account the number of steps needed to execute multi-omic experiments successfully. The incorporation of scalable and easy-to-search databases, machine-learning techniques, and technological application programming interfaces (APIs) are exciting ways to rapidly address the needs of existing and future "omic" data processing and integration pipelines. For Big Data scientists, one critical topic to consider (soon) is:

- How can some 300 billion data points be translated into objective quantitative evidence for diagnosis, therapy, and new insights into public health, illness, and treatment?
- Which methods are best? Will the inference methods commonly used continue to play certain roles? For example, should it be experimental versus computational; the hypothesis is driven versus data-based approaches; conventional statistical modeling versus data mining approaches, and artificial intelligence.

Then, some common questions are turned into useful socially supportive solutions, such as "How to turn Big Data into good research problems/questions/hypotheses." The exponential growth of EHRs, mHealth, eHealth, Smart, and Connected Health, for example, and the convergence of telehealth devices with social, behavioral science, genomics, and economics have led to the development of emerging technologies and the introduction of personalized medicine and patient care with more individualized health care systems.

Notice that it is now possible to refine 300 billion samples/data using modern methods, but it was not possible to refine billions of samples in the previous decade. Hence, this section addresses many open problems and threats against biomedical data plagiarism report science and biomedical engineering. This section also offers some valuable opportunities for future researchers. Lastly, several interesting issues, challenges and opportunities (for future researchers) toward healthcare applications can be found in[22, 23, 24, 25, 26, 27, 28].

14.10 CONCLUSION

Usually, proteomics gives us a greater understanding of an organism than genomics. In this, the transcription level of a gene only provides a rough approximation of its level of expression into a protein. Numerous research holes, problems, and concerns have been included for the future in this work. This article deals with various topics related to biomedical imaging, biomedical data analysis, with varying degrees of balance between biomedical data models and their use in applications in the real world. In the 21st century, biomedical imaging will go beyond testing typical morphological and metabolic parameters, such as tumor size and metabolism of glucose, and validating new data sources in imaging that will improve patient care and correlate with clinical outcomes. As our molecular understanding of the disease increases and individualised therapies are implemented, the imagery must also progress to the molecular level to provide a tailor-made approach to each clinical, it's associated study scenario. Notice that the advancement of imaging technologies has been continuing, such as higher field MRI, volumetric CT scanning, hybrid imaging (PET-CT, PETMR), and a growing role of radiological imaging in the diagnosis, management, and direction of patient therapies.

The data-rich environments of medicine and biomedical sciences require sophisticated methods and approaches to analyze the ever-growing "big" and "complex" data sets that have been gathered so far. The availability and free distribution of such data is, sadly, still inconsistent. In reality, although "omics" data repositories strongly support fundamental developments in molecular medicine, there is still a vast scarcity of clinical data that can be used in comparable ways, with notable exceptions being projects like eMERGE. Important issues related to data ownership, privacy rights, and national and international law slow down the "phenotype" trend, which can have an impact on biomedical science (i.e., omics). To invent new algorithms, new methods, and new technological tools, simple, curiosity-driven research will still be crucial. For effective and realistic solutions, researchers need to share the source code of their (innovative) data analytics methods and algorithms with international research groups and communities.

In addition, the establishment of large and multidisciplinary research teams with the objective of selecting, tailoring, developing, and ultimately delivering solutions tailored to the specific needs of the intended consumer will become increasingly important when moving from basic to applied research. Moreover, to develop new innovative strategies, such teams would also need to remain mindful of current approaches and techniques outside of biomedical informatics. Biomedical computer scientists will, therefore, become more and more like technology developers, who can combine various techniques, methods, and resources for healthcare and biomedical research purposes.

KEYWORDS

- **genetic**
- **genomic**
- **security**
- **privacy**
- **data analysis**
- **biomedical imaging**
- **biomedical data analysis**

REFERENCES

1. Tewhey, R.; Bansal, V.; Torkamani, A.; Topol, E. J.; Schork, N. J. The Importance of Phase Information for Human Genomic. *Nat. Rev. Genet.* **2011,** *12* (3), 215–223. doi:10.1038/nrg2950
2. Hounsfield, G. N. Computed Medical Imaging. *Med. Phys.* **1980,** *7* (4), 283–290. doi:10.1118/1.594709
3. Bryant, A.; Charmaz, K. Introduction: Grounded Theory Research: Methods and Practices, 2011.
4. Bryant, A. Re-grounding Grounded Theory. *J. Inform. Technol. Theor. Appl.* **2002,** *4,* 25–42.
5. Webb, A.; Kagadis, G. C. Introduction to Biomedical Imaging. *Med. Phys.* **2003,** *30* (8), 2267–2267. doi:10.1118/1.1589017
6. Shendure, J.; Ji, H. Next-generation DNA Sequencing. *Nat. Biotechnol.* **2008,** *26,* 1135–1145. doi:10.1038/nbt1486
7. Liu, B. H. *Statistical Genomics: Linkage, Mapping, and QTL Analysis*; CRC Press, 1997.
8. Clow, D. An Overview of Learning Analytics. *Teaching Higher Education* **2013,** *18* (6), 683–695. doi:10.1080/13562517.2013.827653
9. Whirl-Carrillo, M.; McDonagh, E. M.; Hebert, J. M. et al. Pharmacogenomics Knowledge for Personalized Medicine. *Clin. Pharmacol. Therap.* **2012,** *92* (4), 414–417. doi:10.1038/clpt.2012.96
10. Tyers, M.; Mann, M. From Genomics to Proteomics. *Nature* **2003,** *422,* 193–197. doi:10.1038/nature01510
11. Fiehn O. Metabolomics—the link between genotypes and phenotypes. *Plant Mol Biol.* 2002 Jan; *48*(1–2):155–71. PMID: 11860207.
12. Mazziotta, J. C.; Phelps, M. E.; Plummer, D.; Kuhl, D. E. Quantitation in Positron Emission Computed Tomography: 5. Physical–Anatomical Effects. *J. Comput. Assist. Tomogr.* **1981,** *5* (5), 734–743. doi:10.1097/00004728-198110000-00029
13. Schulze, D.; Heiland, M.; Thurmann, H.; Adam, G. Radiation Exposure during Midfacial Imaging Using 4- and 16-slice Computed Tomography, Cone Beam Computed Tomography Systems and Conventional Radiography. *Dentomaxillofacial Radiol.* **2004,** *33* (2), 83–86. doi:10.1259/dmfr/28403350
14. Picard, R. W.; Kabir, T. Finding Similar Patterns in Large Image Databases. In *IEEE International Conference on Acoustics Speech and Signal Processing*, 1993. doi:10.1109/icassp.1993.319772
15. Obenshain, M. K. Application of Data Mining Techniques to Healthcare Data. *Infect. Control Hosp. Epidemiol.* **2004,** *25* (08), 690–695. doi:10.1086/502460
16. OpenNeuro.org
17. Maldjian, J. A.; Laurienti, P. J.; Kraft, R. A.; Burdette, J. H. An Automated Method for Neuroanatomic and Cytoarchitectonic Atlas-Based Interrogation of fMRI Data Sets. *NeuroImage* **2003,** *19* (3), 1233–1239. doi:10.1016/s1053-8119(03)00169-1
18. Bishop, J. M. The Molecular Genetics of Cancer. *Science* **1987,** *235* (4786), 305–311.
19. Wak, N. K.; Chong, C. H. Input Feature Selection for Classification Problems. *IEEE Trans. Neural Netw.* **2002,** *13* (1), 143–159. doi:10.1109/72.977291
20. Shruti Kute; Amit Kumar Tyagi; Rohit Sahoo; Shaveta Malik, "Building a Smart Healthcare System Using Internet of Things and Machine Learning," in Big Data

Management in Sensing: Applications in AI and IoT , River Publishers, 2021, pp.159–178.

21. Martinez, M. G.; Walton, B. The Wisdom of Crowds: The Potential of Online Communities as a Tool for Data Analysis. *Technovation* **2014,** *34*, 203–214.
22. Shamila M, Vinuthna, K. & Tyagi, Amit. (2019). A Review on Several Critical Issues and Challenges in IoT based e-Healthcare System. 1036–1043. 10.1109/ICCS45141.2019.9065831.
23. Madhav A. V. S., Tyagi A. K. (2022) The World with Future Technologies (Post-COVID-19): Open Issues, Challenges, and the Road Ahead. In: Tyagi A.K., Abraham A., Kaklauskas A. (eds) Intelligent Interactive Multimedia Systems for e-Healthcare Applications. Springer, Singapore. https://doi.org/10.1007/978-981-16-6542-4_22
24. Nair M. M., Kumari S., Tyagi A. K., Sravanthi K. (2021) Deep Learning for Medical Image Recognition: Open Issues and a Way to Forward. In: Goyal D., Gupta A. K., Piuri V., Ganzha M., Paprzycki M. (eds) Proceedings of the Second International Conference on Information Management and Machine Intelligence. Lecture Notes in Networks and Systems, vol 166. Springer, Singapore. https://doi.org/10.1007/978-981-15-9689-6_38

PART 4

Future Research Directions for Intelligent and Automated Healthcare Environment

CHAPTER 15

Healthcare 4.0 in Prospective of Respiratory Support System and Artificial Lung

MOUPALI ROY[1*], ARPAN DAS[1], BISWARUP NEOGI[2], and PRABIR SAHA[3]

[1]*Narula Institute of Technology, Kolkata, India*

[2]*JIS college of Engineering, West Bengal, India*

[3]*National Institute of Technology, Meghalaya, India*

[*]*Corresponding author. E-mail: moupali.roy@nit.ac.in*

ABSTRACT

The terminology Healthcare 4.0, which originated from Industry 4.0, is an emerging area of research and innovation. On converging to this Healthcare 4.0, the respiratory support system representation and artificial lung-related chronological research are introduced by this chapter. An informative analysis of this area of research concentrating on the product design approach has been elaborated. The challenges in this particular area of the lung support system need to be explored. Several products and models related torespiratory support systems have been analyzed and recognized on the basis of application aspects and commercial perspective. Most significantly, the artificial intelligence, Internet of medical things, advanced data science, and machine intelligence concepts connected to respiratory system analysis have been introduced with a product-based manner. Moreover, all the cutting-edge technologies and research challenges have been covered under this lung-oriented research toward the aim of technological products.

Intelligent Interactive Multimedia Systems for e-Healthcare Applications. Shaveta Malik, PhD
Amit Kumar Tyagi, PhD (Eds.)

15.1 INTRODUCTION

Healthcare 4.0 is fundamentally a part of Industry 4.0. The number "4.0" is nothing but a version, whichmeans Healthcare 4.0 has been upgraded a lot till date. Therefore, at first, it known as what Industry 4.0 is. It indicatestoward the revolution of the Industry.[1] The concept of Industry 4.0 was first made known in the year2011 as the fourth industrial upheaval.[2] The initial industrial upheaval commenced in the 18th century. Industry 1.0 was about steam power and mechanization of construction. After that, it led to the second industrial upheaval in the 19th century. In this revolution, electricity and gathering line production were introduced. Then, the industry extends one step more in the year 20th century. Industry 3.0 was about partial automation using memory programmable control and computers.[3] Finally, at present, we are implementing Industry 4.0. So, in this way, we have got the entire revolution up to Industry 4.0.

Now, it has come to the period of Healthcare 4.0. It is a cooperative stretch of digital, physical knowledge introduced as smart and wireless technology, mobile, online, e-Health, medical IT, and tele-health.[4] In the decade 1840–1860, we got the first healthcare revolution, which was Healthcare 1.0.[5] After this decade, due to continuous revolution, what we are seeing today in the medical world is Healthcare 4.0. It is the combination of three main paradigms—IOT, Big Data, and Cloud Computing.[6]

Lung replacement is the most viable treatment for patients suffering from the advanced stages of lung disease.[7] Healthcare 4.0 is highly connected with lung transplantation. Artificial lung isone of the huge examples where Healthcare 4.0 is being used. This is like a device that can be involved in those patients with progressive lung disease.[8] There is one type of lung disease called chronic obstructive pulmonary disease (COPD), which encompasses chronic bronchitis and emphysema that is produced by airflow barrier. One major characteristic of COPD is that it develops gradually over a period of time.[9] This ailment has become the third greatest common cause of death in the world. Those suffering in this disease have to opt for lung transplant. But it requires a few months to receive a lung transplant. During this waiting period, patients need a temporary lung replacement. For this purpose, Healthcare 4.0 is being used to develop artificial lungs, which is used as a replacement of a lung.

Numerous respiratory system-oriented challenges are there to be researched, such as identical mutual ventilatory illnesses, like bronchitis and emphysema, with regard to COPD. In this context, scarce electrical models are there that are equivalent to the physiological edifice of the human respiratory system. Lung cancer is a big challenge in the present times; in this aspect also, Healthcare 4.0 is of great help. To addressing the delinquent that healthcare conveyance schemes in the world-wide are progressively tackled with the contests to arrangement with the budding petition for high eminence services despite the fact the possessions are getting threatened..

Artificial Intelligence (AI) mentions the intelligent procedures or structures, which are much similar to human abilities, such as the capability to motive, determine sense, or learn from practice. This emergent technology has presently been functional in assorted submissions in progressive healthcare and medicinal schemes, and this is on the growth cracking a diversity of difficulties for hospitals, patients, and the global healthcare manufacturing. It is close to human perception capability in the investigation of composite health or medical statistics for the anticipation or treatment practices, such as analysis procedures, treatment decorum, personalized medicine, drug development, and patient monitoring and maintenance.[5]

A strict characterization for Healthcare 4.0 for monitoring pulmonary systems such as lung plantationartificially with the help of AI, IOMT, and ML is needed. Healthcare is fluctuating from outdated hospital-centric attention to a more cybernetic, scattered care that profoundly leverages the state-of-the-art technologies everywhere, like AI, data analytics, robotics, genomics, deep learning, and home-grown healthcare. Respiratory system relates to the upcoming instruments, the H 4.0will be supported as a unified continuum of care in the clinic-centered point-of-care. Using the H 4.0 technologies, a techno-commercial business model on respiratory system is perceived as a worthwhiledirection to innovate. For this, wearable beams are being included in the database about air quality from a particular area, body area network (BAN), actions of respiratory rate, and tidal volume. Wearable activity trackers are designed by using various types of biomedical sensors.

15.2 INDUSTRY 4.0 TOWARD HEALTHCARE 4.0 IN FUTURISTIC APPROACH

Industry 4.0 is the revolution in the direction of computerization and statistics exchange in modern technologies. These are included in many areas of technologies like CPS, IOT, IIOT, IOS, cloud computing, cognitive computing, and AI. Industry 4.0 referred physical invention and operations with smart digital technologies, IOMT, ML, and Big Data.[10]

Cyber corporal systems or CCS is the incorporation of networking, computation, and physical process. Internet of things is a perception that is connecting "things" like matters and machines to the internet. It is a platform where embedded devices can accumulate and exchange information with each other over the internet. Internet of Services (IOS) is a next-generation block chain concept. It provides network infrastructure to hold up a service-oriented ecosystem.[5] The impression of cloud computing is to customize hardware and software to transport the computing resources from submission to data storage as desired on a pay-for-habit basis over the network. Google's Gmail is one example of this.[11] The subdivision of the computer sciences that emphasizes the growth of astuteness machines, discerning and occupied like humans is known by AI.[12] All these areas triggerthe Industry 4.0 vision.

Figure 15.1 signifies the relation of revolution between Industry and Healthcare. At the initial Industrial revolution, mechanization of manufacture was found. At the same time, a new disease was discovered called pulmonary tuberculosis. It is a bacterial infection of the lung. Chest pain, severe coughing, and breathlessness are some common symptoms of this. This disease can be life threatening if propertreatment does not happen. In this way, the first healthcare revolution has come.

After two decades, the next industrial revolution happened. In this revolution, electrification was introduced. Healthcare revolution happened simultaneously and cobalt therapy was developed. Cobalt therapyis the therapeutic use of gamma rays after the radioisotope cobalt-60 to treat circumstances such as cancer. Cobalt-60 was extensively used in exterior beam radiotherapy tackles, which fashioned a beamof gamma rays, which was engaged into the patient's body to destroy tumor tissue.

After that, automation had come, which indicates another industry revolution. At the time another diseasecame called COPD. COPD is a chronic provocative pulmonary sickness that reasons obstructed inhale exhale for the lungs. Warning sign encompass breathing trouble, mucus construction, cough and wheezing. While treatment was needed to rescue those patients from this disease, the healthcare reached its next revolution.

Finally, industry 4.0 has come with the use of cyber physical system. Followed by this healthcare 4.0 has also come and this revolution still continuing its journey in today's world. In healthcare 4.0, Incentive Spirometer is being used. An inducement spirometer is a handheld expedient that helps our lungs recovers afterwards a surgery or lung sickness. Our lungs can develop weak after prolonged disuse. Using a spirometer assistance keep them dynamic and free of fluid.

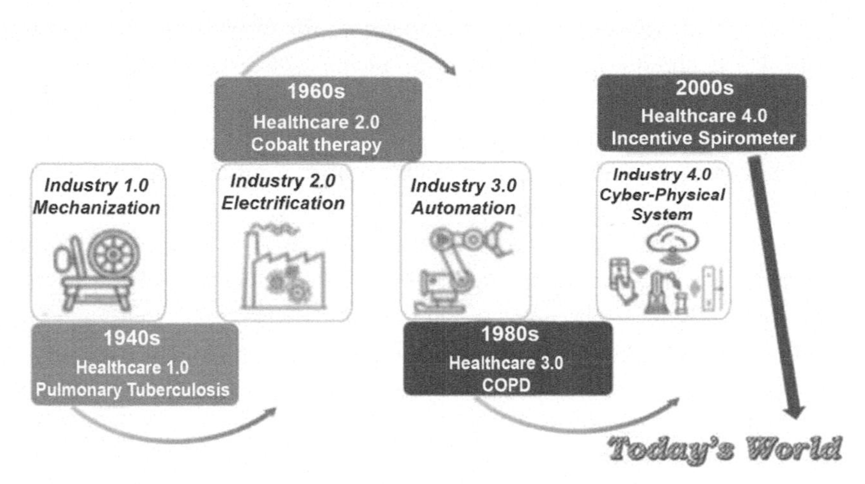

FIGURE 15.1 Lung related biological engineering innovation across the twice decade (history of industry-healthcare 4.0).

This Industry 4.0 vision helps to create intelligent and inter-connected world. It is used by some smart factories to connect between physical production and digital networks. Industry 4.0 has a major influence on our civilization. Also, in business, education, environment, manufacturing sector and all the sectors of economy, it is being used. Industry 4.0 is assembly it informal for establishments to work in partnership and segment statistics among clienteles, constructcrs, traders and other revelries in stock hawser. It progresses throughput and attractiveness, empowers the changeover to a digital currency, andaffords prospects to accomplish economic development along with sustainability.[13]

This insurgency, while it happening as an idea for novelty in industrial, is pleasing pedigrees in all characteristics of human natural life. For occurrence, so long as healthcare to a budding elderly human inhabitant is part of it.[5] This health domain is also an example of human activities which is affectedby Industry 4.0. The concept of Healthcare 4.0 is to discover more and more technologies by the usingof Industry 4.0 that is existence adapted in Health Services.

At the stage of the first upheaval (I 1.0 and H 1.0), approximately recent medical gears were conceivedand pragmatic in health center, Piston syringe, Stethoscope flexible tube, and clinical thermometer transferable such as few medical instruments introduced in the early 18th century set by step. Without power source, these health apparatuses familiarized in this era are submissive strategies meeker active strategies announced advanced; the manufacture tools requests sophisticated automatic project and dispensation. Through H1.0, the automatic processing proposal it established procedures conceivable.

Once later upheaval (I 2.0 and H 2.0) originated, additional intricate medical apparatus was developed in health centers. During late-18th to early-19th century, few medical imaging instruments were introduced likely initial sphygmomanometer, X-ray, and electrocardiograph. All those novel remedial machineries are frequently engaged in difficult electronic and electrical production in adding to the mechanical too. The energy supply announced by innovation fetched remedial apparatus into the epoch of electrification.

After that, in the next upheaval phase (I 3.0 along with H 3.0), the progression empowered in the computer, robotics, biomedical, and microelectronics subdivisions to origination and acceptance of additional multipart therapeutic schemes, as intensity mode middle 19th century ultrasonography, X- ray, computed tomography and implantable pacemaker, artificial heart, MRI and PET, and so on. These health preparations crucial cultured innovativeness of computer software, control algorithms and mechanics, and electronics. High-precision dealing out and quality regulator is a precondition to harvest such systems. Deprived of the progressive industrial technologies familiarized by Industry3.0.[5]

After Healthcare 4.0, the complete healthcare section has accomplished momentous growth in the direction of the eventual apparition of 8-P medical-care: participatory, prognostic, preventive, personalized, patient-adjusted, pervasive, accuracy, and preventative healthcare.

15.2.1 PRECAUTIONARY AND PERSISTENT MEDICAL CARE

In this system, many more shrewd and inconspicuous feelers have arranged for all physique existence or for specific atmospheres concluded in the advancement of active substructure exclusively before one turn into gruesome.The inter-operability of remedial strategies to spread extra precise and comprehensive statisticsgroups that simplify improved facility or analyses is significant characteristic of the organizations of H 4.0. Disseminated and associated strength history repositories and movement of actual statistics after wearable diplomacies to information analytics current with additional opportunities in the possibility, performance, and excellence of medical-care amenities and competences. Intellectual actuators particularly automatons will be developed along with the prevalent ones, which will provocatively lead to a growth of the usefulness of telehealth. So that, unparalleled actual and comprehensive statistics and inclusive attention ofliving circumstances container be on condition that to authorities for supplementary preventiveand pervasive medical-care.

15.2.2 PRECISION AND PERSONALIZED MEDICAL-CARE

Additionally, greater association adventure the limitations of dissimilar officialdoms will be permitted by digitalization of originalities. The actual statistics will be effortlessly merged and evaluated composed with all genetical statistics, particular healthcare chronicles, and additional ancient statistics apprehended by innumerable organizations. All these statistics can be quite accompanying with the conforming separate with guaranteed concealment conservation. Thus, the analysis and treatment will be developed moremodified and accurate.

15.2.3 CENTERED PATIENT MEDICAL CARE

Once the blockades of evidence stream all over the comprehensive medical-care provision stream is fragmented dejected, medical-care 4.0 can afford patients with unified integration of enduring movements and all-inclusive arrangement ofmedical-care procedure optimization.

Cultivating human fitness administration within emotional, psychological, physical, and socialcontexts can be accomplished by means of the knowledge pouring I 4.0. That would have stretched the methodical promises as of isolated expanses toward patients and evolvingbiosphere to have admittance to specialists in progressive fitness amenities over the sphere. Nevertheless, principled, permissible, and dogmatic fences strength are hindrances that should be overawed in instruction to improvement to assistances from this probable fitness upheaval. The profound inter-disciplinary conjunction of automation, computing, biomedical, and strength informatics is important for forthcoming investigate in university and trades.

15.3 RESPIRATORY SYSTEM-ORIENTED CHALLENGES AND RESEARCH TOPIC

In Figure 15.2, there are numerous types of respiratory system challenges in regular life. Each breath origins an inhalation of around 7 mL of air volume per KG of body weight. A child who weighs 30 KGinhales approximately 210 mL of air volume (210X30). In the period of a minute, some 4200mL of air capacity enters and are excluded from the lungs. Athletes breathe somewhat deeper and slower. With each breath, they inhale roughly 10 mL of air per KG.

Thus, in Figure 15.3, an athletic child who considers 30 KG will only breathe 15 times in the period spaceof a minute. Each breath will entail some 300 mL of air capacity. In the planetary of a minute, 4500 mL of air capacity will enter and be ejected from lungs. It can be understood from this that athletes express their airways in a much more effective way.[14]

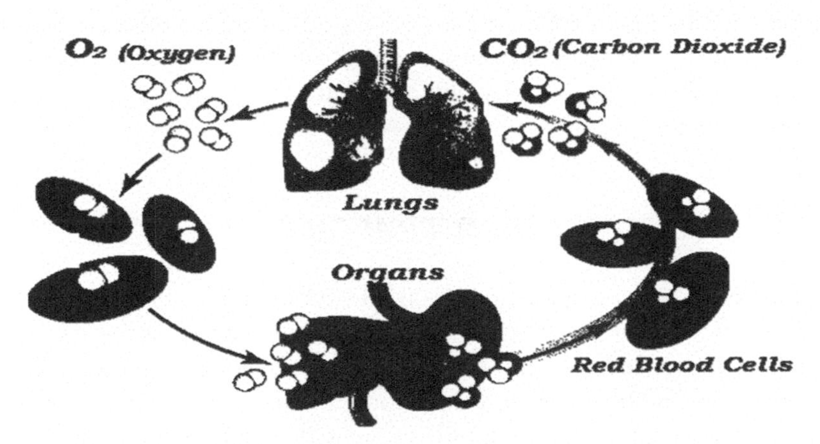

FIGURE 15.2 Lung cell airflow processing.[14]

15.3.1 PROCESS FOR TEST

- Patient is inquired to inhale as profoundly as possible.
- Patient is queried to exhale persistently into the spirometer.
- Patient is requested to stay to exorcise air for a few seconds, regardless of the solid urge tobreathe in.

a. In their case of the pulmonary transfer encouraged oxygen onto mixing hemoglobin deprived of substantially fraternization air as well as blood.
b. Make the most of transmission effectiveness while preserving the veracity of the air-tissue- blood boundary.
c. Deliver feedback apparatuses that contest respectively period of the handover process to one additional and to entire physique metabolic stresses,

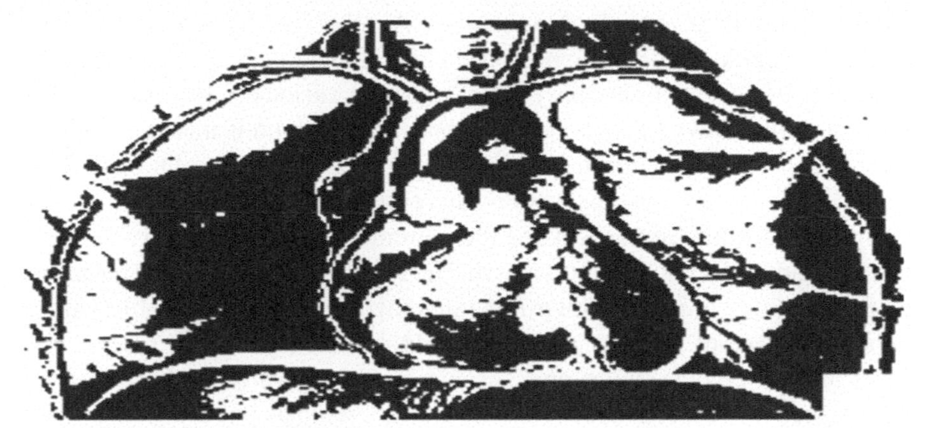

FIGURE 15.3 Human lung architecture.[14]

d. Integrate sufficient operational and purposeful investments and the aptitude to speedily recruit investments as mandate growths, and
e. Afford multipurpose adaptive machineries to recompense for the unpredicted loss of dimensions, and reservation or re-establish gas-exchange purpose in the face of exterior abuses or disease.[15]

In Figure 15.4, all-over India, there are various types of lung diseases and respiratory research-related projects are going on. Few of these are enlisted there:

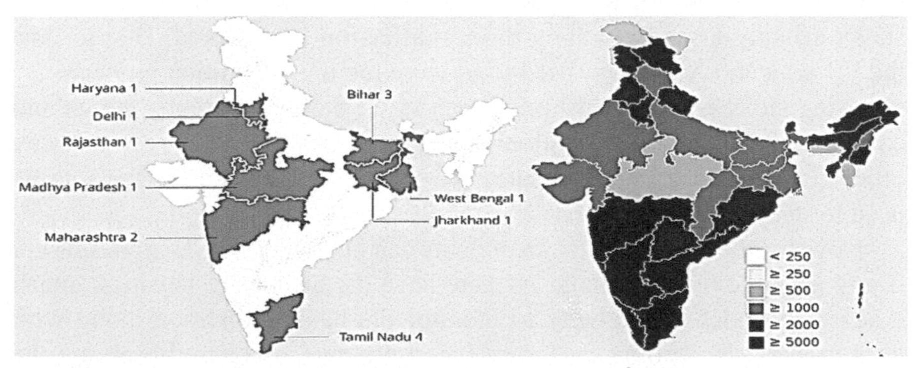

FIGURE 15.4 All India lung related project progress process.[8]

One development was on respiratory sputum cysteine acetyl effect. On the other hand, chronic bronchitis has an effect on emphysema of lung, which has been classified in various researches. Another project proposal is there on passive smoking for women exposed to cotinine in urine. In a laboratory, immunity of cell for pulmonary tuberculosis in various stages is studied; a clinical study showed in a project that bronchial asthma is traditional and is prevalent in school children too. In Chandigarh, another evolutional function said sports persons are healthy in ventilatory functions; smoking—an attributable health care aspect for the community—is also a very relevant research project topic. In the case of HFA budesonide, a project proposal works on blind clinical formoterol in randomized order for the stable asthma patients in case of inhaling or exhaling. In contrast, an ICMR research on adult asthma epidemiology and chronic bronchitis can also be discussed and projected; a project under RNTCP for tuberculosis patients preservedin generic utility and health quality instruments outcomes for disease-specific research have been done;another project works are there where lung tuberculosis is a popular occurrence and intensified to finding the cases between the high-risk nominated clinics with the attending of public secondary facilities of health care and Internet of Medical Things (IOMT).[8]

In Figure 15.5, the initial steps of lung healthcare system to measure lung capacity can be shown. So, there is need to observe the lung condition. As an instance, Oximeter was first used to observe the condition of lungs. After checking the condition, it may be happening that the capacity of lung is less. Then, the use of Spirometer was introduced. It will help to check the capability of lung by observing how much oxygen can be inhaled and exhaled as well as the sustainable power. ECG patch can also observe the relation between occupation of heart and lung along with these. In the follow-up stage, actually how much lung disease affection is observed. Due to this, the CT scan and X-ray was used to observe the lung condition properly. By checking the X-ray plate of lung, it can say if there is presently cough and cold or can check more condition of lung after observing CT scan that any other diseases like COPD or cancer or else. All of those processes can be used to study the lung properly.[5]

Now, for the second steps, treatments can be done using H4.0 in this era if any negative result of lung. Asthere have the healthcare logistics, attach a sensor bracelet in the body to monitor the health condition along with lung properly as well as can sense blood pressure. Because observing the blood pressure is also dynamic object. If any unstable condition detected then immediate hospitalization with all super speciality is the main steps to the patient. Lastly, after completion of the primary treatment, if the lung condition keeps on same then surgery is the final one to contemplate about artificial lung. The concept, as the heart transplantation same to implement an artificial lung.

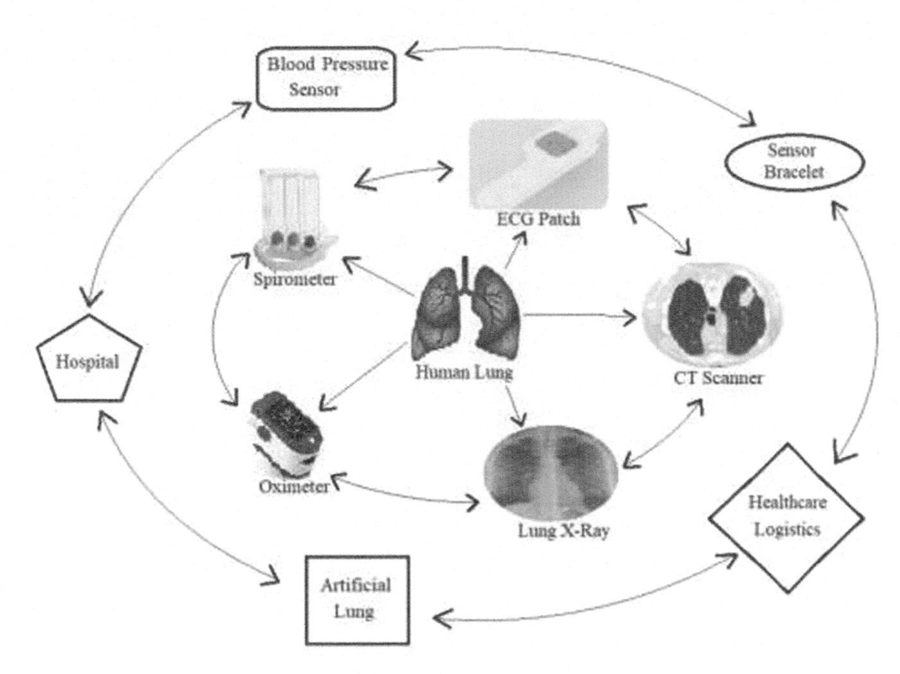

FIGURE 15.5 Respiratory supported healthcare system.

15.4 FUTURISTIC INSTRUMENTS RELATED TO RESPIRATORY SYSTEM

2025 onward, all the continuing concept will be implemented electronically for the product shape towards forthcoming technological terms symbiosis, systemic, informatics, cybernetics. The relationship between CPS and systems science and Systems science discipline depends on the cyber systems, Engineering complex cyber-physical systems toward core system science with Physical Systems. Fourth Industrial Revolution (I4.0) has an interdisciplinary relation with industrial information and industrial integration, with systems engineering, science and theory.[16]

Considering science and engineering as a system, this system can be described in others systems like system theory, cybernetics, system approach etc. System science is directly related to this flow. By thissystem, the nature container was studied from simple to complex. This system can be used in human respiratory system, also. System engineering is a combination of constituent that accomplishes collectively a convenient function. This system is also present in a rummage-sale in healthcare system. System theory is the learning about any system. To patterned lung condition, there are numerous instruments present. For producing, these many technologies are there. Cybernetics is another system that can be considered here. This system is an approach to exploring regulatory system. Data storage,cloud computing, and machine learning are used in this field. This technology is being used in healthcare 4.0. In this field, already they used in so many systems. So, there must be a huge data. Here, system analysisis needed. After analysis of those data, lung condition can be judged more accurately. After that there is another system called system methodology.

There are many systems which can be use is shown in Figure 15.6. But which is suitable that thing need to be expressed. For this purpose, system methodology is needed. In system approach, any new approachcan be done. During lung related invention, there must be needed so many tastings. So, system approachis also important in Healthcare 4.0. For using system science and engineering, primary thinking may be needed. For that reason, system thinking may use. Like that every system is very much important duringcreate the lung related instruments.

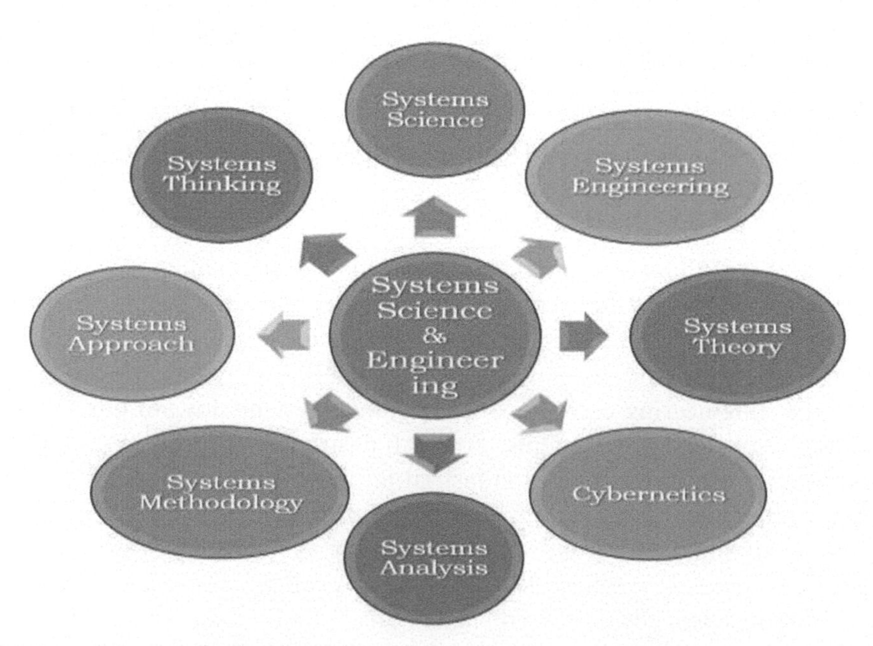

FIGURE 15.6 Systems involvement to industry 4.0.

The combined habit of the feasible arrangement model (FAM) and the fractional slightest squares pathway demonstrating approach to measure the supportable organization carrying out commercial actions. A systems-thinking tactic assimilating systemic organizations in three segments:

a. The FAM was initially used to progress an abstract model of the organizational delinquent;
b. PLS-PM was used to suggest a hypothesis to outline an explanation and statistically authenticate the associations planned in the intangible model; finally,
c. The contacts between performers were reconsidered in instruction to encourage supportable presentation.[17,18]

There are many wearable devices which can be use as instruments. Wearable represents whatever can be wear as jackets, pants, hats, sweaters, glasses, watches, socks. The main motive to make device wearable is not to bother the daily activities and also to monitoring for healthcare. Today these wearabledevices are very much trending for sports and fitness.

Pulse-oximetry is important for intensive care the tolerability of freshening, calculating arterial hemoglobin inundation alone is not sufficient and it motionless agonizes from few of hands-on downsides, as well as concentrated perfusion at the position of dimension, as in hypothermia or shockwave. Respiratory ventilation is the merchandise of pulmonary rate and the inhaled and exhaled volume of air with each breath. One-to-one care pulmonary freshening earnings intensive care these two essential constraints, and all the restrictions that can be derivative once encouragement and termination are distinguished, namely expiratory period, responsibility cycle or inspiratory time, whichis the ratio between total respiratory cycle time and inspiratory time or mean inspiratory flow which one is the percentage among encouraged time and supportive volume and also uncaring expiratory stream is expressed by the percentage perished time vs. expiratory capacity.

So, restrictions are important to judge respiratory supporting occupation for the patient. Pulmonary percentage characterizes an energetic symbol used to display the headway of sickness and an uncharacteristic pulmonary rate is an essential extrapolative influence and a significant indication of solemn illness. Pulmonary rate has been exposed to be intelligent to envisage opposing clinical measures, such as cardiac apprehension or charge to the ICU, and to differentiate between patients at risk improved than pulse and constant patients and blood pressure dimensions. Many lung sicknesses, such as affect pulmonary rate, pneumonia, in the clinical environment, in particular breathing rate, and respiratory activity, should be monitored uninterruptedly over time in penetrating cases. Specialist care ventilation during snooze, in footings of hypopneas and apnoeas whereas enumerating the manifestation of epochs with truncated or no aeriation, is a stipulation in the broadcast for sleep sicknesses for opportune treatments and diagnosis.[19]

Now-a-days, there are some irresistible number of newly wearable devices, not all of those are proficientof determining reliable databases about health status. Equally, there are adequately of sensors that quantity physiological limitations but are not in a wearable arrangement. Wearable biomedical instruments are consequently the subdivision of strategies that are able both to quantity biological limitations and to be damaged.

Frequently, wearable technology is constructed on predictable electronics, either inflexible or bendable,powered by conformist batteries. This comprises mobile phone peripherals (devices, boundaries or sensors associated to the phone). In contrast, wearable expertise is more disruptive, and embraces apparel and fabrics with distributed meanings, into which electronics are warmly combined. So far, theprogress is not understandable because strategies have to be stretchable, washable and foldable, and occasionally printable or crystal clear.

In connectable lung will be premeditated to become patients up and affecting in the interior the hospitalscenery, which is significant for equally patient repossession and enlightening a patient's position erstwhile to a replacement of respiratory system. The pulmonary replacement device will accompany recent exertions by possibly enlightening the effectiveness of the assignment of oxygenand carbon dioxide and Federspiel explicated, snowballing biocompatibility.[20]

A prosthetics respiratory system, an adequate has been supported in a haversack to be exposed to exertion in few animals. Quite a few devices are there presence established that could convert the breathes of individuals with respiratory system failure, it's presently reliant on large apparatuses. The novel expedient still involves a gas boiler is turned everywhere, even though those models are experienced.

People with respiratory failure are generally be associated to an appliance that pumps blood concluded anda gas exchanger to afford oxygen and eliminate carbon dioxide—but this frequently confines them to bed. The extended they are bed-ridden, the weedier their muscles become, and the less probable they areto convalesce.[21]

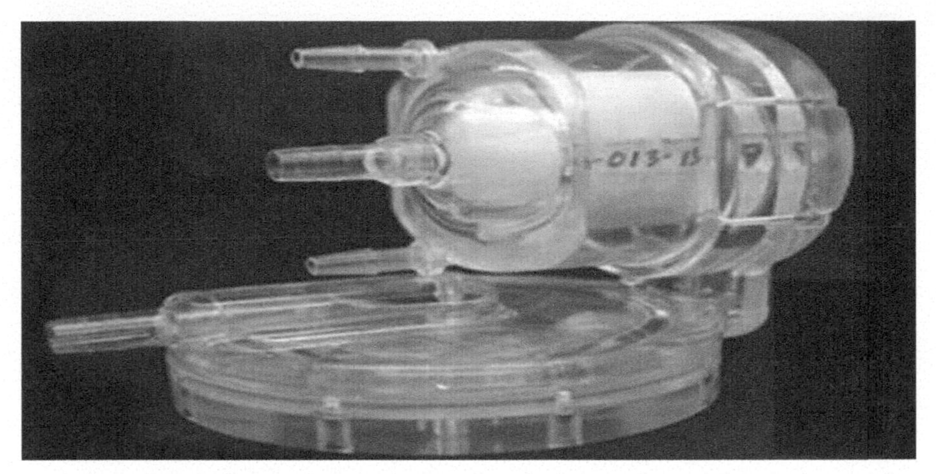

FIGURE 15.7 Artificial lung mechanical model.[20]

In Figure 15.7, artificial respiratory support systems have stretched to help sickening child until a respiratory replacement is obtainable. Whereas, this strategy will be cooperative in providing oxygen for progenies anguish from pulmonary hypertension, cystic and respiratory fibrosis, during the wait for lung transplants, and they control motion. A researcher investigating behaviors to generate additional compressed Pulmonary support expedient for kids. The exertion was productive, through the Supportive Respiratory directing to assist as per a connection toward replacement or retrieval in broods with criticaland continuing respiratory letdown. Invention such as, tiny in the size, will be premeditated for incessantrespiratory sustenance.

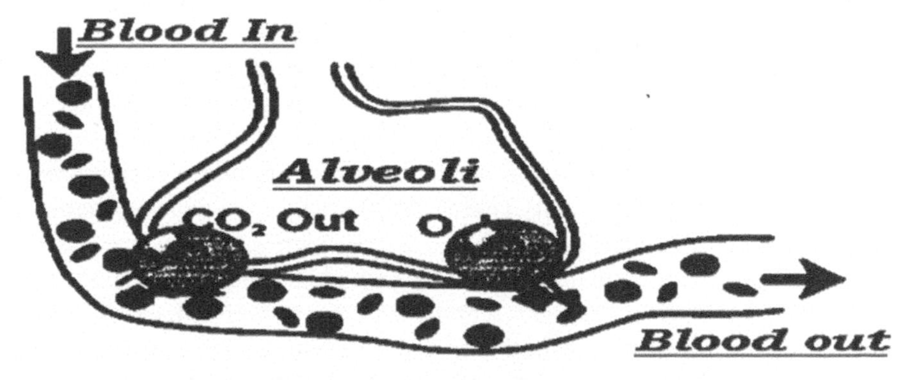

FIGURE 15.8 Alveoli: Blood in-out process for lung.[20]

In Figure 15.8, conceivably most prominently nevertheless, the expedient has assistance patients come out of engross in movement rapidly conceivable subsequently a procedure. Trainings will be exposed in ambulation throughout the coming up retro for pulmonary replacement suggestively expands after transplantation. Researches said the upcoming step for the artificial Respiratory support system is to discovery traditions to further interpret the development toward experimental custom. The investigatorsare employed with Institute, in partnership with sciVelo, to assistance development the commercial transformation.[22]

15.5 ARTIFICIAL LUNG ON THE ASPECT OF H 4.0

In e-healthcare, a lot of statistics to doctor IoT campaigns are providing, also serving doctors, to recognize specialized medicine conversation, or for remembering countless certain activities of a patients it's also helps doctors, etc. Equally, IoT make communication easier between human beings by conversation submissions in online social networking. These are with submissions in this real ecosphere and associated on a huge measure and combined strongly through Internet substructure. IoT devices ortenders are assembling a statistic from each tracking human's movement. These statistics is deposited as big numbers and vended or used to other organization to make revenue for analytics development.[23,24]

Researchers have industrialized an artificial pump lung (APL) destined for damaged on the persistent and tolerate for peripatetic respiratory or cardiorespiratory sustenance. The Flexible Extracorporeal Respiratory Assist System participates a pivot-bearing adjustive pump with a package and is proposedto be ramshackle on the continuing all over long-term sustenance. The segmental expedient design empowers its conformation with Human Forces movement packages of fluctuating size, by this meanspermitting this system to be used in a variability of Lung promotion submissions.[7]

In Figure 15.9, consider there is a citizen. He has a mobile, jacket, bracelet, and some measuring instruments.There have also a system, which is mapping with the researchers and the providers. Researchers can direct access the system. Medical clinic and hospital are the providers. They are using the citizen's data. The provider and the researchers are connected with the citizen by this system. So, what happensin this system? Researchers use healthcare 4.0 technologies in their instruments. And provide jackets, bracelet which gives lung related data by using mobile health data. This data obviously depends upon environment. According to the locality of the citizen, this data can be different. Environment data needto check is the areas is polluted or not, heal area or not, highly populated or not etc. In the different environment, the healthcare data of citizen can be different.

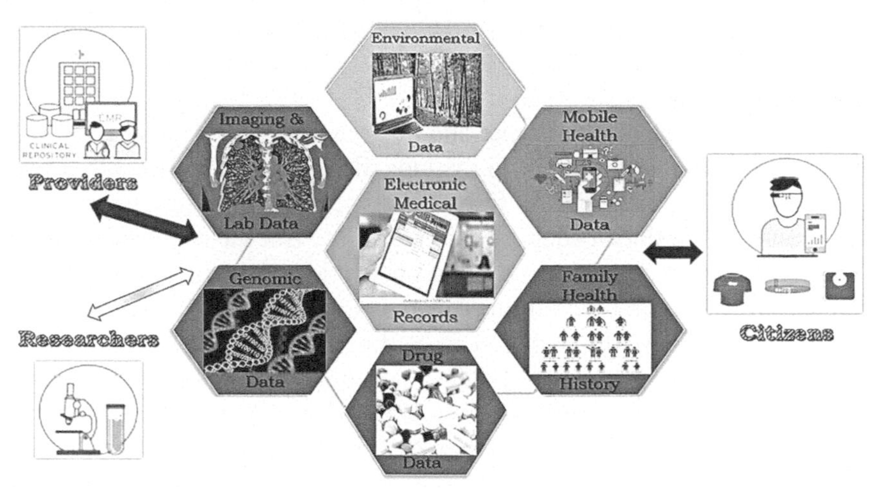

FIGURE 15.9 H4.0: System chain with researchers, providers and citizens.

Here artificial monitoring system is being used to collect all this data[R]. In this way after collecting allthe data of citizen's, researchers can make a database. By this database, citizen's family healthcare history can be added. The electronic medical record also has to add in their system depending on their family healthcare history. Simultaneously drag data need to add in this system. If we consider the family healthcare history, then obviously genetic data also should be there. Finally, researchers need the lung related imaging data and lab data using all these data. Researchers have access to all this data which they can use for further research. Citizens can also go to the clinical stores with this data to get the clinical support.[25]

15.6 RESPIRATORY/ LUNG RESEARCH OBJECTIVE ON THE PLATFORM OF H 4.0

Botnet is an assemblage of cooperated Internet hosts, it mounted with inaccessible control software established by spiteful users to exploit the profit accomplishment illegal happenings like Phishing, DDoS, and Spamming etc. occurrence on online network. Commonly, the development to HTTP instigated with advances in exploit kits Botnet but in far ahead due to etiquette fluctuating; Alternating communications; Announcement encryption; firewall approachable of Botnet, Botnet subgrouping Foundation concealment and HTTP Botnet is the utmost interest of the research municipal. There a new universal HTTP Botnet recognition framework for physical period network by means of Artificial Immune System (AIS). Mostly, a novel bio-inspired model (AIS), which spread over to resolving numerous problems in statistics sanctuary.[26]

Nowaday's, many technologies and instrument are used under Healthcare 4.0. In the respiratory systemor in the lung also, this Healthcare 4.0 is being used as shown in Figure 15.10. Using these new technologies, many instruments can be introduced such as oximeter. This is a tiny device. Usually it slides over fingertips or can be clips on ear lobe. Using this instrument, the measurement of oxygen in the lung has been done. Many types of body sensor also can be used which can sense the motion of the body. Using this sensor, the body movement is being monitored. We check the connection between heart andlung and blood circulation ratio also have been observed. Consider a wrist band. It is like a smart watchwhich includes pulse Ox sensor. It helps to monitor blood oxygen saturation level. It can also monitor the energy level of the body. Ear buds connects the brain to the heart and lung. Oxygen is required in our brain. So, measuring the oxygen is also needed. Again, consider shirt or jacket. It wraps the entire body. So here also healthcare 4.0 can be used. All these technologies are representing healthcare 4.0 which is being used in human lung.[19]

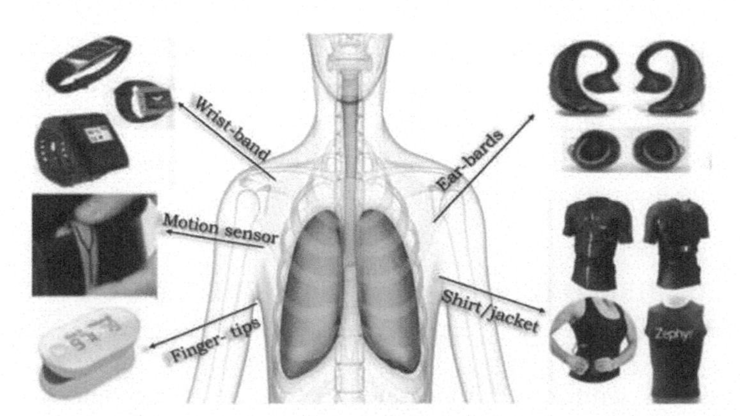

FIGURE 15.10 Lung related wearable devices related with H 4.0.

In topical expansion in clever/ computing or electronic devices, Big Data (a lot of data) is being produced. A highly useful Big Data is used for expecting future admiration to a request after having a appropriate and effective investigation. So, numerous glitches are there, challenges and issues have been corrected which are essential to overcome in forthcoming era. Hence, it stretches perception to how and where big data container be castoff in health-care structure, also afford quite a lot of encounters and problems with veneration to collected statistics genomic analytics.[27]

These all the instruments are industrial related. So here also many technologies are being used related to industry 4.0. Like IOT mapping can be done by using these instruments. That means all these data related to lung condition can be saved in the cloud. This data can be read any time when anyone wants.It can be accessed by our mobile or the monitor of the instruments.

The people famine to adore wireless facilities everywhere like in hotels, colleges, etc. LBS, the computer sequencer subdivision of level amenities castoff in various arenas and funding, the submission are approximately classified as Navigation and Maps, Tracking package, Information service, Vehicularnavigation, Social networking, Games and Advertising etc.[28] Position privacy for mobile users is now a days important for various types of mobile applications which are mainly related with the physical analysis and determined into the data conversion with this again stored in paid or free cloud linked with the mobile. Hence the persistence and involvement of various confidential and challenges issues in LBSs growing in future that have not been invented till in the any research.[24]

15.7 TECHNO-COMMERCIAL BUSINESS MODEL ON RESPIRATORY SYSTEM TOWARD H 4.0

Business model is a term, which mentions to a concern's strategy for assemble revenue as specifiedin Figure 15.11. It categorizes the models or amenities the professional strategies for vend, its recognized goal marketplace, and somewhat anticipated incidentals. Business models are imperative for both innovative and traditional businesses.

When any novel business model or strategy is being developed, then always essential to go through theproper steps. Everyone should spread over that scientific technique to progress and promotion the product properly.

In the beginning, discover is the problem. Discovery is approximately determining that the explanation or the model is working properly or not. Check the problem statement first. According to the problem, try to make the solution. It may be an impression or a business model. Then, exertion on this model and justify the explanation. Check the market validation to know the importance of the model. Make the prototype. Chartering a designer or a developer if needed. Finally, start testing the model. In every single step, always require to crisscross the result. If anywhere change to adverse result, then replicates that step again.

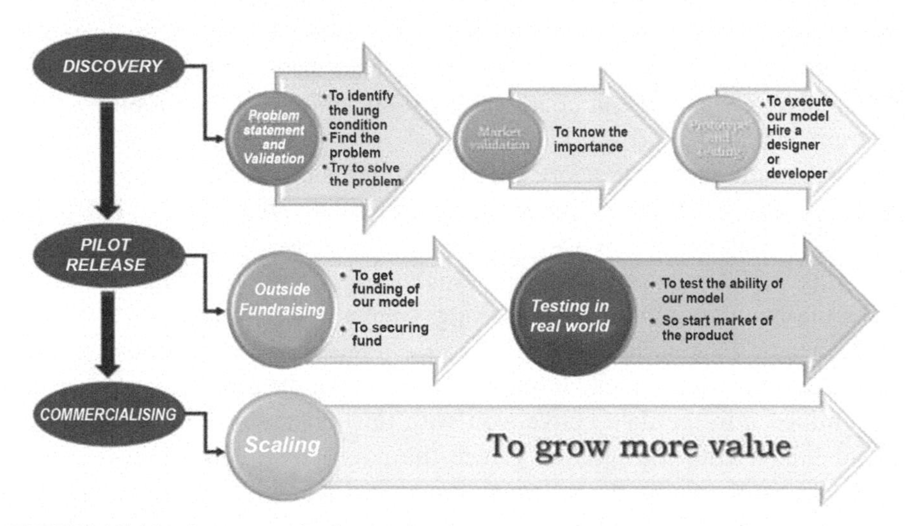

FIGURE 15.11 Startup production roadmap.

Then go to the next step, which is pilot release. In this step, the business model or proposal is going to grow real. Here the marketing skill is needed very much. Try to learn what works and what does not. Also need funding in this step to grow fast. Look for someone different who willing to invest the problem of this solution. At the same time also need to examine the capability of this model. Check the solution is work for many people or not. Test the model in real world. Start marketing of the invention by using this solution.

Identification and the validation of the problem have been done properly. A prototype of this model is also ready. Now need to commercialize. This is engrossed on scaling and growing supplementary worth. Check the products are meeting to customer needs or not. Grow the entire team and client base. These are the whole process to make a plan or business model successful.[29]

This issue is related to about respiratory system. So, contemplate a spirometer because it is an apparatusfor inspection human lung condition. How the entire progression has been done. First, we need the problem statement and validation to identify the lung condition. Find out the problem and try to solve the problem using healthcare 4.0. In this way, an idea or model can be prepared. Find out the market validation to know the importance. Then make the prototype to execute the model. Hire a designer or adeveloper to make this and start testing the model. Try to get funding for this model. Finally start testingin real world to test the ability of this apparatus as well as start market of this instrument. Try to scalingto grow more value of this product. This is the techno-commercial aspect for this particular apparatus.

15.8 CONCLUSION

To conclude this chapter, this aspect was cleared that upcoming future engineering and medical field research will be overlapped each other. The interdisciplinary nature of research in I 4.0 to H 4.0 most specifically toward human respiratory system was depicted with fine tuning with this chapter. This summary is not proposed to be a comprehensive survey on this matter, though a sincere determination was made to concealment all the recent the whole kit and caboodle as much as probable and any oversight of additional works is purely accidental. The roadmap of H4.0 on specific to the human respiratory system toward the future was the specific contribution of this chapter. From past every 20 years gap, the Technology has been rebooted and in case of Healthcare the lung specific innovative researches forecast. After 25 years in case of Industry 5.0 some probable research topics such as "symbiosis, systemic, informatics, cybernetics" was introduced. In case of repertory system research. all these new research themes will be integrated. Most significantly, the techno-commercial aspects on respiratory system research on the platform of H 4.0 was introduced toward futuristic aspiration. This summary is not envisioned to be a comprehensive investigation on those areas; however, a genuine determination was finished to cover all the modernistic research works as much as probable and any exclusion of other researches is innocently unintended.

KEYWORDS

- **healthcare 4.0**
- **artificial intelligence**
- **respiratory system**
- **artificial lung**
- **wearable devices**
- **techno-commercial business model**

REFERENCES

1. M. M. Nair, A. K. Tyagi and N. Sreenath, "The Future with Industry 4.0 at the Core of Society 5.0: Open Issues, Future Opportunities and Challenges," 2021 International Conference on Computer Communication and Informatics (ICCCI), 2021, pp. 1-7, doi: 10.1109/ICCCI50826.2021.9402498.
2. Luenendonk, M. Industry 4.0: Definition, Design Principles, Challenges, and the Future of Employment. *Cleverism*, Sept 23, 2019. https://www.cleverism.com/industry-4-0/#:~:text=The%20term%20also%20refers%20to,competitiveness%20in%20the%20manufacturing% 20industry (accessed Jul 16, 2020).
3. Tyagi A.K., Fernandez T.F., Mishra S., Kumari S. (2021) Intelligent Automation Systems at the Core of Industry 4.0. In: Abraham A., Piuri V., Gandhi N., Siarry P., Kaklauskas A., Madureira A. (eds) Intelligent Systems Design and Applications. ISDA 2020. Advances in Intelligent Systems and Computing, vol 1351. Springer, Cham. https://doi.org/10.1007/978-3-030-71187-0_1
4. Jayaraman, P. P.; Forkan, A. R. M.; Morshed, A.; Haghighi, P. D.; Kang, Y.-B. Healthcare 4.0: A Review of Frontiers in Digital Health. *WIREs Data Mining Knowl Discov* Dec 25, **2019,** e1350.
5. Pang, Z.; Yang, G.; Khedri, R.; Zhang, Y.-T. Introduction to the Special Section: Convergence of Automation Technology, Biomedical Engineering, and Health Informatics toward the Healthcare 4.0. *IEEE Rev. Biomed. Eng.* Jul 26, **2018,** *11*, 249–259.
6. Aceto, G.; Persico, V.; Pescapé, A. Industry 4.0 and Health: Internet of Things, Big Data, and Cloud Computing for Healthcare 4.0. *J. Indust. Info. Integr.* **2020,** *18*, 100–129.
7. Orizondo, R. A.; Cardounel, A. J.; Kormos, R.; Sanchez, P. G. Artificial Lungs: Current Status and Future Directions. In *Thoracic Transplantation* Kobashigawa, J.; Patel, J.; Eds.;Nov 11, 2019; pp 307–315. https://doi.org/10.1007/s40472-019-00255-0
8. Pitt-Developed Artificial Lung Shows Promise in Pre-Clinical Trials. *The McGowan Institute for Regenerative Medicine*, Apr 13, 2017. https://mirm-pitt.net/news-archive/pitt- developed-artificial-lung-shows-promise-in-pre-clinical-trials/ (accessed Jul 17, 2020).
9. Bagchi, S.; Chattopadhyay, M. Electrical Modelling of Respiratory System and Identification of Two Common COPD Diseases through Stability Analysis Technique.

In *2012 IEEE International Conference on Advanced Communication Control and Computing Technologies* (ICACCCT), Aug. 23–25, 2012, India.

10. Moore, M. What Is Industry 4.0? Everything You Need to Know. TechRadar, Nov 5, 2019. https://www.techradar.com/in/news/what-is-industry-40-everything-you-need-to-know (accessed Jul 20, 2020).
11. Nouri, S. M. R.; Li, H.; Venugopal, S.; Guo, W.; He, M. Y.; Tian, W. Autonomic Decentralized Elasticity Based on a Reinforcement Learning Controller for Cloud Applications. *Future Gen. Comput. Syst.* May **2019,** *94*, 765–780. DOI:10.1016/j.future.2018.11.049
12. Saeed, F. 9 Powerful Examples of Artificial Intelligence in Use Today. *IQVIS*, Feb 16, 2020. https://www.iqvis.com/blog/9-powerful-examples-of- artificial-intelligence-in-use-today/ (accessed Jul 25, 2020.
13. Yuan, X.-M. Impact of Industry 4.0 on Inventory Systems and Optimization, 2019. IntechOpen, Mar 18, 2020.
14. O'Malley, R. J.; Rhee, K. J. Contribution Of Air Medical Personnel To The Airway Management of Injured Patients. *Air Med. J.* **1993,** *12* (11–12), 425– 428. DOI: 10.1016/S1067-991X(05)80138-5. PMID 10130326
15. Hsia, C. C. W.; Hyde, D. M.; Weibel, E. R. Lung Structure and the Intrinsic Challenges of Gas Exchang. *Comprehen. Physiol.* Apr **2016,** *6*, 827–895.
16. Xu, L.D. The Contribution of Systems Science to Industry 4.0. 2020 John Wiley & Sons July/Aug **2020,** *37* (4), 618–631. https://doi.org/10.1002/sres.2705
17. Sánchez-García, J. Y.; Ramírez-Gutiérrez, A. G.; Núñez-Ríos, J. E.; Cardoso-Castro, P. P.; Rojas, O. G. Systems Thinking Approach to Sustainable Performance in RAMSAR Sites. *Sustainability* **2019,** *11* (22), 6469. DOI: 10.3390/su11226469
18. Grinin, A.; Grinin, L. Cybernetic Revolution and Forthcoming Technological Transformations (The Development of the Leading Technologies in the Light of the Theory of Production Revolutions). In *Cybernetic Revolution and Forthcoming Transformations*, Jan 2015.
19. Aliverti, A. Wearable Technology: Role in Respiratory Health and Disease. *Breathe (Sheff)* 2017, *13* (2). e27–e36.
20. Mishra S., Tyagi A.K. (2022) The Role of Machine Learning Techniques in Internet of Things-Based Cloud Applications. In: Pal S., De D., Buyya R. (eds) Artificial Intelligence-based Internet of Things Systems. Internet of Things (Technology, Communications and Computing). Springer, Cham. https://doi.org/10.1007/978-3-030-87059-1_4
21. Wilson, C. Artificial Lungs in a Backpack May Free People with Lung Failure. *NewScientist*, Mar 21, 2017. https://www.newscientist.com/article/2125422-artificial-lungs- in-a-backpack-may-free-people-with-lung-failure/#:~:text=An%20artificial%20lung%20that's%20small,currently%20dependent%20on%20large%20machines (accessed Sept 4, 2020).
22. Amit Kumar Tyagi, Dr. Meenu Gupta, Aswathy SU, Chetanya Ved, “Healthcare Solutions for Smart Era: An Useful Explanation from User’s Perspective”, in the Book "Recent Trends in Blockchain for Information Systems Security and Privacy", CRC Press, 2021.
23. Tyagi, A. K.; Rekha, G.; Sreenath, N. Beyond the Hype: Internet of Things Concepts, Security and Privacy Concerns. In *Advances in Decision Sciences, Image Processing, Security and Computer Vision*, Jul 13, 2019; pp 393–407.

24. Tyagi, A. K. Building a Smart and Sustainable Environment Using Internet of Things. In *Proceedings of International Conference on Sustainable Computing in Science, Technology and Management* (SUSCOM), Feb 26–28, 2019; Amity University Rajasthan, Jaipur-India, 2019.
25. Placing a Bet on Platform Business Model in Healthcare & Winning. *Science Service Dr. Hempel Digital Health Network*, May 20, 2019. https://www.dr-hempel- network.com/digital-health-startups/placing-a-bet-on-platform-business-model-in-healthcare/ (accessed Sept 6, 2020).
26. Tyagi, A. K.; Nayeem, S. Detecting HTTP Botnet Using Artificial Immune System (AIS). *Int. J. Appl. Info. Syst.* **2012,** *2* (6), 34–37.
27. Shamila, M.; Vinuthna, K.; Tyagi, A. K. A Review on Several Critical Issues and Challenges in IoT Based e-Healthcare System. In: *2019 International Conference on Intelligent Computing and Control Systems (ICCS)*, May 15–17, 2019; Madurai, India: IEEE, 2019; pp 1036–1043.
28. Tyagi, A. K.; Sreenath, N. Future Challenging Issues in Location Based Services. *Int. J. Comput. Appl.* Jan 1, **2015,** *114* (5).
29. Marsiglia, M. The Startup Product Roadmap—Maturing Your Software Idea into a Scalable Product. *Atomic Objective*, Sept 4, 2018. https://spin.atomicobject.com/2018/09/04/startup-product-roadmap/ (accessed Sept 6, 2020).

CHAPTER 16

Lifestyle Revolution: The Way to Healthcare, Case of India

RAJU K. KURIAN*, JYOTSNA HARAN, and SAURABH OJHA

Royal College of Arts Science and Commerce, Mira Road, Mumbai, India

Corresponding author. E-mail: rajukkurian@gmail.com

ABSTRACT

Do you know what people understand by health? It is never physical, but it seems physical. "Health is the outcome of physical, mental, and social well-being. If you are not sick, it does not mean that you are healthy. The enjoyment of the highest attainable standard of health is one of the fundamental rights of every human being without distinction of race, religion, political belief, economic, or social condition."[21]

Note that Health is the basic requirement of life. It is the real wealth.

Here our focus is to discuss all those factors which enrich health, not from a cure perspective but on prevention background. We want to bring out the relevance of the popularly known saying "prevention is better than cure." The present COVID-19 pandemic teaches the same. To understand and practice prevention, one must go deeper in various changing phases of lifestyle, particularly in India. This work takes one on a journey through various stages of lifestyle since ancient India.

We have to emphasize all aspects of life, for example, eating habits… what and when, sleeping and working habits, division of time and work, occasional irrational behavior and many more.

Intelligent Interactive Multimedia Systems for e-Healthcare Applications. Shaveta Malik, PhD
Amit Kumar Tyagi, PhD (Eds.)

There is scope of health education and its awareness in schools and at community levels too.There is a need to think of parenting too in this regard and counseling of our youth.

Mental and psychological aspects of health based on common logic and spiritual thought process are also looked into to find the solution.

Realizing the importance of lifestyle practices in real life in real terms would bring down thecost of healthcare and would resolve other problems associated with it. This may have many positive effects.

We have based our research work on secondary sources, observations, and cases.

The chapter observes the significance of various lifestyles giving emphasis on healthy lifestyle and educating people for the same. This work also discusses the interplay of Big Data and healthcare. Big Data size in healthcare is constantly increasing. While Big Data in healthcare is considered to be a boon, this paper evaluates Big Data in healthcare in both positive and critical manner by making some practical observations especially in the context of the COVID-19 pandemic and suggests how economies avoid investment in healthcare infrastructure including Big Data in healthcare. This may bring down the cost of healthcare in public as well as private sector. This shift in expenditure may have many positive effects like room for expenditure on education, infrastructure, and many more. In totality, all the factors will have cumulative and multiplier effects taking the family, society, and the economy ahead.

16.1 INTRODUCTION AND SCOPE OF THIS WORK

An individual's lifestyle plays a very critical role in his/her health prospects, both mental and physical. It has been observed that a person who owns a healthy lifestyle is in good health. Lifestyle is the way of life a person leads. It is the sum total of different activities a person undertakes such as food habits, exercise, sleep patterns, and such more. A good "way of life" is crucial to health. Lifestyle has always played a central role in our lives. In the current scenario of COVID-19 pandemic, this discussion has gained momentum. This pandemic situation has forced everyone to take a closer look toward lifestyle patterns and amends to unhealthy practices. We experience a lifestyle revolution as we realize radical changes aroundus. Lifestyle and health are embedded together and have action reactions affecting each other.The purpose is to bring forward a holistic view on healthcare through the lenses of lifestyle, historical events, health awareness and COVID-19. The growing concept of Big Data in healthcare, how is it related to lifestyle? Can better lifestyle improve quality of Big Data?

We observe the relation between the quality of Big Data and better lifestyle. Health and well-being has entered the self-tracking phenomenon through data via sensors and apps that allow our health to be improved on the move. Big Data can thus provide information of well-being and health to people via app by tracking their habits and thereby helping to improve their health.

16.1.1 RESEARCH PROBLEM STATEMENT

How to take care of health is the problem? To keep self-healthy is the most important and crucial issue. Though it is individual, but it matters a lot to the family, so to the society and ultimately to the country. Just imagine a country where most of its people are healthy. Health is wealth. Is healthy employee not an asset to the company? Yes, it is.

This work has the focus to opt for those steps in one's daily routine as a pillar to improve one's health in order to lead a "3 H life" (healthier, happy, hearty) so the family, society, and country at large. This is expressed as modest, disciplined, purposeful, and regularized (MDPR) lifestyle.

Quick technological advancements have been made in clinical investigations, strategies for examining enormous amounts of information, and then gathering wide range of data, known as Big Data. Healthy lifestyle in practicality would improve data quality for comparing health conditions and provide guidance to healthcare service providers and data collection service providers.

This study reviews a transition of lifestyle practices, originating a century ago. The work is based on secondary sources of data and observations. Some personal interviews were conducted to support our assumptions.

STEPS:

1. Survey and review current lifestyle practices and health conditions.
2. State present lifestyle conditions with the past and historic practices.
3. Review the current health practices and lifestyle conditions in the light of COVID-19.
4. Propose an alternative which focuses more on prevention rather than cure.

5. Identify gaps in current and the proposed alternative.
6. Develop guidelines/models for addressing and filling the void between the present and the proposed alternative.

Novelty: The outcome of the research is a set of provisional guidelines/theories/models addressing lifestyle and healthcare needs. The authors come out with a model of lifestyle supporting health along with its implementation suggestions.

16.1.2 FURTHER RESEARCH INSIGHTS

To realize and understand practical difficulties in real life and to find ways to resolve.

16.2 LITERATURE REVIEW

In recent decades, lifestyle as a crucial issue of health has shown increased interest in researchers. The United Nations agency has connected different factors to individual health and quality of lifestyle which relate to health. Many individuals follow an unhealthy fashion.

"In today's modern era, various changes have occurred in the lifetime of many people. Deficiency disease, unhealthy diet, smoking, alcohol, overwhelming, drug abuse, and stress project unhealthy lifestyle habits that are dominant in today's life. Moreover, the lives of individuals face new roadblocks, as an example, upcoming new technologies inside IT like the web and virtual communication infrastructures, lead our world to a serious challenge that concerns the physical and mental state of masses. In some nations, high use of medicine doses could be a major unhealthy lifestyle issue. Asian nations are among the nations that use the most medications. Pain relievers, eye medications, and antibiotics have the foremost usage in Asian nations, whereas self-medications like antiviral drugs have a negative result on the body system, if the individual would be littered with infection, antiviral will not be effective in treatment. Overall, a small percentage of those who are self-medicated can face severe complications like medicine resistance."[8]

"A healthy lifestyle adds life to your years and not just years to life. With the increasing instances of lifestyle diseases in India, one out of four Indians is at risk of dying from noninfectious diseases, such as diabetes, cardiovascular conditions or cancer in the late 60s or early 70s. Practicing bad habits, such as poor and unhealthy diet, lack of exercise, irregular sleep pattern, and resorting to excessive smoking or alcohol use might put you in with 38 million people, who are limited due to one or more typical health conditions.[22]

"There are various merits of lifestyle factors that can promote good health if one wants to live a long and healthy life. It is not possible to alter the genes or much of the environment around us, but making educated and intentional choices when it comes to food, activity, sleep, alcohol use, and smoking can minimize health hazards and potentially add years to our life."[7]

16.3 IMPORTANCE OF HISTORY

India has a rich, extraordinarily old legacy of medical and health sciences. The methodology of the old Indian clinical agenda was one of the all-encompassing treatments. The historical scenery of medical services in India can be followed to the Vedic grounds (5000 BCE), inwhich a portrayal of the Dhanwanthari, the Hindu lord of prescription, rose. Atharvaveda, oneof the four Vedas, is judged to have formed into Ayurveda, a regular Indian type of all-encompassing medication. The way of thinking of Ayurveda, "Charaka Samhita" (the renowned masterpiece on Medicine arranged by Charaka), and the careful skill formulated by Sushrutha, the father of Indian medical technique, bear declaration to the outmodedconvention of logical medical services among the Indian public. Verifiably, the most extraordinary clinics in India were those worked by King Ashoka (273–232 BCE). Medicationexperts on Indian clinical standards were educated in the Universities of Takshashila and Nalanda.

The Vedic scriptures speak in detail about the ancient practices in India. This can also be witnessed in our immediate environment. For example, Many grandparents had this practice of drinking water early morning in copper vessels, sleeping early at night, practicing yoga early morning, eating a healthy breakfast and not resorting to medicines for basic health issues. Modern science now proves that taking medicines repeatedly impairs the ability of the immune system and makes the body more susceptible to diseases. This does not carry old practices that were perfect without shortcomings. The history has examples of pandemic in every century which challenges not only Indian age old health practices but also the health practices across the globe. The Spanish Flu of 1920 can be cited as a good example in this regard. The Flu had caused massive deaths and the traditional health practices had proved fatal. A strong need of antibiotics was felt in those times which led to the invention of penicillin in 1928. Hence, no matter what, there is always a scope of improvement in every aspect of life and healthcare is no exception.

In the 18th century, as historical peep continues, a cholera pandemic in various regions of Asia, Africa, Europe, etc. extended for 8 years (1852–1860), there are records of small pox in the late 18th century and early 19th century and also references about bubonic plague in the 13th century. This plague was so dreadful that it was referred to as "Black Death" by historians. All of these had occurred in the past and the likely blame was such an enormous loss of human life is to be attributed to health practices prevalent in those times. Hence, it proves that at no point in history, were humans ready to face a health crisis and the health system at respective times had proved to be fatal and inadequate.

16.4 ANGLES OF LIFESTYLE

Lifestyle can be viewed from different angles. It is the style of living. It explains what one does. How does one do it? And why so? There can be as many types of lifestyle as many numbers of people. It is really very challenging to come out with a standard lifestyle, which has general acceptability. We can observe multidimensional angles and the beauty is all right from their own perspectives. These are presented in triangular form on the next page for clarity. These are on the basis of different perspectives such as profession, culture, religion, tradition, health, society, country, region, modern era, technology, spirituality, economic conditions, we can go on incorporating many aspects respecting different limitations of different people. There is no competition of these styles, for no style is superior or inferior to others. There is no comparison, though people do it even in spite of understanding it. But, it is natural for being human.

Without going deeper in various lifestyles, here we would put forth a lifestyle which may be accepted and practiced with little difficulty at the outset and then will become an integral part of everyone's lifestyle (Fig. 16.1).

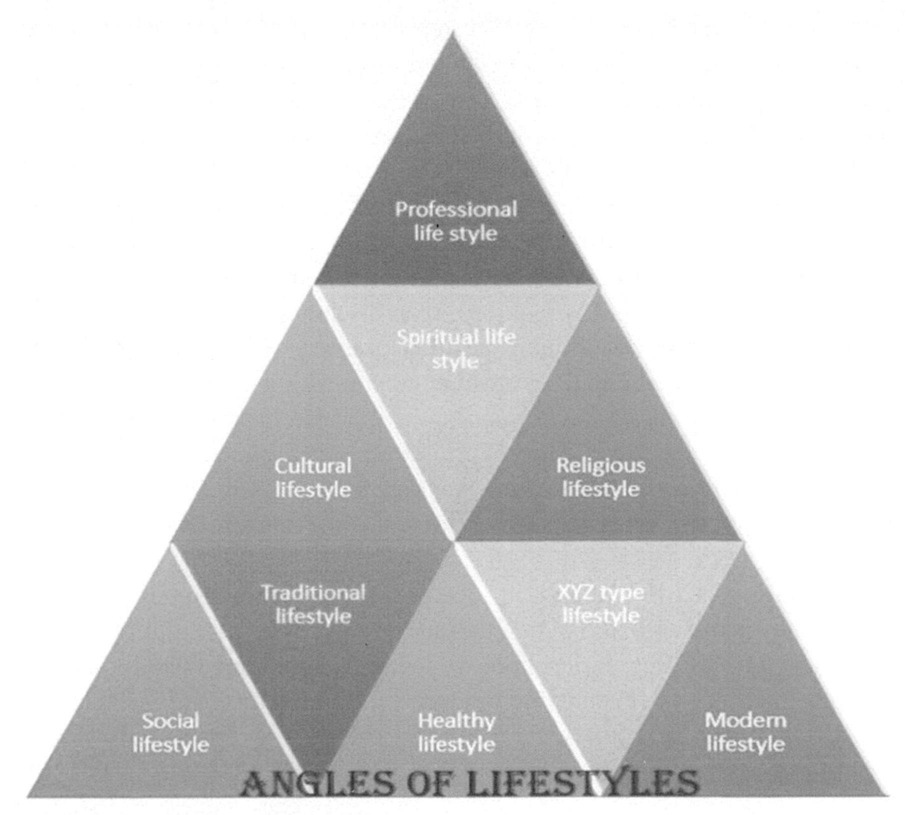

FIGURE 16.1 Angles of lifestyle.
Source: Prepared by authors

16.5 MDPR LIFESTYLE

MDPR means modest disciplined purposeful regularized lifestyle (Fig. 16.2). It is easy to adopt. First, it is to be understood with its pros and cons. Obviously, pros being far higher than cons, so the the end it is beneficial and brings its holder in a win-win situation. MDPR what does it mean?What do authors understand? What they want to convey? What is their objective? Here it is what do we mean-

Modesty in Life: The meaning of modesty is the worth or state of being unassuming in the assessment of one's abilities. This implication is to be understood, not read, and known to everyone, but to be practiced by all. It is like saying "honesty is the best policy," it is little difficult to adopt, but knowing its outcome and satisfaction, it is easy, and at the end, it wins, so why not to adopt in the beginning itself. It requires a bit of self-control. Modesty can be maintained at individual levels irrespective of profession, caste, religion, society, economic standards, family culture, and other factors. It is too noncomparable and it varies with the background, but it is to be understood, if objectives are clean then no issues.

Discipline in Life: Discipline is learned in educational institutions. It is assumed to leave it there as one steps out of it. No, it is an integral part of life. To this discipline, no external monitor is required. It is self-discipline. Even in some cases, if it is required, then discipline has to be followed for a short duration. This discipline is required at every step of life and must be self-controlled, so that one does not feel its imposition.

Purpose in Life: Whatever may be the purpose in life, it should not be above humanity and basic ethics. In the journey of life, we must keep the recall of the purpose in life and the ways are to be properly justified. A well-defined purpose and being focused on it, may make one'slife meaningful and purposeful.

Regularity in Life: Regularity does not mean a smooth journey in life. Life journey is full of disturbances, ups and downs. It can be social, economic, and others as well. Regularity, here we mean, is maintained handling of the situations, which may even come randomly with prudence and wisdom.

MDPR lifestyle is not very ideal and difficult to practice. It is very common to all and can be at mass level. Of course, it requires educating people and taking support of notable examples in each family and society. Narrating real stories of such lifestyle followers.

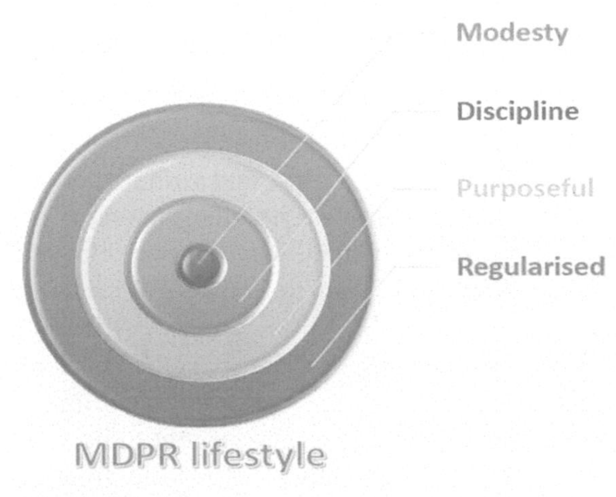

FIGURE 16.2 MDPR lifestyle.

It is an integrated and a package of lifestyle. This is not confined to any of the angles discussedabove. This may have many beneficial effects in the long period, not at all ignoring short timespans. Actually, summation of all short periods is a long one. So pros would be realized in every aspect of life. This lifestyle has the concept of "prevention is better than cure." Two dudes take a long walk, one takes to control diabetes and another takes to avoid it to come. Both have to walk, then the latter option is better.

When we consider healthcare, we are concerned about sleep, exercise, food, work, mind pressure, mediation, and such more. These are of great significance and have a positive impacton body and mind. There is a need to detail the manner of eating, sleeping and working, exercise, meditation, thought process in daily routine. It is observed that the proper style is missing in the majority of people. They feel as if they are right, but not, because we find many examples of illness, immunity, weakness, which would not have been, if routine would have been the so-called proper. There is no work life balance properly maintained. Sleeping habits are just the opposite of what it should be. People have forgotten the working of the phrase "early to bed and early to rise makes man healthy, wealthy, and wise."

16.5.1 HOW PETS SUPPORT HUMAN HEALTH?

Here, we discuss a few unpopular things which help us deal with our mental health. One of these methods is about owning a pet. This concept is extremely discussed in the west especially in the US.

In India too, this trend is getting momentum. "India is home to approximately two crore pet dogs. The value of food they consume is to the tune of ₹1200 crores annually, and it is growingat a healthy pace, according to Jiyaul Hoque, Country Manager, Pedigree, an animal feed. About 19 million other domestic pets are also currently kept in homes. Annually, about 600,000more are being adopted. The market is growing at 13.9% annually and will be worth roughly $430 million or ₹278 crores by 2020."

"Pets, particularly dogs and cats, can minimize anxiety, tension, as well as melancholy, fill seclusion, they empower workout and strength, and even they increase the overall well-being of an individual. The idea is that an innocent animal can help kids with growing up safe and sound and vibrant. Pets additionally give vital companionship to more established grown-ups. A pet adds real euphoria along with unrestricted happiness to our day."[14]

16.5.2 HOW PETS ALTER HUMAN WELL-BEING?

While people with pet animals often experience various health advantages, a pet does not necessarily be some particular animal or species like dogs or cats. Even observing fish in an aquarium can help minimize muscle tension and ease the pulse rate.

"Studies have shown that[14]

People with pets are less likely to bear from depression than those without pets.

Individuals with domestic animal have lower blood -pressure than those without it.

- One research study even found that when individuals with borderline high blood pressure adopted dogs from a refuge, their blood pressure dropped considerably in 20 weeks or so.
- Playing with a dog or cat can improve levels of serotonin and dopamine, which facilitates ease and cool.
- People with household animals have lower triglyceride and cholesterol levels (signs of heart illness) than those without them.
- Heart patients with pets live a lengthier life.
- Older people (above 65 years of age) visit their doctors less by 30%[14]

What do you think about pets? It is a more western trend these days. But it has been verycommon in India since its ancient times. It is a costly affair. Those who can and have interestshould prefer for two major perspectives, one is self-health, and another is animal care. Yes, of course, data speak about its growing trend in middle-income class in India. There are multiple other reasons to count upon, but here we limit ourselves.

16.6 YOGA—A WAY TO HEALTHY LIFE

Yoga has been the backbone of ancient Indian healthcare practice. In the ancient Indian scriptures, there are many yogic poses, culture, and how priests practice divinity through Yoga. Yoga is not a physical activity; it is mingled with the internal journey of life. Yoga is an Indian concept and philosophy, but we feel with passage of time, it could not get placed inone's daily routine life. How so ever, because of its golden virtues, Yoga is gaining momentum and efforts are in process to bring this in normal routine life. It imbibes the ability to treat problems, such as diabetes, heart issues, blood pressure, joint pain, digestion issues, and much more. Yoga has become a subject to teach offline and online both in today's tech world.Yoga has received a lot of support from the government in the form of policy framework, branding, especially the celebration of International Yoga Day, tax exemptions, etc. Yet, a lot needs to be done in order to make sure that the benefits of these reach to the maximum. Various NGOs are now working with the government in order to spread the healthy practice of yoga to each corner of the country. All these efforts in itself are the evidences of its value in the real life of a person.

It is crystal clear that Yoga is a way to a healthy life. Not only it helps to save on medical bills and expenditure, it also helps us to build a sustainable life as Yoga is believed to sync the mind and body. Popular Yoga asanas like Pranayama have been known to improve blood circulation in the body. In fact, many health experts have suggested that Pranayama might be helpful to improve oxygen levels in humans thereby reducing the risk of respiratory diseases including COVID-19. Yoga also helps in treating mental health conditions like depression, anxiety, panic attacks etc. Yoga is a pillar to the immune system, especially various breathing techniques. One might raise the question of any widely accepted scientific theory, but mass observation and experience do not require proof of Yoga benefits.

To support these assertions, we interviewed a few people who are regular yoga practitioners. A Mumbai resident narrated her success story of yoga result outcome. Teacher by profession,she was suffering from Gastritis and indigestion for a long time. Despite being treated in the best hospitals of the city, she could not scale down her problem and suffering continued. A yoga practitioner her colleague advised to practice a few Yoga asanas. She could feel the difference. Her health improved tremendously within a span of 6 months and she is now completely free of medicines. One can find many such stories in their neighborhood. The important issue is yoga is an experience and rather difficult in the beginning, its regular practice requires high level determination on the part of the individual. People find it easy totake medicines and even people prefer injections to get rid of problems without realizing the negative effects in future. They go with the saying "jo hoga so dekha jayega abhi to problem resolve karo." This very short temperament and lack of patience aggravate their problems. Here, there is a need for real education.

We are all aware of the benefits of yoga and would substantially reduce the cost of healthcare.Yoga helps us in curing various diseases, the spread of

Yoga as a way of life will help millionsof India to live a healthy life. Yoga is the preventing tool for all types of ailments. The ultimateoutcome of healthy life is reduced healthcare cost. This may bring down the size of Big Data concerned to healthcare and quality and direction of data too will bring a positive change. Data would give rosy picture when reviewed in perspective of intertemporal and interspatial comparisons. So, in a nation like ours the government must invest in yoga to cut down the cost of health expenditure . There is a need for reallocation of investment. In the beginning, invest in yoga promotion and practice and this will release funds from medical expenditures resulting in better health of people in both quantity and quality. Yoga can surely help the poor to stay healthy and reduce their dependence on expensive and unaffordable medicines. Government will also be able to rationalize the healthcare budget. However, the question that arises here is what is stopping us to adopt and embrace yoga into our daily life? The answer is simply our lifestyle. There is an urgent dire need for a lifestyle revolution.

Revolution works in cumulative and in circular flow having multiplier and accelerating effects. To make it clear, yoga helps health, and health helps yoga. Good health cuts medical expenses in the family budget and redirects in health promoting heads. It has spiral effects. Alternate lifestyle, by alternate authors, means "unhealthy lifestyle" leads to poor choices. It too has spiral effects. It is similar saying wealth makes wealth, health makes health. There is a need to break the vicious circle that affects health. To break this vicious circle of ill health requires lifestyle revolution, Yoga is one element. Life style is a package program, discussed in later pages of the chapter.

16.6.1 HOW TO PROMOTE YOGA?

We need to build systems around which a healthy lifestyle can be created. Institutions like Schools, Colleges, Universities, Government places, and offices should either conduct Yoga sessions or encourage people to adopt Yoga as a way of life. Incentives can be provided in various forms. Organizations often provide Gym membership fees to their employees or establish gymnasiums in office, similar efforts can be made for Yoga. Organizations can provide space for Yoga, make arrangements for Yoga mats, which in comparison to gymnasiums will be much less expensive while delivering the same end results, that is, healthy lifestyle.

16.7 COVID-19 AND CHANGES IN INDIAN LIFESTYLE

How COVID 19 has changes the lifestyle of Human being, such changes will be explained in this section.

Mental Health: The pandemic COVID-19 has taught things differently. The world opted for lockdown, to protect physical health, India too followed suit. It has an impact on the mental health of people in large numbers around us. A psychiatrist in our region is extremely busy in attending to patients all day. She believed that the lockdown had caused sudden lifestyle changes and people were not mentally ready to cope up and accept sudden shock. Restrictions in movement have made people feel helpless and stressed. People are staying at home, but effectively their minds are craving for the normal life. The problem is more acute to real job losers, and also symptoms for clinical depression, anxiety, narcolepsy, stress disorders were noticeable. India being in ancient culture of yoga and meditation, has moved away from its ancient practices and the value of mental health. The healthcare professionals too are facing mental health issues due to sudden upsurge in their nature of work.

Physical Health: Currently, people have forced a new lifestyle, many could not accommodate themselves say for instance in their sleep pattern, food pattern, exercise at home, yoga, work from home. Maybe, we moved away from our ancient health practices of yoga and exercise at home. Gymnasiums, swimming pools, sports clubs, parties, get-together,night-outs, all these are part and parcel of life for the majority. Lockdown had forced these institutions to shut down. It deeply impacted the fitness of many people around us. Obesity has now become a common site in every society, colony, or neighborhoods. Though there hasbeen a lot of push for yoga and fitness at home, still it has not made much progress as some habits cannot be changed so easily. Besides this, the large spread of COVID-19 and the increasing deaths have only proved that our immune system was not ready or strong enough to fight such diseases and the recent surge in sale of immunity boosters are a resounding testimonial for the same.

16.8 BIG DATA AND LIFESTYLE

The recent health crises made Big Data an integral part of our healthcare system. This in turn has increased the cost of healthcare in India. The only way to solve this is through health education and awareness among the people. Big Data, though it is an efficient tool, but it is expensive to take care of health issues. The best example is the US, known for one of the best healthcare systems in the world. The healthcare cost in the US is exorbitant. It is almost mandatory for every household to have an insurance policy without which it would be difficult for them to treat even basic disorders, such as fever, cough, common flu, etc. Amidst this COVID-19 pandemic, the value of Big Data has become more critical as it is with help of Big Data, coupled with supercomputer technology. The data of coronavirus is stored and studied, and different simulations are generated for the invention of vaccines and medicines. Countries are making fast progress in this field and a lot of success is to be attributed to Big Data. Countries with heavy investments in Big Data in healthcare are well placed than the others which are not. Comparison between The Developed West and The Poor African nations would be a perfect example to explain this preposition. While most European nations and other developed nations like the UK, Australia, New Zealand, Japan, etc. are opening up their economies, countries in the African continent & also other countries like Iran, with low investment in Big Data in healthcare are struggling with the COVID-19 pandemic. However, this issue of high investments in Big Data in Healthcare can be reduced by increased health awareness or formal health education.

Health education a source for better lifestyle: Health schooling are often defined as an idea by which individuals and groups of individuals learn to behave in a manner encouraging to the promotion, upkeep, or renewal of health (en.wikipedia.org).

Importance of healthiness in one's life needs no explanation. But, where does it exist in textbooks and its practical aspects. It is taught in some institutions in the south part of India. But it is not to the required level of importance. It is not a part of the major curriculum. We spoke to students graduating from different colleges in order to know if at any point in their curriculum were they taught about health. We also tried to access the course curriculum across different streams, and we were shocked to know that we did not find any curriculum which speaks extensively about health awareness or health education in a formalized manner. This is where the real problem lies. In olden days, when Gurukuls system was followed in India, disciples were taught the Vedas, Upanishads which also covered health education and knowledge. However, in modern education, this does not find any place. Hence, when individuals face minor health issues, they resort to incorrect self-medication, especially in Asian nations (as mentioned in the Literature review) which ultimately affects the immunity. Such practices coupled with unhealthy eating habits and sleep patterns severely impact human health, again proving the fact of positive correlation of lifestyle and health and being consistent with the literature review above.

However, the COVID-19, pandemic has forced us to slightly alter these habits. People have reduced junk eating and resorted to healthier choices, though not in absolute terms, but as lockdowns have forced food-joints to shut, people do not have much choice of gobbling outside food except the limited food deliveries available. But how far will it go?

Though these habits are good, the pandemic has economic impacts also and significantly affects employment levels in India. However, this sector might see some renewed approach wherein more emphasis might be given to nutrition and hygiene aspect rather than taste but only time will speak how things would get shaped.

Healthcare Revolution in India: In India, there is a lifestyle revolution happening in health space as well. The government has recently announced a National Digital Health Mission wherein the healthcare solution in India is set to be revolutionized. The primary focus of the government is to digitize healthcare solutions in India via Digital Health Card. The Digi Card will store all the details of a citizen. The details shall include every data of a citizen which are generated at the time of visit to a doctor, getting a medical test, medicine prescriptions, and every other tiny detail. Every citizen will have a health card and will be required to show the same on every visit relating to healthcare at multiple places as mentioned above. This is nothing but an indirect way to Big Data in healthcare. India being a country with a vast population, and DigiCards being required to be presented at every medical visit, there will be a lot of data generated. Since different people might have different health profiles, the data diversity will be largely scattered and speedy. Hence, the 3Vs are present which are Volume, Velocity, and Variety, and therefore, a healthcare revolution in India has arrived which is based on Big Data. This healthcare revolution in India comes under National Digital Health Mission (NDHM).[12]

What is NDHM?[12]

The National Digital Health Mission was an idea brought forward in the year 2018 by NITI Aayog, the government's think-tank for improvements in various sectors. Under this, all the health-related facts of a person will be stored at one place.

What is the goal of the national health mission?

The objective of the National Digital Health Mission is to build a national digital health infrastructure,providing inexpensive and reliable health service to the citizens of the nation. This operation will also be a major milestone on the way to the United Nations Sustainable Development Goal 3.8 of Universal Health Coverage.[12]

Key highlights of the NDHM

Below are the six key highlights of the NDHM by which the individuals will gain a timely, secure, and low-cost healthcare approach.

1. Health identification
2. Digi Doctor
3. Health facility records office
4. Individual health files
5. Online pharmacy
6. Telemedicine

Health identification credentials and its formation

A Health identification credential will act as a Health Check Account of every person. This consists of all the information with regards to the tests, diseases, doctors visited, medicines, reports, and diagnosis. Every fact related to the health of a person will be included in the health identification card. The health identification card is made with the help of the details, such as Aadhaar and phone number of an individual, generating a unique code for every person. Also, the health identification card is not mandatory, and a person will be given the treatment if he does not want a health identification card. National Health Authority (NHA), Ministry of Health and Family Welfare has given the green signal for the implementation of the National Digital Health Mission in the nation. As per Union Minister of Health and Family Welfare, Harsh Vardhan, The plan will be executed via a test launch in the Union Territories of Chandigarh, Ladakh, Dadra, Nagar Haveli, Daman and Diu, Puducherry, Andaman and Nicobar Islands, and Lakshadweep. With the initial outcomes in the Union Territories, the scheme will be implemented in the States too.[12]

We have various examples for the healthcare revolution in India. This revolution took place in heart health space and we present before you a practical case study as below.

The most severe problem confronting India's healthcare system today is the high number of patients demanding immediate medical help. Since most Indian public healthcare centers are small and unequipped, they simply do not have enough cots to meet the challenge. As of now, an estimated 20-lakh beds are required in various healthcare centers throughout the nation. This means that the responsibility of providing good healthcare service is passed on to the nonpublic healthcare institutions, which often charge exorbitant fees that most of the masses cannot pay. In a nation with widespread deprivation, analysts expect Indian healthcare industry incomes to top $155 billion by 2017. As of now, it ranks 14th in terms of value, and third by volume. The problem, then, is making private healthcare less expensive. One area that needs attention is heart care. Heart surgeries by their very character are one among the costliest operations in healthcare, requiring highly trained cardiologists. But though exclusive hospitals charge close to $4000 for heart surgery, most have significant waitlists of patients awaiting spots. There is certainly a big difference between supply and demand.[15] However, it also means there is enormous opening.

This is the problem that Dr. Devi Shetty, a cardiac specialist working at NarayanaHrudayalaya, a private organization, took on 11 years ago when he started the first heartcare health center in Bangalore. Dr. Shetty is as close as India gets to a celebrity doctor, having pioneered the use of microchip cameras during open-heart operations. He also had the opportunity of operating on Mother Teresa in her final years.

The results at these new healthcare centers have been encouraging. Narayana currently undertakes more heart operations than most other healthcare centers throughout the world, and the most across India by a huge difference. Over 12% of the country's heart surgical procedures are carried out at Shetty's healthcare centers in Bangalore and Kolkata. And the death rate is lower than that of the best healthcare centers in New York. Most important of all, the costs of surgery are almost 40% low than those of other private healthcare centers in India—a statistic that has drawn the attention of healthcare industry in the West as they strive to bring down their own costs. Narayana hospitals levies approximately $2000 for open-heart operation, compared with $20,000–$100,000 in the United States of America. Furthermore, their sliding scales for medical charges mean that rich patients subsidize poorer ones, providing close to 80% with some form of discount or other.[15]

Learnings from Real Life: For the purpose of this research work, we tried to reach out to a few people in order to know their views on healthcare, lifestyle, and health awareness. We have asked various questions like- how lifestyle affects health? Practical difficulties in following a healthy lifestyle, etc. We were surprised to know that some of those respondentswere actually well aware about the good lifestyle practices but were unable to follow the same.We have classified the same into three as below:

(a) **Busy Work Life:** Despite knowing the best health practices, the respondents were not able to adopt the same due to a busy work life. This means that not only health education should be a part of the curriculum but also work life. Companies should place high emphasis on employee's health thereby reducing the health issues, investment in healthcare, investment in Big Data, both for the company and the nation as a whole. Though, we might find some companies that have started paying attention to employees' mental well-being, especially during lockdown, a lot still needs to be done. Merely giving insurance and medical claim to employees cannot be a solution. Steps should be taken so that even the smallest of the smallest organizations take health seriously and this can be made possible only through health education and awareness.

(b) **Competitive Study Environment:** While we might wonder that students/children should be able to adopt good health practices, the same seems unlikely as even students these days have an extremely tight schedule. The students are busy with school, coaching classes, homework, assignments, preparing for some entrance exam, etc. so much so that many of them sleep for less than 8 h a day. The discussion of anxiety and stress is not uncommon among students and that we also come across extreme incidents about how students suffer. This is where the government can play a role. Through policy changes, efforts can be made to make healthy, yet productive changes in the overall study curriculum. We cannot expect any radical changes by merely "Health education" as a subject in the curriculum. A more holistic, pragmatic, and practical approach is required to ensure that the benefits of health education are felt.

(c) **The Cycle of Evil Habits:** Habits once developed are difficult to change. In fact, it takes days, sometimes even months to develop a new habit. The more we delay our good habits, the more difficult it becomes to adopt them. As rightly said, "Practice makes man perfect," no matter what we practice, we eventually become good at it. As we keep repeating the evil habits, we become so good at it that it requires high motivation and a strong reason to develop new habits. Again, this is only possible for health awareness. As people will move toward a healthy lifestyle and experience its benefits, it will encourage other people to adopt similar healthy choices thereby motivating to be fit and healthy. Some efforts can be seen in some places like community gyms, public parks, yoga centers, where people from a particular locality come together and exercise together. Events like Mumbai Marathon, Run Powai Run, etc. are also good initiatives. It motivates people to train their body to follow healthy choices and that we need more of such events on regular basis, at every district which can help India to become a fit and healthy nation.

(d) **Notable Exemplary:** Some well-known examples to strengthen our concept and ideology We have extensively discussed healthy lifestyle, what constitutes a healthy lifestyle, how the same can be spread and taught, its impact on Big Data. However, we will now discuss a few real-life examples and scenarios which will prove our assumption about a healthy lifestyle.

We all know about famous Indian actor Akshay Kumar. He is a famous name across India. He fits accurately into our concept of healthy lifestyle. He is an early riser, eats healthy, practices yoga, and goes to bed early. The result of these healthy habits are very evident to all of us. He is one among the fittest actors in India. Even after being a mid-aged person, he is able to perform a lot of stunts on his own. We have gone through many of his interviews and different articles where he has reinstated the secret of his success to his healthy lifestyle. Thereare various other examples too. Actress Malaika Arora Khan is known to start her day with aglass of water, which is one of the healthy habits discussed earlier in this work. Many popular celebrities who start their day with warm water. Also not only Indian celebrities but also international Hollywood celebrities have understood the importance of starting the day with warm water. One such popular example is Meghan Markel.

Another celebrity who has been vocal of Yoga and healthy diet is Shilpa Shetty. Yoga is a very big part of her life. In fact she has been doing it for the past 18 years for 3 hours a week.Also, it would be worth noting that she had started Yoga in order to cure cervical spondylosis which further proves our premise that Yoga is a way to affordable healthcare. She promotes Yoga on various platforms and believes that Yoga is a way to healthy life.

In this research work, we have also made references to the ancient Indian healthcare system of Ayurveda. Coconut oil pulling is one of the many practices in Ayurveda, a method by which one swishes coconut oil inside our mouth, teeth, and gums for close to half an hour on an empty stomach to eliminate harmful mouth bacteria. Famous Hollywood celebrity Gwyneth Paltrow has been reportedly known to practice oil pulling.

Also, all of the aforementioned celebrities include a significant portion of fruits and leafy vegetables in their diet, which further proves our point that fast food is not a way to healthy life and how the world is slowly understanding and adopting Indian ways of health and food. All of these celebrities are reportedly known to be fit and are not known to suffer from any serious or life-threatening diseases. Also, a healthy lifestyle leads to better immunity which helps our body to fight diseases and reduces our dependencies on medicines. It is probably because of these reasons that India has low COVID-19 cases per million compared with most developed nations in the world. The secret to this lies in our eating habits. While Indians are known to be foody and there are few Indian foods that are not a healthy option, still as compared with the West's culture of fast food or quick service restaurants, it would be safe to conclude that India is placed much better when it comes to food. To support this, there was a recent research which said that *Dal-chawal* is healthy and helps our body to fight various diseases by improving our immunity. One might not find such extensive reports for other food items. Also, in India, people do not frequently visit restaurants compared with the West. In the West, going to restaurants regularly is a common food culture, whereas in India, most of the Indians prepare food at home and go to restaurants less frequently than the West. However, this culture is slowly changing in urban areas as people who work for long hours frequently eat outside rather than cooking at home.

We preferred to take examples from the film industry, for being very popular. We have observed MDPR lifestyle followers among common people like us, but the percentage is verymeagre. So there is a need to make it a Janta Andolan and bring the revolution in lifestyle. It will become an Indian example before the world. This would surely bring a remarkable twist in the quantum and quality of Big Data too.

16.9 CONCLUSION

Health education as a part of a curriculum and other walks of life would go a long way towardpromoting a healthy lifestyle. Social catalyst like teachers, TV players, film industry, social reformers, leaders, parents, doctors, gym centers, coaches, common exercise places, real friends, and advertising agencies have big and crucial responsibility in boosting MDPR lifestyle. A health educated individual would make better lifestyle choices and hence would be less susceptible to health hazards. A well-designed health curriculum as basic education may help in lifestyle revolution and evolution. This would in turn help economies to save billions in healthcare costs, especially in Big

Data as it reduces all 3Vs and the resultant money can be invested in other important areas.

Alternatively, money that will be invested in Big Data healthcare in the future should rather be (fractionally) invested now in health awareness. It might save humanity from future healthcare crisis, after all "Prevention is better than cure.

KEYWORDS

- **lifestyle revolution**
- **healthcare revolution**
- **education**
- **big data**

REFERENCES

1. Amringer, C. F. *The Health Care Revolution*; University of California Press: London, 2008. Retrieved Dec 5, 2020.
2. Sheth, H.S.K., Tyagi, A.K. (2022). Mobile Cloud Computing: Issues, Applications and Scope in COVID-19. In: Abraham, A., Gandhi, N., Hanne, T., Hong, TP., Nogueira Rios, T., Ding, W. (eds) Intelligent Systems Design and Applications. ISDA 2021. Lecture Notes in Networks and Systems, vol 418. Springer, Cham. https://doi.org/10.1007/978-3-030-96308-8_55
3. Catalyst, N. *Healthcare Big Data*, Jan 1, 2018. Retrieved Oct 20, 2020, from catalyst.nejm.org: https://catalyst.nejm.org/doi/full/10.1056/CAT.18.0290
4. Das, P. *Lifestyle Changes during Covid-19*, Apr 20, 2020. Retrieved Dec 1, 2020, from timesofindia.indiatimes.com: https://timesofindia.indiatimes.com/blogs/melange/lifestyle- changes-during-covid-19/
5. David Grossman, M. M. *American Indian Health: Innovations in Health Care, Promotion, and Policy*, Oct 2001. Retrieved Dec 5, 2020, from jamanetwork.com: https://jamanetwork.com/journals/jamapediatrics/article-abstract/191056
6. Dhani, U. *11 Amazing Health Benefits*, May 4, 2020. Retrieved Oct 15, 2020, from food.ndtv.com: https://food.ndtv.com/health/12-amazing-healing-benefits-of-drinking-water-in-a-copper-vessel-1658134
7. Dyer, K. A. *6 Positive Lifestyle Factors*, Jan 27, 2020. Retrieved Oct 17, 2020, from verywellhealth.com: https://www.verywellhealth.com/lifestyle-factors-health-longevity- prevent-death-1132391
8. Farhud, D. D. https://www.ncbi.nlm.nih.gov/pmc/articles/PMC4703222/. *Impact of Lifesyle on Health, Iran J. Public Health*, Nov **2015,** 1442–1444. Retrieved Nov 1, 2020, from https://www.ncbi.nlm.nih.gov/

9. Futurist, T. M. *How Could Digital Technology Make An Impact On Primary Care?* Mar 22, 2018. Retrieved Dec 2, 2020, from medicalfuturist.com: https://medicalfuturist.com/digital- technology-make-an-impact-on-primary-care/
10. https://www.who.int/about/who-we-are/constitution. (n.d.). Retrieved from https://www.who.int/
11. Jarus, O. *20 of the Worst Epidemics and Pandemics*, Mar 20, 2020. Retrieved Oct 1, 2020, from livescience.com: https://www.livescience.com/worst-epidemics-and-pandemics-in- history.html
12. Javaid, A. *Health ID Card*, Aug 17, 2020. Retrieved Oct 2, 2020, from https://www.jagranjosh.com/: https://www.jagranjosh.com/general-knowledge/national- digital-health-mission-1597647525-1
13. Jawahar, D. S. *Healthcare Scenario in India*, Winter 2007–2007. Retrieved Oct 3, 2020, from healthcaremanagement.org: https://healthmanagement.org/c/icu/issuearticle/healthcare- scenario-in- india#:~:text=Historical%20Background&text=The%20history%20of%20healthcare%20in,In dian%20form%20of%20holistic%20medicine
14. Kai Lundgren, L. R. *The Health and Mood-Boosting Benefits of Pets*, Sept 2020. Retrieved Dec 1, 2020, from helpguide.org: https://www.helpguide.org/articles/mental-health/mood- boosting-power-of- dogs.htm#:~:text=Pets%2C%20especially%20dogs%20and%20cats,valuable%20companions hip%20for%20older%20adults.
15. Kaul, R. *India's Health Care Revolution*, July 2012. Retrieved Oct 4, 2020, from https://www.thesolutionsjournal.com/: https://www.thesolutionsjournal.com/article/indias-health-care- revolution/#:~:text=That%20means%20that%20the%20burden,reach%20%24155%20billio n%20by%202017.
16. Khadijah Breathett, M. S.-A. *Cardiovascular Health in American Indians and Alaska Natives: A Scientific Statement from the American Heart Association*, May 28, 2020. Retrieved Dec 6, 2020, from ahajournals.org: https://www.ahajournals.org/doi/10.1161/CIR.0000000000000773
17. S C Tiwari 1, N. M. *The Indian Concepts of Lifestyle and Mental Health in Old Age*, Jan 2013. Retrieved Dec 1, 2020, from https://pubmed.ncbi.nlm.nih.gov: https://pubmed.ncbi.nlm.nih.gov/23858270/
18. Shaukat, N. *Physical and Mental Health Impacts of COVID-19 on Healthcare Workers: A Scoping Review*, July 20, 2020. Retrieved Oct 5, 2020, from intjem.biomedcentral.com: https://intjem.biomedcentral.com/articles/10.1186/s12245-020-00299-5
19. Szasz, G. *A History of Pandemics*, Mar 3, 2020. Retrieved Sept 25, 2020, from bcmj.org: https://bcmj.org/blog/history- pandemics#:~:text=In%20the%2017th%20and%2018th,poor%20moral%20and%20spiritual%20condition.&text=A%20cholera%20pandemic%20originating%20in,killing%20over%20a%20million%20people
20. Tang, S. Y. *Alexander Fleming (1881–1955): Discoverer of Penicillin*, July 2015. Retrieved Sept 20, 2020, from ncbi.nlm.nih.gov: https://www.ncbi.nlm.nih.gov/pmc/articles/PMC4520913/#:~:text=He%20named%20the% 20'mould%20juice,first%20antibiotic%2C%20or%20bacteria%20killer
21. WHO. Oct 20, 2020. https://www.who.int/governance/eb/who_constitution_en.pdf. Retrieved from https://www.who.int/governance/eb/who_constitution_en.pdf: https://www.who.int/governance/eb/who_constitution_en.pdf
22. Zahid. *Healthy Lifestyle. National Health Portal*, 2016. Retrieved Sept 26, 2020.

PART 5

Shifting of Healthcare Systems Toward Emerging Technologies

CHAPTER 17

Internet of Things-Based Cloud Applications: Open Issues, Challenges, and Future Research Directions

SIDDHARTH M. NAIR[1], R. VARSHA[2], AMIT KUMAR TYAGI[3,4*], and S. U. ASWATHY[5]

[1,2,3]*School of Computer Science and Engineering, Vellore Institute of Technology, Chennai Campus, Chennai 600127, Tamil Nadu, India.*

[4]*Centre for Advanced Data Science, Vellore Institute of Technology, Chennai 600127, Tamil Nadu, India.*

[5]*Department of Computer Science and Engineering, Jyothi Engineering College, Cheruthuruthy, Thrissur, Kerala, India*

[*]*Corresponding author. E-mail: amitkrtyagi025@gmail.com*

ABSTRACT

In the last decade, several developments have taken place, including the famous cloud computing. Cloud computing is typically the on-demand availability of services, such as operating systems, data storage, and computing power. Cloud computing offers those services without direct user activation on the Internet (anything). Infrastructure-as-a-Service (IaaS), Platform-as-a-Service (PaaS), and Software-as-a-Service (SaaS) are commonly categorized into three groups. There is also another recent "Internet of Things" breakthrough, mainly in all possible applications, such as defense, agriculture, development, healthcare, etc. Such things are referred to as Information and Communications Technology (ICT) or intelligent things used in the creation

Intelligent Interactive Multimedia Systems for e-Healthcare Applications. Shaveta Malik, PhD
Amit Kumar Tyagi, PhD (Eds.)

of smart apps. Such smart devices produce a lot of data known as Big Data, which is stored in the cloud (connected on the Web/via the Internet). However, a number of serious concerns (issues) have been raised here, such as the protection of stored data, the safety of data communications, and confidence in using our data with other (unknown users). So many researchers often focused on security and privacy aspect, but here in this chapter we address the common confidence issues in cloud-based Internet of Things or Internet of Things-based cloud in detail, that is, with many opportunities, potential research directions (or research gaps) in detail.

17.1 INTRODUCTION

The term Internet of Things (IoT) initially referred to the devices maintaining a connection with the Internet. But more recently, IoT devices refer to those devices which can communicate with each other. It is comprised of a variety of devices like sensors, laptops, smartphones, etc. By linking these connected devices with artificial intelligence and mechanical systems, it is possible to collect data, and use them to perform specific tasks. IoT is responsible for linking private networks so the devices could "talk" to each other.[1] These connected networks of devices have proven to be instrumental in many industries, making manufacturing, scaling and other processes very efficient. Although setting up such networks on a moderate or large scale requires an initial investment, in the long run, it enables the saving of time, money, and industrial effluents as well. With IoT, different information required like the data of sound, power and other scientifically relevant data can be obtained by using RFID method and other sensor-like devices. So IoT has managed to interface such devices with the real world for human interaction and the betterment of lifestyle.

In any event, the assignments with high computational complexity and the vast amount of data stored in the IoT ecosystem are constantly taken care of by the asset-rich cloud worldview due to the asset limitations of IoT gadgets, which increases their effectiveness extensively. IoT gadgets, for instance, generate enormous data measures that place colossal strains on the IoT. The cloud can be used to calculate and store IoT gadgets' large-scale information, which will enhance the overall effectiveness of cloud-based IoT settings.[2] Figure 1a delineates cloud-based IoT engineering. We have the ability to increase the use of the open innovation that is offered in cloud conditions through the mix of IoT and the cloud.[3] Nevertheless, as with various modern technologies, there are a few difficulties in advancing the cloud-based IoT environment[4–6] and the IoT climate.[7,8] Security (e.g., real layer security and access control of executives) and confidence (e.g., malignant hubs and knowledge abuse) are two of the challenges for the cloud-based IoT setting. The appropriation of the cloud-based IoT worldview therefore transfers the IoT's security and confidence problems to the cloud. The protection of the IoT setting can be guaranteed through a reliable cloud to address this problem, as shown in Figure 1b. By the way, there is little writing on the cloud-based confidence evaluation of the IoT environment, while the latest traditional writing on IoT security addresses remote organizations.[22,23]

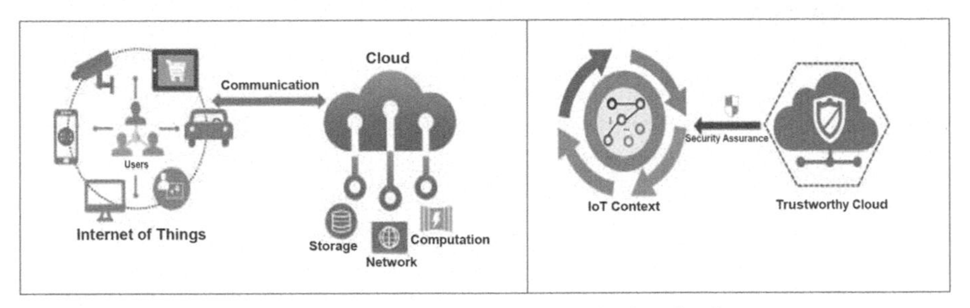

FIGURE 17.1 (a) IoT based on cloud, (b) ensures security by cloud.

In such interconnected networks, the challenges of maintaining security and trust exist. Trust is the belief that reliability and security of the system is maintained. It is the assurance of privacy in technological systems. For example, in the case of home automation systems, the sensors and other devices collect data from our day to day lives. The daily routines and patterns of peoples' lives are "watched" by these systems. Since these systems are to be linked with human interaction, the data are usually stored in the cloud. Such data are private and must not fall into the hands of a third party. Regarding security, information obscurity, privacy, and uprightness need to be ensured, just as verification and approval instruments so as to forestall unapproved clients to get to the framework.[2] Though, concerning security prerequisite, the integrity in the storage and access of data must be maintained and clients' individual data classification must be guaranteed, since gadgets may oversee potentially private information.

Hence, this chapter can be organized as: Section 17.2 discusses about the related work in detail. Section 17.3 discusses our motivation behind writing work toward this emerging area. Further, several issues faced in the Cloud and IoT (in the past, present, and future), such as security, privacy, trust, scalability, and standardization are discussed in Section 17.4. Challenges in IoT-based cloud applications have been discussed in Section 17.5. Later, Section 17.6 discusses about several future opportunities toward/identified in IoTs-based cloud applications. Further, Section 17.7 discusses some solutions and techniques, required to secure a system (in general). Then, an open discussion is provided based on user's and business perspective in Section 17.8, that is, thought about using IoTs devices/smart devices in future. Lastly, Section 17.9, concludes this work with including several useful remarks for future researchers.

17.2 RELATED WORK

A. A. Khan, M. H. Rehmani, and A. Rachedi, in their work, "Cognitive-Radio-Based Internet of Things: Applications, Architectures, Spectrum Related Functionalities, and Future Research Directions (2017)," had paid attention to the fact that IoT applications are meaningless if they are not equipped with cognitive radio capability. A new research dimension opens when cognitive research is associated with IoT.[3] In continuation, the work done by Atlam, Hany F., Alenezi, Ahmed, Alassafi, Madini O. and Wills, Gary (2018) in "Blockchain with Internet of Things: benefits, challenges, and future directions"[4] had emphasized on the importance of IoT on how it could associate billions of articles which impacts on people's life and how it can tremendously change everything in a large way. In their work, they have emphasized on how the centralized server/client has posed a problem in adopting the IoT in the real world as it can create a single-point failure if even one connection collapses. They have referenced how versatility and security issues emerge because of exorbitant number of IoT protests in the system. To resolve these issues, they associated IoT with Blockchain. Blockchain is a decentralized system and this helps in solving the security issues of IoT.

In "Internet of Things (IoT): A vision, design choices and potential directions (2013)," Gubbi, Buyya, Marusic, and Palaniswami have paid attention to the detail in which wireless sensor network (WSN) innovations have increased the efficiency of modern day living and this integrates seamlessly with the world around us in combination with IoT and the knowledge is exchanged to build a common activity. Technologies, such as RFID tags, sensors, and actuator nodes have increased the performance of the Internet. The transition from www to web2 to web3 has shown how the need for data has been increasing significantly. M. Abomhara and G. M. Køien in the paper "Security and privacy in the Internet of Things: Current status and open issues,"[5] in the year 2014 had stated that reliable, inexpensive security and privacy solutions are necessary to ensure accurate, restricted, and authenticated information which is passed on between different objects, because without trust and privacy of information, it is absolutely worthless having IoT to send information as it will be mistreated and misused.

Issues in the present power systems which arise due to one-way direction of information flow, wastage of energy and growing demands of soon becoming extinct energy has given rise to smart grids. These smart grids (SGs) provide a bi-directional stream of vitality between specialist partnerships and clients, including power production, transmission, appropriation, and frameworks for usage. SGs use numerous gadgets, transmitted to power plants, circulation focuses, and in a large number of shoppers' premises for the surveillance, inquiry, and control of the framework. A SG therefore requires networking, computerization, and other gadgets to follow. With the assistance of IoT, this is accomplished as IoT encourages SG structures to support various device functions in the era, transmission, transmission, and use of vitality by joining the IoT gadgets (sensors, actuators, and keen meters, e.g.), as well as by providing the network and computerization. The above explanations were provided in the paper written by Y. Saleem, N. Crespi, M. H. Rehmaniand R. Copeland in "Internet of Things-Aided Smart Grid: Technologies, Architectures, Applications, Prototypes, and Future Research Directions,"[6](2019).

Tsai, C., Lai and C. And Vasilakos, A.V. Future Internet of Things: Open Issues and Challenges (2014)[7] centered on the development and fusion of creative processes into IoT and is actually known as the Future Internet of Things. IoT is used to delete the data and also is used to migrate the "data" from the layer detection to the main problem of the application layer. The work carried out by Adat, V., Gupta, B.B in "Security in the Internet of Things: Problems, Difficulties, Scientific Classification, and Engineering" in the year 2018 was uncommonly taken into account in discussing how today's network progress is unpreventable.[8] The number of Internet-related devices, those with a modernized character, is rising step by step. The IoT has become a noteworthy part of human life with the advancements in progress. It is not, however, especially presented and guaranteed. By and by, for an unquestionable IoT situation, different security problems are considered as a major issue. With the proposed frameworks and the advances made by the establishment of the IoT, there are a number of security problems. In order to guarantee the IoT situation, some lucrative and promising security frameworks have been developed, but there is a lot to do. The challenges are continuously changing and the plans for the game must continually be changed.[11,12] presents a study of enabling technologies, problems, and open-ended research concerns. Another worldview of the IoT provides a lot of new administrations for the ensuing rush of mechanical innovations. IoT implementations are infinite, thus allowing the digital world to blend consistently with the physical world. Be that as it may, regardless of the enormous efforts of standardization agencies, partnerships, corporations, scientists, and others, there are still numerous issues to manage in order to achieve the full IoT potential. From various points of view, these topics should be considered, such as empowering innovations, applications, action plans, social, and natural consequences. Open problems and concerns approached from a creative point of view are at the heart of this article. We only highlight numerous visions that remain behind this world view for clarification, so as to promote a superior understanding of the highlights of the IoT. In addition, this detailed summary offers bits of information to the best in the IoT empowerment and growth progress class. The most important among them are influenced by such subtleties.

A new strategy and trust system based on the cloud model hypothesis was suggested by Yang et al.[13] This tool takes confidence, expenses, and time into account and utilizes the strategy of logical progression measure to allow CSCs to choose the proper administration of the cloud. A trust evaluation framework was proposed in[14] that utilizes the portion of consistency observation to determine the reliability of CSPs (Cloud Service Providers). Nonetheless, cloud administration QoS data are difficult to access and frequently scattered. The QoS data of cloud administrators can also be inconsistent. It is therefore difficult to assess the exact reliability of CSPs based solely on the QoS calculation.

Nagarajan et al.[17] promoted a major framework for the preparation of knowledge for the evaluation of cloud administration reliability. By using a cloud representative that fuses the MapReduce framework, it pre-measures the critique evaluations of CSCs. In[18] the input evaluation section and the Bayesian game model were consolidated by a novel confidence assessment technique to interpret malignant CSCs and their assessments of criticism. The former is used to analyzes and recognize counterfeit identities, and the latter is used to differentiate between harmful clients and their criticism. Noor et al.[15] designed and revised the organization of executives with a standing-based trust. This framework will measure the credibility of vital assessments to protect malignant CSCs from cloud administration. A lightweight standing estimation method was suggested in[16] for cloud administrations based on the cloud model. In order to obtain the standing scores of cloud administrations as per the input assessments of CSCs, this technique uses fluffy set hypothesis. Nevertheless, in the actual cloud world, there are malignant customers and unjustifiable critical evaluations that ultimately affect the status of CSPs. Essentially, it is difficult to attain the genuine reliability of CSPs based only on consumer feedback evaluations.

17.3 MOTIVATION

Today's IoTs are used in many applications, such as agriculture, transportation, defense, manufacturing, communication (e-healthcare, telecom, etc.), etc. These devices are generating a lot of data (called Big Data), which is stored at cloud side which can be accessed anywhere and anytime by the end users. These data can be used for forecasting and predicting useful decision and can be more helpful for serving humanity in a better way. For example, in e-healthcare, these smart devices (interconnected devices) can be used to make communication among other medical devices. These devices can be useful for doctors or health professionals for predicting immune systems about virus (e.g., COVID 19, H1N1, Swine Flu, Spanish Flu, etc.). We need to make useful of these smart devices at a large scale in many applications. By generating big data, we can serve humanity in a better way, that is, can make people' life convenient, easier, and longer to live. Hence, this section discusses our motivation behind writing this article in brief. Now, next section will discuss about the popular issues in Cloud and IoT (i.e., in past, present, and future).

17.4 POPULAR ISSUES FACED IN THE CLOUD AND IOT IN THE PAST, PRESENT, AND FUTURE

Another worldview that blends angles and technologies coming from different methodologies is IoT. Omnipresent encoding, inescapable registration, Internet Protocol, progress identification, advances in correspondence, and inserted gadgets are combined to frame a system in which the actual and computerized worlds meet and are continuously in harmonious union. The main article is the square structure of the IoT vision. They are turned into shrewd articles by putting insight into daily objects, capable not only of collecting data from nature and associating/controlling the physical environment but also of interconnecting with each other through the Internet to exchange knowledge and data. The usual enormous number of interconnected gadgets and the vital measure of accessible knowledge open up new opportunities for the general public, state, economy, and individual citizens to benefit from significant benefits.

There are a wide variety of issues with the amazing skill of the IoT. Since the IoT is focused on the Internet, Internet security problems may also arise in the IoT. There are also three layers in the IoT: the awareness layer, the transport layer and the application layer. The IoT front-closes are responsible for collecting data and oversight of status, while in the cloud server, the generous measure of data is put away and monitored. Accomplishing information security and framework effectiveness in the information procurement and transmission process are of incredible importance and testing on the grounds that the force matrix-related information is touchy and in enormous sum. Information obtained from the terminals will be parceled into squares and scrambled with its comparing access subtree in succession, subsequently the information encryption and information transmission can be prepared in equal. Besides, we ensure the data about the entrance tree with limit mystery sharing strategy, which can safeguard the information security and uprightness from clients with the unapproved sets of properties.[20] The conventional investigation exhibits that the proposed plan can satisfy the security prerequisites of the cloud-IoT in shrewd matrix. The numerical examination and test results show that our plan can successfully lessen the time cost contrasted and other famous methodologies. The IoT regarding client protection issues incorporates:

(a) Person knowledge access.
(b) Improving developments in the field of defense and important recommendations.
(c) Standards in methods and programming to deal with the personality of consumers and goods.

A portion of the problems include the following with respect to classification:

(a) The need for an easy-to-use trade in basic, safe and confidential data
(b) Confidentiality must be included in the configuration measure of the IoT

The growing interest in food, both in terms of quantity and quality, has increased the need for intensification and industrialization of rural areas. The

IoT is a community of advances that is deeply promoting and suitable for providing various arrangements for the modernization of agribusiness. Like the company, rational meetings and discovery organizations are trying to convey more and more IoT goods to the horticultural business partners in a race, and in the long run, set up the systems to provide an unmistakable job when IoT becomes a normal innovation. At the same time, cloud computing, which is now well established, and fog computing provide adequate tools and answers to help, store, and investigate the colossal measurements of IoT gadget knowledge. IoT knowledge administration and analysis ("Big Data") may be used to mechanize types, predict circumstances, and even constantly enhance various exercises. In addition, the principle of interoperability between heterogeneous gadgets has enabled the development of suitable tools that can be used to render new applications and administrations and provide an additional motivation to the information streams delivered at the edge of the device. The rural division was strongly affected by the developments of the WSN and relies on the IoT to benefit equally. Heterogeneity: brought about by a variety of gadgets, working systems, phases, accessible administrations, correspondence results, calculations and power points of view, reliability is needed for critical application, big data will cause transport, stockpiling, access and handling problems, monitoring can take on some security problems and difficulties when it comes to haze.

Security: With the broad use of its architecture, the security concerns of the IoT are clearly defined. Starting with the presentation of IoT security engineering and highlights, the recognition layer, including key administration and measurement, security steering convention, information combination creativity, as well as validation and access control, and so on are particularly expounded among these wellness estimates concerned. Private data is a huge segment of the data executed between IoT gadgets, which must not be listened to or altered stealthily at all. In this way, IoT gadget protection is of primary importance for the further development of innovation. Such devices usually have restricted region and vitality properties, which use great cryptography that is restrictively expensive. Physically unclonable functions (PUFs) are a class of novel native security equipment that guarantees a shift in outlook in various security applications; their normally straightforward engineering can fix a large number of vitality-compelled IoT gadgets' security difficulties. RFID is one of the IoT technologies that empowers creativity. RFID can theoretically enable machines to recognize things, understand their status, and if necessary, convey and make a move to make "ongoing knowledge."

The inescapability of RFID technology has given rise to numerous important issues, including security and safety issues.[9]

Trust: Trust executives carry on a major IoT job for accurate information mixing and mining, professional mindfulness-setting administrations, and enhanced customer safety and data security. It allows individuals gain an impression of insecurity and potential and participates in the identification and usage of IoT administrations and applications by customers.

In any event, not many studies on the IoT trust framework can be found in writing, but we argue that there is an impressive need to apply the trust aspect to the IoT.

We decay the IoT from parts of the IoT framework synthesis into three layers, which are the sensor layer, the focus layer, and the application layer. Each layer is bound by trust for clear explanation by the board: self-created, loaded independently with feeling of teamwork, and multiorganization. Similarly, the organization requester conducts an official decision formation as shown by the assembled confidence data similar to the method of the requester. Finally, to see all of the above trust section, we use ordinary semantics-based and fluffy set speculation, the result of which provides a general basis for the progress of IoT trust models.[10] As per the system strategy, which relies on the data generated by IoT gadgets, the assets should be allocated to gadgets. If there are some gadgets demanding assets, they can report managed vindictive data to get more assets for their own enthusiasm. That is, the brilliant assembling system could be defenseless because of narrow-minded shrewd assembling gadgets' practices. This lessens the effectiveness of the whole framework and besides stops the plant-wide procedure. While many exploration commitments identified with the trust processing target distinguishing vindictive hubs. Sensors or sensor-installed things may build up direct correspondence between one another utilizing 6LoWPAN convention. A trust and notoriety model are perceived as a significant way to deal with shield a huge disseminated sensor arranges in IoT/CPS against malevolent hub assaults, since trust foundation instruments can invigorate cooperation among circulated figuring and correspondence elements, encourage the discovery of dishonest elements, and help dynamic procedure of different conventions. The result after various tried methods shows that the fuzzy methodology for trust-based access control ensures adaptability, advertisement, it is vitality effective,

Privacy: In light of the previously mentioned security defects, numerous other security and protection issues present themselves in IoT. IoT uses the Internet as a main platform for interconnecting different topographically extended IoT hubs, and cloud is therefore used as a primary supporting base for the back-end. The set of IoT hubs and the cloud is all called an IoT cloud in writing. Tragically, the IoT cloud has various drawbacks, such as the colossal inactivity of the device as the amount of knowledge prepared within the framework increases. The concept of haze registration is proposed to mitigate this problem, in which fog computing is located between the IoT hubs and the cloud system to process a lot of local information locally. In comparison with the first IoT cloud, the correspondence inertness will basically decrease in the haze figuring upheld IoT cloud, which we will allude as IoT mist, just as the overhead at the back-end cloud base. Thus, a few essential administrations that were difficult to convey via the traditional IoT cloud. A couple of them are as follows:

- Burglary of delicate data like bank secret phrase.
- Simple openness to individual subtleties likes contact address, contact number, and so on.
- It might prompt open access to classified data like money-related status of an establishment.
- An assault on any one gadget may bargain the respectability of the various associated gadgets. In this manner, the interconnectivity has an immense disadvantage, as a solitary security disappointment can disturb a whole system of gadgets.
- The dependence on the Internet makes the whole IoT engineering powerless to infection assault.

The vision of the IoT, which uses open Web quantifies to achieve knowledge exchange and article interoperability, began a notable forward leap in beating every distinction between virtual and physical worlds. In addition, the Social IoT (SIoT) loosens IoT to enter sharp articles with casual connections and allegedly ties physical and virtual worlds, just as support for participation between physical devices and human beings continued.

17.5 CHALLENGES IN INTERNET OF THINGS-BASED CLOUD APPLICATIONS

As of late, the IoT and distributed computing have generally been considered and implemented in various fields, as they can provide another technique for M2savvy M's observation and association (counting man-to-man, man-to-machine, and machine-to-machine), and individual use and proficient sharing of assets on request. In particular, the accessibility of data at scales and worldly longitudes unheard of up to this stage, combined with another era of wise planning calculations, presents a challenge for the IoT-based cloud applications. Like various parts of the IoT biological system, information placed in the cloud is similarly sensitive. The foundation should have the opportunity to protect data put away in the cloud. Insurance mechanisms include adequate encryption, power, and so on.

Security vulnerabilities consistently exist regardless of how much endeavors we pay to upgrade our item code and equipment. For this situation, we should initially have an arrangement to fix mistakes and rapidly discharge patches, rather than leaving the blunders unfixed for an extensive stretch. Next, the necessity of furnishing clients with an immediate and secure technique to fix blunders. Right now, it is well known to refresh online gadgets over the air, yet we should guarantee that the above strategy itself will not become a security vulnerability. IoT gadgets are frequently situated in open fields, and are unattended and not genuinely ensured. We should guarantee that they will not be malevolently messed with by horrible association, penetrated by programmers, or worked utilizing a level head screwdriver. Likewise, we should ensure information that gets put away on the gadgets in any structure. Despite the fact that it is expensive to insert a security assurance part on each IoT gadget, it is as yet critical to scramble information on these gadgets.

17.6 OPPORTUNITIES IN INTERNET OF THINGS-BASED CLOUD APPLICATIONS

IoT and cloud are an ideal association and have particular working and they are agreeable in nature and work best as unclear accomplices. Distributed figuring assists with taking care of, getting ready, and moving data in the cloud instead of related gadgets. The following are a segment of the preferences got from the intermixing of both the advancements.

- **(a)** Giving establishment: IoT and software integration grants open cloud organizations to help untouchables get to the scheme, which can assist IoT data or device modules that operate over IoT contraptions.
- **(b)** Pay more just as costs emerge (PAYG): PAYG cloud storage is a portion of the technique that enables clients to be unmistakably paid for the information they store.

(c) Expanded execution: In order to quickly interface and partner with different devices, data generated by endless IoT contraptions requires increasingly imperative execution. IoT and cloud consolidation will coordinate and gain rapid significance from their basic nature of exchanging information between devices.

(d) Improved versatility: IoT gadgets require a ton of information-sharing capacity for significant purposes. A portion of the IoT-based cloud stages that customers deliver are cloud administrations, such as Microsoft Azure IoT Suite, IBM Watson IoT Platform, Google Cloud's IoT Platform, AWS IoT Platform, etc.

(e) Cloud administration: It empowers the remote IoT device lifecycle. The executives assume a key job in empowering a 360-degree information perspective on the gadget framework.

(f) Gadget shadowing or advanced twins is another advantage that endeavors through cloud administrations. Engineers can make a reinforcement of the running applications and gadgets in the cloud to make the entire IoT framework exceptionally accessible for shortcomings and disappointment occasions.

(g) Cloud and IoT put together in the health care department encourages an advancement in the act of medication, empowers personalization of treatment and these are the possible alternatives which are focused especially for treating particular conditions and in fulfilling the needs of the individual which also helps to decrease the expense of human services while at the same time improving results.

Therefore, with the use of the IoT and cloud computing, today industries are changing with technology, we have seen several revolutions in the previous decade, and many technology advances will also be seen in the next decade. In several industries, Blockchain technology, cyber-physical systems, and artificial intelligence can play significant roles. Such new technologies can create a smarter atmosphere for the efficient execution of tasks (using IoT devices). In many applications, such as military affairs, including aerospace, military reconnaissance, intelligence grid system, smart transport, intelligent medical, environmental monitoring, industrial control, etc., today's IoT is used. On the other hand, we may argue that big data also has a role to play in making necessary and successful decisions. Using its intelligent research, IoT applications gather data from users or other devices/machines. The framework will provide efficient research results/outcomes to integrate the collected information with the Internet and other networking, which will satisfy the demand for intelligent communications and decision support in many sectors. Big data will therefore also play a major role in evaluating such actions or decisions in the near future (generated with context information technology). These decisions made by deep learning/deep learning on big data will provide a realistic, efficient, and perfect framework of application for public policy, public services and public business, improve the efficiency of government services, and improve the quality of life of the people.

17.7 SUGGESTED SOLUTIONS: TECHNIQUES AND METHODOLOGIES

To overcome these issues such as privacy, various techniques are developed in which these classifications are divided into Techniques and Methodologies.[22,24]

Under the Techniques:

Encryption. To guarantee information privacy, the cycle comprises on utilizing a cryptographic answer for encoding information put away in a cloud by either the information proprietor, by the cloud specialist co-op or by the two of them. This method is viewed as a decent arrangement, yet it presents a few restrictions, for example, the Twitter and Google episode in 2010.

Handling Encrypted Data. It is a method which is used to hide the impediments of the past transactions. This method guarantees security and privacy in a system but this method is quite complex and costly. The cloud specialist co-op does not have to unscramble information for inquiry execution and can execute inquiries legitimately on encoded information.

Jumbling. It' s a cycle of dispersing delicate information prior to sending it to the cloud specialist organization utilizing a mystery key or obscure technique by the last mentioned. In contrast with the encryption measure, the confusion is the most vulnerable technique and it is as of now done by hand or semi-robotized.

Anonymization. This technique n wipes out or removes the actually recognizable data PII from the information recorded prior to sending it to the cloud supplier. At that point, it can measure the genuine information and saving security of information proprietors. This cycle can be bombed when utilizing a connecting assault.[32]

Clingy Policy. It permits to "attach security strategies to information proprietors and drive access control choices and strategy implementation."

Confided in Platform Module. It is an equipment-based arrangement that gives the capacity to tie down activities to ensure client's information privileged insights, yet it is not planned to perform secure information handling.

Information Segmentation. It comprises on putting away unique portion of information in isolated nonlinkable pieces. It thinks about that private information is a delicate information and the relationship between information is likewise touchy.

Confide in Third-Party Mediator. It is a middle person among client and cloud supplier to ensure and look at strategy requirement and convey an examining. At last, these methods adapt to the protection issue and there are likewise steps toward to manage similar issues in.[31–34]

Under Methodologies:

Administration of Identification and Control (AIC). It is essential for any security system to be at the center. It makes it possible to view the customers, administrations, staff, mists, and some other substances. A lot of data/information is connected to a specific element in order to do so. Such specifics are based on the particular situation. Client protection must not be exposed by the character of an element.

Key Management. It speaks to an intend to deal with encryption key administration. As classification is one of the principle objectives of security, encryption which is the primary answer for the secrecy objective must be productive. Encryption calculations have serious issues identified with: how to safely produce, access, store, and trade discharge keys.

Security Management. Given the amount of cloud users, the dependence stack, and the security controls, executives need to function as a CML module to deal with security prerequisites, approach specifics, nature of security controls (as defined by the strategies), and so on.[26–29]

17.8 AN OPEN DISCUSSION: FROM USER'S AND BUSINESS'S PERSPECTIVE

IoT and cloud is making a huge difference in everyone's life and how it is affecting the users, and business has made a huge impact and it is seen in the following way:

From User's Perspective: With the extended use of PDAs, Virtual Assistants, IoT-enabled nuclear family devices like coolers, garments washers, security devices, and some more, the customers are going to see a flood in the encroachment of IoT contraptions in their own lives. These devices, while empowering a pleasant and improved lifestyle to the customers, will in like manner produce a mass of data that ought to be readied, taken care of, and changed over into information huge for business and deliberate use. These contraptions, anyway wise, have compelled dealing with and limit control and should be in a condition of congruity to offer a perfect help. It is basic for relationship to get overwhelmed with the proportion of data that will be made.[34,37,38]

PAAS—(Platform as a Services) the current circulated registering circumstance, is where an outcast provider urges a shrouded stage to run, make, or have your present programming. Tremendous data are a field that oversees such degrees of datasets which are absurdly enormous for standard data planning. They point toward evacuating business-relevant information important for insightful assessment. Encouraging and keeping up big data organizations is a costly and attractive undertaking. Consequently, it is attainable for affiliations that make this kind of big data to make sure about Platforms as a help from cloud figuring dealers which are significant for big data. This further allows the aggregation of Business Data and the data delivered from the IoT devices onto a lone stage opening approaches to dynamically correct judicious examination, dependent on the progressing data. At present, the spread of these IoT devices is prudent, with solitary firms including to the game plans subordinate their spaces. Thus, the shrouded stages for evaluating the course and separate these applications are disconnected. In any case, bit by bit, there will be a need to standardize these key AI or machine learning applications which are reusable and made reachable to the most diminutive of venders. Computer-based intelligence and machine learning programming must be calibrated to fill in according to our requirements and the outcomes that we have to pick up from them. This is a collective undertaking of utilizations from comparative spaces contributing their learnings together.[39]

From Business Perspective: Cloud innovation can empower us to impart these insights to our supervisory crew—all before we even arrive at our office. Organizations started embracing IoT as a component of their information methodology to give ongoing data to these current announcing frameworks, more setting regarding how the business is working, and more prominently perceivable into territories of the business that were not already conceivable. For certain organizations, this is a chance to increment operational productivity and smooth out expenses. For other people, it can open new plans of action and income streams. Since information from conventional undertaking frameworks are layered in with information produced from sensors and associated gear, organizations are finding that IoT information has unexpected attributes in comparison to customary endeavor information. The speed and volume of this sort of information can overpower frameworks that are not set up for it. It likewise requires some re-architecting of information models since it speaks to various sorts of data that might not have been a piece of earlier arranging. This is the place distributed computing comes in. As a result of the cloud's capacity to house a lot of information, organizations can process and store two informations from their venture frameworks and their IoT gadgets in a similar spot. The cloud turns into an incredible conglomeration point for every single dissimilar framework, where organizations can scale their endeavors up or down with not many impediments. Associations would then be able to wipe out the requirement for incorporation and reviews between frameworks that harvest up when their information is put away independently. With cloud innovation making it simpler to store significant archives, recover records from various stages and that's only the tip of the iceberg, the requirement for in-house stockpiling is tossed out the window.[40,41] Lastly, several interesting issues, challenges and opportunities (for future researchers) toward IoT, Cloud and its related technologies can be found in.[42–50]

17.9 CONCLUSION

The IoT begins to show its content. IT and correspondence positions, and more network upgrades. Although the critical concepts and institutions are painstakingly represented and created, it is anticipated that more prominent efforts will release their full potential and integrate and unify frameworks and entertainers in the use of the indifferent fields of their office. The Internet of Vehicles (IoV) and its applications are discussed in this section. It starts with the introduction of the IoV base, concept, and system design and then breaks down the IoV characteristics and the comparison of new difficulties in creative IoV work. A portrayal of empowering advancements is at the center of the section. Macintosh conventions and principles, for instance, IEEE 802.11 and IEEE 802.11p/WAVE, and steering conventions, for instance, AODV and OLSR, are presented in detail. The other center part is the conversation of applications. A total scientific categorization is introduced and various classes are examined, including driving security, effectiveness administration, shrewd traffic the board, and instructive administrations. In summary, today's new industrial revolution is cyber-physical systems revolution. The integration of CPS with IoT device provides a unique environment, that is, put foundation of cyber-physical systems revolution. Further, this revolution can be merged with big data and analytics and blockchain technology, to changes/provides/ biggest shift in business and technology. Hence, Industries 4.0, smart factories, cyber-physical production systems (CPPS), and IoT are the necessity of future technology. All researchers, whoever are working related to above areas (listed in Section 17.6), are invited to continue their research work toward the above-said futuristic problems.

KEYWORDS

- **Internet of Things**
- **cloud computing**
- **cloud-based Internet of Things**
- **trust**
- **security and privacy issues in cloud-based internet of things**

REFERENCES

1. Formisano, C. et al., The Advantages of IoT and Cloud Applied to Smart Cities. In *2015 3rdInternational Conference on Future Internet of Things and Cloud*; Rome, 2015; pp 325–332.doi: 10.1109/FiCloud.2015.85
2. Chen, S.; Xu, H.; Liu, D.; Hu, B.;Wang, H. A Vision of IoT: Applications, Challenges, and Opportunities with China Perspective. *IEEE IoT J.*Aug**2014,***1*(4), 349–359.doi: 10.1109/JIOT.2014.2337336
3. Khan, M. H. Rehmani and A. Rachedi, Cognitive-Radio-Based Internet of Things: Applications, Architectures, Spectrum Related Functionalities, and Future Research Directions. *IEEE Wireless Commun.*June **2017,***24*(3), 17–25.doi: 10.1109/ MWC.2017.1600404
4. Atlam, H. F.;Alenezi, A.;Alassafi, M. O.;Wills, G. Blockchain with Internet of Things: Benefits, Challenges, and Future Directions. *Int. J. Intell. Syst. App.***2018,***10* (6), 40–48, [2030]. doi:10.5815/ijisa.2018.06.05
5. Abomhara, M.;Køien, G. M. Security and Privacy in the Internet of Things: Current Status and Open Issues.In *2014 International Conference on Privacy and Security in Mobile Systems (PRISMS)*; Aalborg, 2014; pp 1–8.doi: 10.1109/PRISMS.2014.6970594
6. Saleem, Y.; Crespi, N.; Rehmani, M. H.; Copeland, R. Internet of Things-Aided Smart Grid: Technologies, Architectures, Applications, Prototypes, and Future Research Directions. *IEEE Access***2019,***7*, 62962–63003.doi: 10.1109/ACCESS.2019.2913984

7. Tsai, C., Lai, C.; Vasilakos, A.V. Future Internet of Things: Open Issues and Challenges. *Wireless Netw*. **2014,** *20*, 2201–2217. https://doi.org/10.1007/s11276-014-0731-0
8. Adat, V.; Gupta, B.B. Security in Internet of Things: Issues, Challenges, Taxonomy, and Architecture. *Telecommun. Syst.* **2018,** *67*,423–441. https://doi.org/10.1007/s11235-017-0345-9
9. Khoo, RFID as an Enabler of the Internet of Things: Issues of Security and Privacy. In *2011 International Conference on Internet of Things and 4thInternational Conference on Cyber, Physical and Social Computing*; Dalian, 2011; pp 709–712.doi: 10.1109/iThings/CPSCom.2011.83
10. Gu, L.; Wang, J.; Sun, B. Trust Management Mechanism for Internet of Things. *China Commun.* Feb **2014,***11*(2), 148–156.doi: 10.1109/CC.2014.6821746
11. Tyagi, A.K.; Rekha, G.; Sreenath, N. Beyond the Hype: Internet of Things Concepts, Security and Privacy Concerns. In *Advances in Decision Sciences, Image Processing, Security and Computer VisionICETE 2019. Learning and Analytics in Intelligent Systems*; Satapathy, S., Raju, K., Shyamala, K., Krishna, D., Favorskaya, M., Eds., Vol. 3; Springer: Cham, 2020.
12. Matheu-Garc´ıa, S. N.; Hern´andez-Ramos, J. L.; Skarmeta, A. F.; Baldinic, G. Risk-Based Automated Assessment and Testing for the Cybersecurity Certification and Labelling of IoT Devices; Elsevier, 2019.
13. Yang, Y.; Peng, X.; Fu, D.A Framework of Cloud Service Selection Based on Trust Mechanism. *Int. J. Ad Hoc Ubiquitous Comput.* **2017,** *25*(3), 109–119.
14. Sidhu, J.; Singh, S. Improved TOPSIS Method-Based Trust Evaluation Framework for Determining Trustworthiness of Cloud Service Providers. *J. Grid Comput.***2017,** *15*(1), 81–105.
15. Wang, S.; Sun, L.; Sun, Q.; Wei, J.; Yang, F. Reputation Measurement of Cloud Services Based on Unstable Feedback Ratings. *Int. J. Web Grid* Services **2015,** *11*(4), 362–376.
16. Noor, T. H.; Sheng, Q. Z.; Yao, L.; Dustdar, S.; Ngu, A. H. H. Cloud Armor: Supporting Reputation-Based Trust Management for Cloud Services. *IEEE Trans. Parallel Distrib. Syst.* Feb **2016,** *27*(2), 367–380.
17. Nagarajan, R.; Thirunavukarasu, R.; Shanmugam, S. A Fuzzy-Based Intelligent Cloud Broker with Map Reduce Framework to Evaluate the Trust Level of Cloud Services Using Customer Feedback. *Int. J. Fuzzy Syst.***2018,** *20*(1), 339–347.
18. Siadat, S.; Rahmani, A. M.; Navid, H. Identifying Fake Feedback in Cloud Trust Management Systems Using Feedback Evaluation Component and Bayesian Game Model.*J. Supercomput*.**2017,***73*(6), 2682–2704.
19. Sahmim, S.; Gharsellaoui, H. Privacy and Security in Interne-Based Computing: Cloud Computing, IoT, Cloud of Things: A Review. ICKIIE, **2017,** 1516–1522.
20. Zhou, J.; Cao, Z.; Dong, X. Security and Privacy for Cloud Based IoT: Challenges. IEEE, 2017.
21. Li, X.; Wang, Q.; Lan, X.; Chen, X.; Zhang, N.; Chen, D. Enhancing Cloud-based IoT Security through Trustworthy Cloud Service: An Integration of Security and Reputation Approach, 2019.
22. Ni, J.; Zhang, K.; Lin, X.; Shen, X. S. Securing Fog Computing for Internet of Things Applications: Challenges and Solutions. *IEEE Commun. Surveys Tuts*. **2018,** *20*(1), 601–628, 1stQuart.

23. Zhang, K.; Ni, J.; Yang, K.; Liang, X.; Ren, J.; Shen, X. S. Security and Privacy in Smart City Applications: Challenges and Solutions. *IEEE Commun. Mag.* Jan **2017,** *55*(1), 122–129.
24. Stergiou, C.; Psannis, K. E.;Kim, B.-G.;Gupta, B. Secure Integration of IoT and Cloud Computing. *Future Gener. Comput. Syst.* Jan **2018,** *78*, 964–975.
25. Halabi, T.; Bellaiche, M. Towards Quantification and Evaluation of Security of Cloud Service Providers. *J. Inf. Secur. Appl.* Apr **2017,** *33*, 55–65.
26. Voas, J.; Kuhn, R.; Kolias, C.; Stavrou, A.; Kambourakis, G. Cybertrust in the IoT Age. *Comput. J.***2018,** *51*(7), 12–15.
27. Daubert, J.; Wiesmaler, A.; Kikiras, P. A View on Privacy & Trust in IoT. ICC, 2015.
28. Sicari, S.; Rizzardi, A.; Grieco, L. A.; Coen-Porisini, A. Security, Privacy and Trust in Internet of Things: The Road Ahead.*Comput.Netw.***2015,***76*, 146–164.
29. Poore, R. S. Anonymity, Privacy, and Trust. *Info. Syst. Sec.* **1999,** *8*(3), 16–20.
30. Bao, F.; Chen, I. Trust Management for the Internet of Things and Its Application to Service Composition. In *2012 IEEE International Symposium on a World of Wireless, Mobile and Multimedia Networks*, WoWMoM2012, San Francisco, CA, USA, June 25–28,2012; *IEEE Comput. Soc.***2012,** 1–6. http://dx.doi.org/10.1109/WoWMoM.2012.6263792
31. Leister, W.; Schulz, T. Ideas for a Trust Indicator in the Internet of Things. In *SMART 2012, The First International Conference on Smart Systems, Devices and Technologies*, 2012; pp 31–34.
32. Hochleitner, C.; Graf, C.; Unger, D.; Tscheligi, M. Making Devices Trustworthy: Security and Trust Feedback in the Internet of Things. In *Fourth International Workshop on Security and Privacy in Spontaneous Interaction and Mobile Phone Use (IWSSI/SPMU)*; Newcastle, UK, 2012.
33. Habib, S. M.; Varadharajan, V.; M¨uhlh¨auser, M. A Framework for Evaluating Trust of Service Providers in Cloud Marketplaces. In *Proceedings of the 28th Annual ACM Symposium on Applied Computing, SAC '13*;Shin, S. Y.,Maldonado, J. C., Eds.; , Coimbra, Portugal, Mar 18–22, 2013, ACM, 2013; pp 1963–1965.
34. Iliev, A.; Smith, S. Protecting Client Privacy with Trusted Computing at the Server. *Security Privacy,* IEEE Mar **2005,** *3*(2), 20–28.
35. Kirovski, D.; Drini´c, M.; Potkonjak, M. Enabling Trusted Software Integrity. In *ACM SIGPLAN Notices* **2002,** *37*(10),108–120.
36. Georgakopoulos, D.; Jayaraman, P. P.; Fazia, M.;Villari, M.; Ranjan, R. Internet of Things and Edge Cloud Computing Roadmap for Manufacturing. *IEEE Cloud Comput.* Jul/Aug **2016,***3*(4), 66–73.
37. Pellicer, S.; Santa, G.; Bleda, A. L.; Maestre, R.; Jara, A. J.; Skarmeta, A. G. A Global Perspective of Smart Cities: A Survey. In *Proceedings of the 7th International Conference Innovative Mobile Internet Services Ubiquitous Computing (IMIS)*, 2013; pp 439–444.
38. Tyagi, A. K.; Abraham, A. Internet of Things (IoTs): Future Challenging Issues and Possible Research Directions. *Int. J. Comput. Info. Syst. Indust. Manage. App.***2020,** *12*, 113–124. ISSN 2150-7988.
39. Tyagi, A. K.; Nair, M. M. Internet of Everything (IoE) and Internet of Things (IoTs): Threat Analyses, Possible Opportunities for Future. *JIAS* **2020,** *15* (4).
40. Tyagi, A. K.; Rekha, G.;Sreenath, N. Beyond the Hype: Internet of Things Concepts, Security and Privacy Concerns. In Satapathy S., Raju K., Shyamala K., Krishna D., Favorskaya M. (eds) Advances in Decision Sciences, Image Processing, Security and

Computer Vision. ICETE 2019. Learning and Analytics in Intelligent Systems, vol 3. Springer, Cham. https://doi.org/10.1007/978-3-030-24322-7_50.

41. Tyagi, A. K.; Agarwal, K.; Goyal, D.; Sreenath, N. A Review on Security and Privacy Issues in Internet of Things. In Sharma H., Govindan K., Poonia R., Kumar S., El-Medany W. (eds) Advances in Computing and Intelligent Systems. Algorithms for Intelligent Systems. Springer, Singapore. https://doi.org/10.1007/978-981-15-0222-4_46.
42. Nair, M. M.; Tyagi, A. K.; Sreenath, N. The Future with Industry 4.0 at the Core of Society 5.0: Open Issues, Future Opportunities and Challenges. 2021 International Conference on Computer Communication and Informatics (ICCCI), 2021, pp. 1-7, doi: 10.1109/ICCCI50826.2021.9402498.
43. Tyagi A. K.; Fernandez T. F.; Mishra S.; Kumari S. Intelligent Automation Systems at the Core of Industry 4.0. In: Abraham A., Piuri V., Gandhi N., Siarry P., Kaklauskas A., Madureira A. (eds) Intelligent Systems Design and Applications. ISDA 2020. Advances in Intelligent Systems and Computing, vol 1351. Springer, Cham. 2021, https://doi.org/10.1007/978-3-030-71187-0_1
44. Varsha R.; Nair S. M.; Tyagi A. K. Aswathy, S. U.; RadhaKrishnan R. The Future with Advanced Analytics: A Sequential Analysis of the Disruptive Technology's Scope. In: Abraham A., Hanne T., Castillo O., Gandhi N., Nogueira Rios T., Hong TP. (eds) Hybrid Intelligent Systems. HIS 2020. Advances in Intelligent Systems and Computing, vol 1375. Springer, Cham. 2021. https://doi.org/10.1007/978-3-030-73050-5_56
45. Tyagi, A. K.; Nair, M. M.; Niladhuri, S.; Abraham, A. Security, Privacy Research issues in Various Computing Platforms: A Survey and the Road Ahead. *JIAS* **2020,** *15* (1), 1–16.
46. Madhav A. V. S.; Tyagi A. K. The World with Future Technologies (Post-COVID-19): Open Issues, Challenges, and the Road Ahead. In: Tyagi A.K., Abraham A., Kaklauskas A. (eds) Intelligent Interactive Multimedia Systems for e-Healthcare Applications. Springer, Singapore. 2022. https://doi.org/10.1007/978-981-16-6542-4_22
47. Goyal, D.; Tyagi, A. A Look at Top 35 Problems in the Computer Science Field for the Next Decade. 2020. 10.1201/9781003052098-40.
48. Nair, M. M.; Tyagi, A. K. Privacy: History, Statistics, Policy, Laws, Preservation and Threat Analysis. *JIAS* **2021,** *16* (1), 24–34.

CHAPTER 18

Genomics and Genetic Data: A Third Eye for Doctors

M. SHAMILA[1], AMIT KUMAR TYAGI[2,3*], and S. U. ASWATHY[4]

[1]*Gokaraju Rangaraju Institute of Engineering and Technology, India*

[2]*Centre for Advanced Data Science, Vellore Institute of Technology, Chennai, 600127, Tamil Nadu, India*

[3]*School of Computer Science and Engineering, Vellore Institute of Technology, Chennai, 600127, Tamil Nadu, India*

[4]*Department of Computer Science and Engineering, Jyothi Engineering College, Thrissur, India*

[*]*Corresponding author. E-mail: amitkrtyagi025@gmail.com*

ABSTRACT

In the recent decade, data have received attention from sectors, organizations, and industries. Numerous data are generated by many smart devices or machines (through communication). Genomic data, genetic data are examples of these data or Big Data. In general, Genetics and Genomics play a vital role in healthcare sector. It should be noted that Genetics and Genomics sounds similar, but they are not, that is, genetics is considered as a subset of genomics. Genetics deals with the study of single gene whereas genomics is the study of group of genes (which is known as genome) and their interrelationship. Genetics is concerned about how the genetic traits are transmitted. Genetics is a familiar term, but genomic is a new field which has become popular in the last few decades due to the advancement in the field of

Intelligent Interactive Multimedia Systems for e-Healthcare Applications. Shaveta Malik, PhD
Amit Kumar Tyagi, PhD (Eds.)

computational biology. The reduction in DNA sequencing cost has increased the usage of genomic data which can help to explore useful information regarding human life. The advancement in genetics or genomics field leads to a transition from traditional medicine to personalized medicine. Apart from the accidental death, genomic factors play an important role in the causes of human death. A better study of genomic data helps the human being to prevent some chronic diseases (like cancer, HIV, etc.) and hence to improve the health of patients and can help to healthcare sector to receive more useful innovations. Currently, many existing tools are available to extract relevant information from genomic data, also to preserve useful/sensitive or genomic privacy. Use of genomic data in research or other clinical purposes may cause leakage of information, which is a serious issue. So it is necessary to find some efficient techniques to overcome such raised challenges or issues. Hence, many researchers tried to provide efficient mechanism, but failed to protect due to various reasons like different preferences of users. Hence, in this article, we give insight to various topics, such as importance of genomic data, genetic analytics, existing tools, limitations, and raised issues, challenges (including identified research gaps) in analyzing this genomic data, or preserving user' information, etc.

18.1 INTRODUCTION

Everyone is familiar with the term genetics but not much with genomics. Genomics is the new field of genetics, which has a wide spectrum of applications in healthcare, such as genome-based disease prediction, genome-based personalized medicine, etc. The analysis of genetic or genomic data not only protects us from ill health but also supports to maintain a good health. Genetics can be treated as a study of single gene, but genomics deals with the study of group of genes and their interrelationships. We all know that human body consists of set of cells. Each cell contains 23 chromosome pairs. A double-helical structure called DNA is retained by each chromosome (deoxyribose nucleic acid). We can tell that DNA is inside genes and gene is inside chromosomes. As per studies 99.9% of the human being holds same set of genes and the 0.1% variation makes the people unique in terms of hair color, skin tone, eye color, etc. DNA is made of four molecules called adenine, thymine, cytosine, and guanine. The genetic code can be considered as set of rules for translating information encoded in DNA into proteins with the help of RNA in genes. A genetic code consists of following features.

The code is nonoverlapping

- They are unambiguous
- The code is comma less
- The code has polarity
- The code is degenerate
- The code is universal

Different programs have been organized by the government to support the genetic data collection and analytics. For example, "All of US program." The genetic data have been gathered from around 1 million US people from different background. Theses collected data are combined with different other records like health records and are utilized for personalized medicine, to improve the health condition, etc. Analyzing these records with different features provides efficient good results in respective (Medicare, biomedical imaging, etc.) applications. In the above applications, a broad variety of different methods, including model systems (e.g., CRISPR-Cas9), can be used to clarify the functions of genes and proteins and to re-examine the relationship between the genotype and the phenotype.[1]

Today, due to the reduction in the cost of sequencing genes, the importance of genomic data and genetic analysis has increased. Admirable progress has been made by the genome privacy group. It took, for instance, more than 13 years and an estimated cost of US$3 billion in 2003 to sequence the first human genome.[2] The reduction of sequencing costs has been improved with genomics as a research discipline with clear application possibilities. As a result, with the improvement of sequencing technology, it has been increasingly integrated into the study of variant interactions. In order to reduce the cost of genome sequencing (GS) to $1000, it is necessary to increase the cost and performance by another four to five orders of magnitude. Therefore, the question of economic feasibility ultimately focuses on whether the new method can achieve such a huge improvement. It should be noted that few advantages of whole-GS include any gene can be converted into digital data for analysis by allowing the entire genome of an individual to be sequenced (refer to genetic data analysis in Section 18.5). While this results in a large number of data, in the near future, there will also be great opportunities for other research communities.

Genomic data are increasingly used in a range of fields, including healthcare (such as personalized medicine), biomedical sciences (such as modern genome–phenome interaction recognition), and direct-to-consumer (DTC) services (such as disease risk test) and evidence collection (e.g., criminal investigation). For example, doctors can now prescribe "the right medicine at the right time" (for some medicines) based on the patient's genome composition. By accelerating the clinical research and drug growth, personalizing treatment regimes, enhancing patient outcomes, and reducing the cost of care, genomics may offer major advantages to healthcare systems. Genomic data, the most personal of all human data, are part of the overall health data revolution, allowing computing capacity to expand exponentially, and wearable devices, such as smartphones, fitness trackers, and heart rate monitors are to be commercialized. Managing these data has led to a change in the way such information is gathered, stored, analyzed and used with privacy, reliability, and security posing prickly challenges. Health and genomic data can now be shared with healthcare professionals and disparate partners.

In addition to the many advantages, there is a paradoxical dark side to the capacity to gather and exchange intimate, confidential information. This can be seen in the form of daunting privacy standards and cyber security risks, which often delay analysis, reduce investment financial returns, and harm the credibility of the brand. The relative benefits of sequencing have been known for a long time. Sequencing, in comparison to array analyses:

(a) It does not include the pre-identification of variations.
(b) It is easier to adapt to more complex variants than single nucleotide changes and very brief additions or deletions.
(c) It does not need to concentrate on variants that are widespread in large populations relative to uncommon or special variations.

Moreover, of the above-listed applications/sector, now a big question is "how Life Sciences companies can take control of data to build a competitive edge," because life science can provide answer to many unsolved questions. Hence, now the remaining part of this work is organized as:

- Section 18.2 discusses work related to genomic, genetic analysis research, and medicine.
- Section 18.3 discusses motivation behind writing this article (or this information).
- Section 18.4 discusses scope or importance of genomic, genomic data, genomic data analysis and genetic work (in past, present, and future).
- Section 18.5 explains about Genetic Data Analysis in detail.
- Section 18.6 tells why we need to protest Genomic Privacy, that is, a type of privacy need to be protected in this smart era.

- Section 18.7 discusses a threat model for Genomic data and Genetic data analyses.
- Section 18.8 discusses various tools and methods existing/available for genomic, genomic data, genomic data analysis, and genetic work.
- Section 18.9 discusses various open research issues, challenges faced in biomedical imaging and opportunities for future as including research directions.
- Finally, Section 18.10 concludes this work with various research gaps and future enhancements.

Hence, in this work/article, our main goal is to fill many identified research gaps in current era through providing an effective literature on genomic research and medicine.

18.2 RELATED WORK

In the past decade, IT technologies, such as whole-GS (WGS) and Digital Genome Database have transformed the complex process of recording a person's entire genetic code from decades, multibillion dollar multinational companies to 1 week, $1000 in operations. We can see dramatic changes in gene sequencing today, that is, in the genomic region. This is because the main objective of scientists/researchers in previous years was to classify all functional components, including regulatory components, in both coding and noncoding regions. In the work of Nambiar et al.,[3] we can find out that "how genomic big data is too useful for life-saving medical innovation." At the same time, an abundance of individual DNA (deoxyribose nucleic acid) knowledge to be handled is also provided by clinical genomics.

Usually, genomic research is a key CSCWW initiative (Computer Supported Cooperative Work). The establishment of the first human genome map set a precedent for genomics from 1990 to 2003. Genomics is a field of scientific research cooperation involving scientists, organizations, and funds around the world. This effort involves participants from different geographic locations and disciplines who use technical infrastructure to facilitate collaboration and arduous large-scale data analysis in order to achieve results that would otherwise be impossible. The spirit of the Human Genome Project itself reflects the value of the research model: sharing these valuable data through Internet databases can be obtained for free, thereby bringing greater benefits to the society.

In the Human Genome Project (HGP), American and British project scientists met in neutral Bermuda to discuss the progress of the project. One of their main questions is how to ensure that genomic information can be in the public interest, not a private property. They developed a set of principles that require that all DNA sequence data are copyright-free and must be distributed to the open access network within 24 h after it is generated.[4] This is in contrast to the traditional scientific practice of releasing experimental data only after publication. The subsequent policy initiatives of the "Bermuda Principles" inspired contemporary open access scientific practice and the concept of presenting information as a global knowledge resource. From the efforts of Reeves and Hrischuk,[5] we concluded that "biological and computational are currently indivisible" for today's genomic science, while we live in an age of smart things, supercomputers, and processing units of high graphics, etc. We (researchers, doctors, or scientists) need to do many things/solve many issues or cure diseases in the near future by GS (properly and efficiently).

18.2.1 PROGRESS IN GENOMICS FROM 19TH CENTURY TO 21ST CENTURY

Since the end of the 1990s, the emergence of the Internet and related information technologies has enhanced the networking potential of genomic research, allowing it to enter a broader field of medical innovation. Today, the integration between biology and computers has transformed genomics from wet laboratory science into a 21st century big data project. Another transformative moment is the arrival of clinical genomics. The cost of the diploid human genome sequence has been reduced from approximately US$70 million to US$2000. Advances in genomics have allowed clinical researchers to correlate mutations and susceptibility to various types of diseases (including cancer)[6] and response to certain therapies. These advances promote the modern age of precision medicine, in which diagnosis and treatment can be personalized to patients on the basis of their genomes, thereby allowing healthcare to be more cost-effective as well as effective.[7]Moreover, the cost of whole-GS (that is, to determine the complete DNA sequence of a person) is now about 1000 US dollars,[8] and doctors have a greater opportunity to diagnose and treat patients with abnormal genetic diseases.[9]

18.2.2 IMPROVEMENT IN GENOMIC DATA SHARING

We have observed and reported many attacks on genomic databases in the past decade. Privacy security tools for managed access, data disruption (especially in the context of differential privacy), and cryptographic solutions have been discussed and classified (with providing brief overview for each category). Wang et al. also analyzed the impact of genome privacy in the United States on health, technology, and ethics in,[10] and studied the various options of genome privacy that can be used to disclose record-related genome research results and ethical and legal options with informed consent, influences. Nayeri and Aghajani[11] introduced an overview of genome privacy work from the perspective of computer science, they have discussed some known privacy threats and available solutions, and solved some of the known challenges in the field, such as genome databases usually not controlled by the health system, or privacy security may affect its use. They also interviewed 61 biomedical experts, discovering that most understand the importance of genome privacy research and the risks of intrusions into privacy. Naveed et al. [12] divide the genome privacy research into three categories which include privacy protection data sharing, secure computing and data storage, and privacy query or output. They discuss the solutions and unresolved problems in these fields, aiming to help practitioners to better understand its performance and protection, and compare it with the use cases and constraints of available cryptographic primitives. Finally, in the work of Greenbaum et al.,[13] the authors focus on research in bioinformatics where it is appropriate to reveal private information. They consider scenarios of genomic data query and sequence alignment, for example, when surveying available solutions for each of them.

In addition, due to legal or policy restrictions, a large amount of research can now be used to support genomics-related applications, which are difficult or impossible to support. For example, genetic and health data do not easily cross national borders, which make international partnerships extremely challenging. In this context, those constraints can be mitigated by mechanisms that demonstrably guarantee the privacy-friendly processing of genomic data and will enable significant progress in science. Scientists and bioinformatics scientists can now conduct large-scale genome data analysis through technological improvements and developments, but there is still a lack of consensus on the social consequences of predictive analysis, otherwise, we must remember that in order to successfully carry out genome/genome work, we must avoid the question of the last mile. Therefore, this section addresses the reviews of already completed genomic/genome work (including their analysis).

18.3 MOTIVATION

Genomics could now be regarded as a big data field and the discovery of this field is the biggest innovation of the century (together with the advent of Blockchain Technology) today in both scientifically and culturally (both). Sequencing a genome was a very complicated task in the previous century. Generally speaking, a single human genome is a complex structure composed of 6 billion information databases. The size of a single genome file ranges from approximately 700 MB of raw data to 200 GB of metadata and variant annotations. GS is no longer a process that relies on test tubes and pipettes, but on databases and IT. The ability to evaluate DNA sequences is beginning to exceed the ability of researchers to store, distribute, and especially interpret data. For example, former actors (in professional sports) are using new analytical methods and collecting new data types to understand player evaluations and team motivation. This is the next example of clinical practice in genomics. Genomics and genetics have become a new research area of partnership today, with new analytical methods that have shaped the advancement of both disciplines. We need to include information related to genomics and its sub-related terminology in a specific way to provide more and more clarity and productive service to people/curing disease. Therefore, our main thing is to save a human life or demonstrate useful service to human beings, so we choose this area/domain/field to write our post. Now the next segment will address the importance of genomes, genomic data, genetic data and their study, including the importance of today and tomorrow in many applications.

18.4 IMPORTANCE OF GENOMICS AND GENETIC DATA IN VARIOUS APPLICATIONS

Genomics has immense potential for re-shaping drug discovery and production in the 21st century, that is, shifting from the blockbusters to niche busters, the Life Sciences industry is leading the drive for personalized medicine. When the human genome was fully sequenced (in 2003), the focus turned to the classification and annotation of its functional DNA elements, including those that regulate genes, for example, the identification of such elements is the most important step toward elucidating pathogenic pathways that affect human health. Some people believe that the same is true for genetic information and health records, but they do not. In short, for many people, genomic data is considered (and treated) no different from traditional health data (e.g., data that may be captured in medical records) or any other type of data. Although genomic data itself may not be an "exception," it has many characteristics that distinguish it from other available data types.

In addition to big data analysis, GS technology is also developing at a rapid growth rate, and now it is possible to generate highly accurate genotypes inexpensively. Different applications may assist in processing and reviewing certain data, including personalized medical services. Although the biomedical community surpasses the benefits of the genomics revolution, the increased availability of such information has a significant impact on personal privacy, especially because the genome has certain important characteristics, including but not limited to:

(i) Connection with features and certain disorders.
(ii) Potential for identification (e.g., forensics).
(iii) Family relationship disclosure.

Today, the genomic data are useful in many applications as given below:

(i) In Research: While a large number of diseases and variable treatment responses have been identified in the genome, new associations are discovered on a daily basis.
(ii) In Healthcare: A mutation in an individual's genomic sequence may affect one's well-being. In general, changes in a single gene may have a negative effect immediately or at any point in the future on a person's health. Genetic testing can be used before delivery to classify various variables that may affect health outcomes.
(iii) In recent years, in DTC services, DTC GS from multiple companies has increased. These services allow individuals to directly participate in the collection, processing, and even research of their genomic data.
(iv) Legal and Forensic: These data have also been used for investigative purposes because of the static nature of genomic sequences.

18.4.1 REVOLUTIONS IN GENOMIC DATABASES OR GENETIC ANALYSIS

In the previous decade after 2003 (done successfully sequencing of first gene), we have seen major advances in genomic databases or genetic analysis, which are included/discussed here as:

- *Reimagining Drug Development and Therapy:* Genomics offers several ways to address some of the Life Sciences' biggest challenges, such as clinical trials, treatment, etc.
- *From the Laboratory to the Laptop:* Data are at the heart of genomics, with researchers using machine learning, a subfield of artificial intelligence (AI) to help oncologists choose the most effective, individualized cancer treatments, by swiftly sorting through research data. New and powerful algorithms may identify which genes are likely to be mutated and predict the most appropriate treatments. The genetic sequencing company Adaptive Biotechnologies is partnering with Microsoft to map the genetics of the human immune system, or immunome, in a bid to help detect early stages of cancers and other diseases. Immunosequencing is designed to help patients' immune systems fight diseases like cancer, by measuring the body's initial response and then using targeted drugs to stimulate a response. The initiative, which uses large-scale machine learning and cloud computing, is attempting to diagnose conditions based upon a simple blood test. And advances such as Illumina's next-generation "massively parallel" sequencing technology enable faster and larger-scale sequencing. These are exciting developments, but one should not get too carried away. Simply finding a problem in a genome does not automatically explain *how* this may affect the person in question. Currently, any personal information needs to be integrated with other relevant clinical data like family history, that is, to improve its predictive power. Successful research increasingly integrates genetic data with phenotypic data (such as medical history).
- *From Insecure Genomic Database to Secure Database:* Here, we discussed that "how life sciences companies can gather, analyze, and use genomic data effectively, to accelerate clinical trials, improve treatment regimens, and demonstrate drug efficacy." But as the amount of data held expands exponentially, it brings with it significant challenges in gathering, storage and analysis, as well as ongoing considerations relating to privacy, reliability, and security. So, since 19th century to 21st century, we have initiated several secure mechanisms which are tamper proof, for example, digital signature, blockchain technology, etc.

This section therefore examines the relevance of today and tomorrow's genome, genomics, genomic data processing, and genetics (i.e., in applications like biomedical imaging). It also shows the revolution of genomics data in many applications which is a major one. The following section will discuss about genomic and genetic data analysis in detail.

18.5 GENETIC DATA ANALYSIS

We defined several features of big data in,[14] but volume, variety, and velocity are among the most useful features. On the other hand, today's human genome has become a more integral part of the Life Sciences/life science stream, and the volume of genomic data continues to increase at a growing pace (massive amount of data). Human genomes are extremely likely to cause pathological conditions, and human health needs functional analysis. We need the unique DNA of the person to do the sequencing process which helps us to find specific characteristics of that individual. This is particularly true for genomic data which, especially with unknown consequences, can be acquired, replicated, or analyzed in unexpected ways with unknown consequences. We need effective instruments, methods, and mechanisms to analyze these massive data, some of which we have in Section 18.8 of this article. GTG banding, FISH, aCGH, Sanger, and NGS are several techniques for functional genome analysis. There are two strategies for the analysis of quantitative real-time polymerase chain reaction (PCR) (qPCR) data: absolute quantification (based on the calibration curve) and relative quantification (based on reference sample comparison). Two distinct concepts include genomic analysis and genetic analysis. Genomic analysis is the detection, measurement or comparison on a genomic scale of genomic characteristics, such as DNA sequence, structural variation, gene expression, or annotation of regulatory and functional elements. Genetic analysis on the other hand, is the study of a DNA sample to test for mutations (changes) that may cause/increase disease risk or affect the way a person responds to treatment.

In the near future, next-generation sequencing or mass spectrometry are mandatory to use for efficient (fast) and accurate sequencing. In addition, reliable scientific findings not only involve laboratory investigation, but also accurate bioinformatics research. These approaches provide an opportunity for precise and systematic functional study covering many fields of research: genomics, epigenomics, proteomics, and interactomics. Notice that the creation of a genomic data strategy that drives more successful R&D and meets regulatory requirements is essential. To boost profitability, research is needed to manufacture new products faster and at a lower cost.

In analyzing genomic data, the biggest challenge is determining the function of genes, gene products (still their interaction is open). Weak spots in big data approaches such as them (methods) are also helpful in terms of what to examine, but they are also open challenges, not why or how we can go about it in the near future, we need to identify trends of social and cultural implications in massive data sets.

Remember that in this smart era, data are the new oil for industries or organizations to get habits of its consumer or his/her events, also to increase profit of their respective businesses. Data analytics is too important for data scientist. In general, data science is the use of advanced analytical techniques and algorithms to train computers "how to use complex data and knowledge from a wide variety of sources to enhance, accelerate, automate, and augment decisions".[15] But generating of vast amount of data also creates several problems like security of data (in motion and rest), privacy of personal information, not having standard tools and worldwide recognized quantification process, etc. Hence, in near future, we will prefer all services with Secure Data Analysis.

For providing security, Blockchain is best ever mechanism in today' era. But, as we know that every technology has advantages and disadvantages according to their features/user's requirement, Blockchain Technology also has some. While encryption key splitting and consent management on a Blockchain may allow for managed and auditable sharing of data, these technologies are unable to protect shared genomic data from intentional misuse. However, by creating a secure computing environment in which data are processed, the privacy of shared genomic data can be protected. Some organizations have realized the idea of using data algorithms instead of sending data to external systems. For example, Blockstack is a company that creates a general-purpose, decentralized computing network that enables users to provide their own computing and storage resources, regardless of the application can be used wherever the data are located. In terms of genomics, the Global Alliance for Genome and Health (GAGH) Beacon Project has adopted a similar concept. A joint ecosystem of connected genome databases owned by different organizations is the Beacon Network. For example, researchers can submit queries about unique genetic variants and then execute those queries on scattered stored data. The research results will be forwarded to the researchers.

By calculating genomic data, many other protected mechanisms (such as digital signatures, homomorphic encryption, etc.) can be used to protect personal/sensitive information from leakage/disclosure to unauthorized users/malicious users. However, data storage providers and computing facilities may also infringe on privacy when data are decrypted for analysis. Absolutely homomorphic encryption and powerful multiparty computing are privacy-protecting technologies that can help solve this problem. These technologies allow data to be encrypted, just like analyzing plain text, but it remains encrypted during the analysis process, so it is safe. Although the implementation of privacy protection technology is hindered by insufficient efficiency, recent developments have made execution time and scalability more and more reasonable. In previous years for example, the privacy-preserving genomic data process was completed by combining multiple privacy-preserving technologies.[16] We have proposed many frameworks that DTC genomics companies can introduce to enhance privacy and thereby strengthen the protection of personal genomic data. However, these mechanisms also constitute self-imposed constraints that run counter to the business model of extracting the maximum value from the generated genome data. Whether private genome companies with a focus on privacy will flourish will depend on whether consumers are compensated by emphasizing data privacy protection. The general trend is that consumers are paying more and more attention to how companies manage their personal data, and even leading large technology companies to introduce stricter privacy policies and more advanced data protection frameworks.

This segment therefore addresses the significance of genomic and genetic information and needs service for it against established vulnerabilities/attacks. Now the next section continues describing genomic privacy and why it needs to be shielded from malicious users/intruders in this smart age.

18.6 GENOMIC PRIVACY: A TYPE OF PRIVACY NEED TO BE PROTECTED

Every day, we learn new knowledge about the genome, whether the knowledge is of a new association with a particular disease or evidence related to previously reported associations. All information about DNA has not been discovered, and it is almost impossible to give precise meaning, so DNA is regarded as a private asset (or public good). Therefore, with the development of the field of genomics, the perception of privacy sensitivity of genomic data will also evolve. In simple words, when the use of genomic data increases in applications, then several concerns regarding the genomic and genetic data also increase. As genomics is integrated into health care delivery, new privacy and data security risks need to be considered. Moreover, there are various types of privacy problems in today's era, discussed in,[17] and included here as

- Information privacy
- Identity privacy
- Data privacy
- Location privacy
- Genomic privacy

Among all listed privacy, genomic is highly essential one, because it is directly connected to user's personal information, such as DNA, RNA, etc. To put this into context: a false-positive on a blood test could lead to unnecessary procedures, while a false-negative might mean a patient dying from an undiagnosed but serious condition.

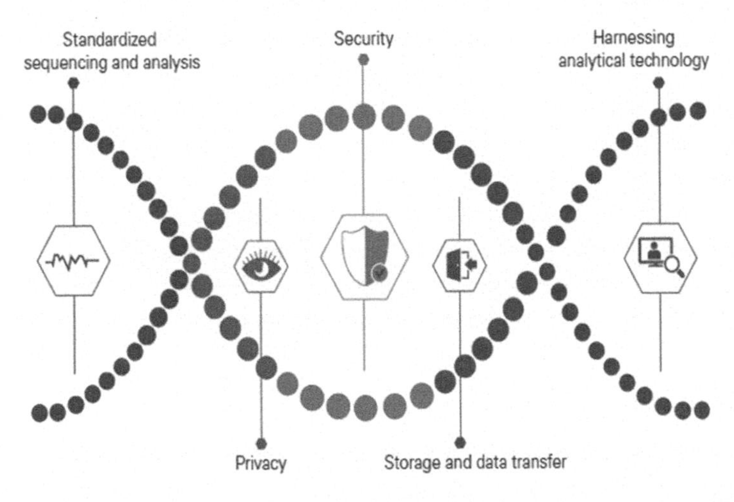

FIGURE 18.1 Security and privacy relationship in genomics.

In short, the privacy issues related to genomic data (Fig.18.1 for the security-privacy relationship) are very complex, especially because such data have a wide range of uses, and the information provided is not limited to the person collecting the data. However, the most important thing is that people may be full of fear of the unknown.

Hence, this section discusses about genomic privacy and need to protect it today and tomorrow. Now, next section will discuss a threat model for genomic privacy, genomic data, and genetic data.

18.7 THREAT MODEL FOR GENOMIC AND GENETIC DATA

Today's open access genome databases are raising some problems. For example, where is the ultimate responsibility for data management and governance? When we bargain, how will personal privacy be challenged? Although genomic data are for research purposes, what happens if a company accesses them for unexpected purposes? As, in the previous years, we have seen a significant shift from on-premises technology into cloud technology, in part to give flexibility of capacity, operational capability and operating costs, and also to provide greater control over genomic data privacy, reliability, and security. In the previous decades, the rate of growth of data generated by machines is increasing exponentially. The existing methods are not efficient enough to process this rapidly generated data especially the genomic data. The main reason is the richness of information that the data hold. Some of the threat models are

(a) Mathematical models are also used to gain insight into the dynamics of transmission and the effects of natural and unnatural emerging and re-emerging infectious diseases that endanger health protection, usually in the form of deterministic systems of nonlinear differential equations.
(b) Other modeling paradigms, such as agent-based and other statistical and stochastic modeling methods powered by data, can also be used to implement the same.
(c) A stochastic risk analysis approach enables model inputs to show a degree of uncertainty. The inputs obey different types of distributions of probability, as opposed to deterministic models. Threat is computed several times by sampling these input distributions. The result of a stochastic risk model is thus a risk distribution, that is, rather than a single value. The main benefit is that it allows for an analysis of the likelihood of certain outcomes in addition to evaluating data.

Various data dangers have the potential to compromise clinical trials in the previous decade. Moreover, networked privacy in clinical genomics is required to be maintained with required efficient security mechanism.

18.7.1 ATTACKS AGAINST GENOME PRIVACY

There has been some re-identification attacks where adversaries/attackers rely on quasi-identifiers, such as demographic information (e.g., links to public records such as voter registers), social media information, and/or the victim's identity. Or search engine records.[18] For example, King and Jobling[19] analyzed the short tandem repeats on the Y chromosome when inquiring the genealogy database, and inferred the personal last name from the (public) anonymous genome data set. Opponents sometimes infer aggregate statistics in the attack based on member statistics to infer whether the target population is part of a research that may be related to the disease. McDowell et al.[20] did this by comparing the target profile with the total number of studies and the reference population collected from public sources. Other methods include the use of leverage interaction statistics for a few hundred SNPs (i.e., single nucleotide polymorphism), while the authors use regression coefficients.[21]By repeatedly sending queries for variants present in the target genome, Raisaro et al. present inference attacks against Beacon,[22] while the attacks are based on microRNA expressions in the paper of Eichmüller et al.[23] More generally, Backes et al.[24] show that even if aggregate statistics are published with significant noise, membership attacks can be successful. For a detailed review of possible/logical inference assaults toward genome privacy, we refer readers to article by Ayday and Humbert.[25]

This section introduces existing work that analyzes security and privacy issues in the context of genome knowledge. Abouelmehdi et al.[26] discuss the different ways that can be used to compromise and protect genome privacy. Possible attacks are divided into three categories: completion attacks, identity tracking attacks and attribute disclosure attacks, and the extent to which they can be overcome by mitigation techniques (e.g., access control, anonymization, and encryption). In their work, Ayday et al.[27] summarize the value of genomics advances and the need for users to protect their privacy when handling their genomic data. In his work, Knoppers[28] discusses about the sharing of genomic data, the potential risks to privacy, and the issues of law and ethics.

Hence, this section discusses the threat model against genomic privacy. The following section will discuss several tools, algorithms, and algorithms (in current) for refining/extracting useful information/making decision from genomic and genetic data.

18.8 TOOLS, MODELS OR ALGORITHMS AVAILABLE FOR REFINING GENOMIC AND GENETIC DATA

In order to understand the genome structure, function, or evolution, it is not enough to obtain DNA sequencing data through next-generation sequencing (NGS). But there is also a need for deep and detailed analysis using bioinformatics approaches. The key road to successful sequence analysis is aligning the sequence of interest with another sequence whose function is known (usually referred to as the reference genome). Notice that the NCBI (National Center for Biotechnology Information) is one of the today's largest repositories of biomedical and genomic information, offering links to many other genomic databases, such as PubMed, Entrez Genome, OMIM, Variation View 23andMe, genomic.org, ancestory.com, etc. we can access the majority of genomic data from many websites, such as 23AndMe, genomic.org, etc. The Personal Genome Project (PGP), generally speaking, is more than just a science repository. The PGP, sponsored by a nonprofit organization called PersonalGenomes.org, is also working to disseminate genomic science and information worldwide, in addition to its publicly available research database, it brings about meaningful and easily accessible advances in the understanding and management of human health and diseases. PGP is also at the forefront of the debate on ethical, legal, and social issues (ELSI) related to large-scale, whole-GS, especially in the areas of privacy, informed consent, and data accessibility. When discussing existing tools to refine genomic data, it is known that many tools are available to refine required data, few names can be included here as:

(a) Bio-weka
(b) Software for quantitative trait loci.
(c) Networks of speech
(d) Gene expression microarray software.
(e) The R Project
(f) Yandell's R Biologists Introduction.
(g) Software Bioconductor.

(h) BioPERL (PERL for Bioinformatics Applications)
(i) Highway of evolution.

General technical advancements made in some next-generation DNA sequencing strategies have a major effect on genetic research. Roche/454 Life Science, Applied Biosystems have produced SOLiD, and Illumina Genome Analyzer which have recently become the most commonly used platforms. So to decrease time complexity and cost, NGS needs to be implemented. Harmanci and Gerstein,[29] in their paper have explored different approaches in order to measure privacy in genomic data. In general, for protein quantification, two well-known techniques/strategies are used based on the purpose of the experiment: Immunoassays or methods of antibody-free detection. Immunoassay, such as the enzyme-linked immunosorbent assay (ELISA), due to its high sensitivity and good specificity, is a commonly used tool. However, when no antibody exists for the protein of interest, researchers may often face the problem. The solution in such instances is antibody-free techniques. On the other hand, mass spectrometry (MS), which calculates the mass to charge (m/z) ratio of ions, is the most effective and systematic analytical instrument for protein detection, recognition, and quantification. MS advancement offers an ability to achieve higher sample density with high accuracy and precision. In addition, for large-scale studies, MS methodology is known to be fast and accurate.

In the past 10 years, we seem to use data mining to discover information for data processing. Generally, data mining refers to methods that try to find trends and definitions from large data sets, of which the Internet is the most obvious example of such databases.[30] Data mining involves "collecting different types of user and consumer information, sometimes when the user knows it, sometimes not collecting this information, and converting it into analytical data points for measurement, classification and classification, in order to achieve various organizations and Institutional goals".[31] But, as these data are increasing with a pace growth by communication of smartdevices, data mining techniques will not be useful. We require efficient and modern tools to refine this large amount of data. To analyze big data, many other tools available today.

- Apache Hadoop
- CDH (Cloudera Distribution for Hadoop)
- Cassandra
- KNIME
- Datawrapper

- MongoDB
- Lumify
- HPCC

Medical researchers and clinicians can use this type of information technology to assist in medical discovery, simplify policies and procedures, and analyze important data for predictive analysis of genomics. These instruments mentioned help a lot in achieving certain (required objectives.

However, as data are decrypted for review, data storage providers and computer facilities can also infringe on privacy. We need some safe and effective mechanism for that. Some of them are identified here as:

18.8.1 PROTECTING PRIVACY AND CONFIDENTIALITY

It is important, as an ethical duty and for the success of the archive, to protect the privacy of individuals who submit specimens for genetic and genomic testing and to preserve the confidentiality of associated health information and research data. Researchers have much to benefit by introducing comprehensive safeguards that in turn, will build an environment of trust and promote recruitment and retention. Two recent attention-grabbing frameworks for strengthening privacy and confidentiality include Confidentiality Certificates and the Genetic Information Nondiscrimination Act. Different methods are described below to protect genomic data privacy.

(a) User anonymity: Anonymous genetic testing allowed by cryptocurrency payments
(b) Data access management: By splitting the encryption key to control multiple parties' access to genomic data.
(c) Document auditability: The immutable storage of data access requests and user permissions on the Blockchain.
(d) Stable data analysis: data analysis augmented by privacy-preserving technologies in managed computing environments.

Other molecular genetic methods can be used for more precise studies, such as microarray-based comparative genomic hybridization (aCGH) or fluorescent in situ hybridization (FISH). However, these techniques have some useful limitations: aCGH does not detect mosaic, balanced translocation and in-versions, while FISH needs specific probes. The main distinction between conventional (i.e., Sanger) technology and NGS is that in massively parallel sequencing technology, the latter is not limited to a single DNA fragment but analyzes millions of fragments. Note that the Sanger sequencing technique was initially established around 20 years ago. If these technologies can achieve a 100-fold improvement, it can be expected to achieve greater improvements by fundamentally changing the chemistry, signal generation, detection sequencing, and instrumentation methods. These methods can integrate chemistry and enzymology, optics and electronics, some major advances in materials science, micromanufacturing, and process control.

Hence, this section discusses available tools, methods in current for extracting useful information from genomic data, and genetic data. The following section discusses several open issues, challenges and provided essential opportunities for future research communities.

18.9 ISSUES, CHALLENGES, AND OPPORTUNITIES IN GENOMICS AND GENETICDATA

Genome science is moving quickly from research laboratories and biobanks to the clinical environment. Sequencing a single human genome takes 24 h and comprises 200 GB of data. It will take 219 years involving 16 million gigabytes to sequence all 80,000 spectators at the 2018 World Cup Final at Luzhniki Stadium in Moscow. In today's age, genomic data safety, reliability, and protection are growing issues, as life sciences businesses aim to comply with strict regulations and verify the outcomes of trials and treatment. The biggest obstacles to address in the next decade are privacy problems and the lack of sufficient professional human-work force and instruments. The truth is that personal data is "public by default, private by effort" in the current sense of network society.

18.9.1 OPEN ISSUES IN GENOMICS AND GENETIC DATA

From academic laboratories and biobanks to clinical environments, genome science has developed rapidly. The resulting big genome data, or large-scale network genetic knowledge, is a pioneering technology. Clinical genomics advances life-saving innovation through precision medicine. On the other hand, the new issue of information exposure to personal privacy comes from the digital database on which it is based. In clinical genomics, privacy issues seem to be a highly specialized field involving a small part of the population. However, as far as digital networks are concerned, whether we are aware of it or not, the protection of health records deeply affects the vast majority of people who use the Internet. For example, a recent study of more than 80,000 health-related websites found that 9 out of 10 visits caused third parties (including online advertisers and data brokers) to leak personal health information. So here was the big question, "So the big question here was," Reacting to some questions.

(a) It is possible to contact and publicly recognize a person's health interests and name.
(b) The purpose of many online databases and algorithm tools is to classify Internet users into categories such as "target" and "waste," and have extremely high insurability, employability, and access to public or retail services.

We may use complex DNA mixtures to correctly re-identify individuals. Using simple allele frequency or genotype counting,[32] genetic data can be de-anonymized from the database. To change the crime scene or medical history, DNA evidence can also be used. However, it is a huge challenge to always protect someone's personal information from intruders. Please note that privacy is not a simple binary file, that is, background issues, personal agency issues and various compromises are increasingly becoming part of the big data innovation transaction. As mentioned in the work of Shamila et al.,[17] different types of privacy protections are known which include identity privacy, data privacy, location privacy, and genome privacy. In order to protect people's data, scientists and doctors need to work together in the near future. In order to continue, genomics not only requires scientists and physicians to cooperate with each other, but also to produce results together in a new fusion space. This collaborative culture presents challenges: even though these groups share a common epistemological background, the organization's culture and priorities are different. For example, the failure of a scientific experiment may indicate progress in the discovery process. The lack of prescribed treatment by oncologists can lead to death. Genomics is a new type of medical literacy that doctors and health practitioners fail to understand and adopt in their work practices.

Many clinicians define the difference between the clinical environment and the research environment as an area of "discomfort." In the current era, scientists and doctors need to have genomic skills, and they can support them even while achieving certain achievements.

(a) New issues, such as using genomic data to cancel the identification of personal data including patients and clinicians.
(b) In view of genomics, it is almost impossible to obtain patient consent under traditional circumstances.

18.9.2 COMMON CHALLENGES IN GENOMICS AND GENETIC DATA

Since the potential use of personal data cannot be determined at the time of data collection, this is also a big problem.

- Data breaches pose another information danger to the privacy of digital genomic data. In the past decade, several individuals have faced many breaches with regard to their health data, as well as government sanctions on intruders for money.
- Rising in data challenge.
- Growing doubt about black markets in particular.

Protection for Medical data also appears today along with data from other applications for example, several counties had attacks on genomic data in the previous decade. The cyber security industry in particular for patient data from clinical trials, emerges as an important one in the Life Sciences sector. The black-market value is much greater than credit card data, genetic data, sensitive health data or the formula for a complex molecular drug. While numerous studies have been performed, scientists still face a major challenge in determining what the sequence means and whether or not a variant detected is pathogenic. A pathogenic variant can contribute to diseases or cause them.

18.9.3 OPPORTUNITIES TOWARD GENOMIC AND GENETIC DATA

Today, we cannot fully protect their privacy, nor can we ensure that the most effective legislation is enacted to protect participants from potential use. Genomic scientists must work with doctors to understand what knowledge is valuable and feasible. Communication between groups in this respect is a crucial technique. Note that the identity person may be verified in public genetic data if the genetic makeup of the person is already identified. We have many genomic and genetic data prospects in the foreseeable future, such as

18.9.3.1 KEEPING DATA SAFE AND USEFUL

By integrating new disruptive forms of personal information into the healthcare system, implementation raises new challenges for doctors, scientists, politicians, and the general public. Not as opposing forces, but as considerations that can be taken into account in the social-technical divide in clinical genomics, medical creativity, and personal privacy are equally important. As the technical structures, societal needs, and developments in medicine clash in convergence regions, today's solutions will not be regarded as adequate. We identified some of the information risks associated with the above genomic data. Finally, we stress the dangers facing medical innovation if there is no progress in sustainable practices in the sharing of genomic data. The information risk of genomic large data is comparable to other forms of personal information in the first instance. The treatment of confidential medical data is a priority for privacy. But genomics model for information-sharing is more applicable than the conventional electronic records. In the near future, genomic data should be considered as a public good and open access database should be developed.

The Life Science sector is also a primary target for IP cyber theft in many other countries including the United States, for popular companies, such as Abbott Laboratories, Boston Scientific and Pfizer. For example, the hack in the computer center of the Maryland Food and Drug Administration revealed confidential data for almost every major drug sold in the United States, including data on drug trials, chemicals, and other data. Moreover, scientists explore how vital genomic data can be used while reducing the likelihood of patients being detected. One method being pioneered is cryptographic "genome cloaking," which claims to hide 97% of each participant's unique genetic information. "Homomorphic encryption" is another route,[33] where scientists are able to decode the final results without ever seeing the source. With genetic data increasingly moving off local servers and onto the cloud, expect more such initiatives to emerge.

As we can find out from the work of Kaul and Prasad ,[34] the cost of cyber attacks continues to rise exponentially each year. Also, with growing data every day, unprecedented exposure to even more nefarious cyber threats and privacy risks will be increased, that is, the chance of breaches arising from insufficient security measures is more. Also, new products will face many ethical challenges, that is, facing difficulties in reaching product to market quicker. In genomic database case, with more parties involved in the clinical trial process, the chance of data being falsified or tampered with increases enormously. Received data can be inaccurate and misinterpreted and cannot be trusted.[36]

Risk specialists, such as internal auditors, have started to explore ways to audit and give assurance over AI considering aspects such as

- Ownership and accountability for AI.
- Corporate values, culture and ethics.
- Hypothesis management.
- Completeness and accuracy.
- Controls over identity and access including the overall Cyber Security controls.
- Logic validation.

18.9.4 TOWARD AFFORDABLE PERSONAL GENOMES

These advancements indicate that technology can be reached and can meet the cost goal of $1000 or less for diploid human genome sequences. In fact, by resequencing individual human genomes in detail, NGS developers have now repeatedly shown that their methods are outdated. The whole genome sequence of at least seven individuals has been reported, and more than one method has been sequenced. Beyond the coverage of the 1000 Genome Project, there are dozens or even hundreds of other unpublished or partially published genomes. Obviously, the era of personal genomics is approaching.

Genetics can offer a quicker market time and a higher success rate, but it could also increase shareholder expectations and place pressure on research and development teams to deliver innovative, competitive products. In the near future, genomics should make it possible for life sciences companies to become more of a fleet in the discovery and promotion of new medicines genomics should soon be able to inspire life sciences companies to extend their footprint in the production and promotion of new medicines. Scientists and physicians are soon to work together in a new area where genomics and genetics converge. For example, through an iterative feedback process, both camps are working in the field of clinical genomics, and colleagues have jointly constructed facts and explanations of research progress. In this section, we emphasize the importance of interoperability and evaluate the effectiveness of such features in a privacy-improving environment. We also call for research on new genome manipulation technologies, such as C RISPR[1] and its possible impact on security (e.g., harmful) and privacy (e.g., editing the genome to recover from data exposure or hinder re-identification).

It should be noted that CRISPR is a family of DNA sequences in bacterial genomes, as well as structures in prokaryotes. Finally, in wider biomedical contexts, the research should be expanded to include users' health and genomic data. We also address numerous useful potential challenges and opportunities in this segment. Now, the following section ends this work with a brief thorough description of this chapter. Lastly, future researchers are suggested to refer[37-42] to know several interesting issues, challenges and opportunities towards healthcare applications/ sectors.

18.10 CONCLUSION

Admirable progress has been made in the privacy of the genome population. For example, the first human genome was sequenced in 2003 for more than 13 years and US$3 billion,[2] but today this could be achieved for a small fraction of the cost within a matter of hours. The cost of a human diploid genome sequence fell from around 70 million dollars to 2000 dollars.We learn more about genome, genomics, and many important problems in the same field, as discussed previously and in the paper of Tyagi et al.,[35] In this research, a systematic approach has been implemented and applied to standardize information about privacy of genomes based on defensive mechanisms that privacy technologies (PETs) offer. In this article, we also request the genome privacy experts to evaluate the importance and difficulty of solving this problem, and provide suggestions on how to minimize it. Today, there are many proven research gaps, such as

(a) High performance technologies, such as quantitative real-time polymerase chain reaction to reactive treatment, next-generation sequences or mass specs have been developed in order to expand current technological limits. These provide opportunities for genome-wide research.
(b) Large quantities of data researchers are still facing with the issue which is typically very time-consuming and complicated. That is why continuous technological advances are required, and more effective analytical tools are created. It is important to remember that complex methodologies that complement each other's shortcomings are key to more detailed rework outcomes.

Briefly, we discovered that the barriers associated with special human genome properties to protect the privacy of the genome may hinder PETs.

For example, since the sensitivity of genomic data does not deteriorate over time, the lack of long-term protection is a major challenge, and it is difficult to solve because the available cryptographic tools are not enough to solve this problem. We also found that most of the technologies proposed for large genomic data sets must choose weaker security guarantees or weaker models. In essence, it is difficult to manage the utility and/or flexibility of the actual function. When combined with assumptions about the format and interpretation of the inspection data, this can cause major difficulties for real-life applications. The interdependence of certain evolving meanings exacerbates these problems. For example, the use of cloud storage for genomic data involves the intervention of a third party, so the increase in usability will be overshadowed by security restrictions. The role of the Internet, digital databases, and IT society in the implementation of public health scientific discoveries.

KEYWORDS

- **genetic information**
- **genome**
- **genomic privacy**
- **genetic analytics**
- **healthcare**
- **gene expression**

REFERENCES

1. Hsu, P. D.; Lander, E. S.; Zhang, F. Development and Applications of CRISPR-Cas9 for Genome Engineering. *Cell* **2014,** *157*(6), 1262–1278.doi:10.1016/j.cell.2014.05.010
2. NHGRI: All about the Human Genome Project (HGP). www.genome.gov/10001772
3. Nambiar, R.; Bhardwaj, R.; Sethi, A.; Vargheese, R. A Look at Challenges and Opportunities of Big Data Analytics in Healthcare. *2013 IEEE International Conference on Big Data*, 2013.doi:10.1109/bigdata.2013.6691753
4. Contreras, J. Bermuda's Legacy: Policy, Patents and the Design of the Genome Commons (SSRN Scholarly Paper No. ID 1667659); Social Science Research Network: Rochester, NY, 2010.http://papers.ssrn.com/abstract=1667659
5. Reeves, G. T.; Hrischuk, C. E. Survey of Engineering Models for Systems Biology. *Comput. Biol. J.* **2016,** *2016*, Article ID 4106329, 12 pages. https://doi.org/10.1155/2016/4106329.

6. Gulland, A. Project to Decode Genomes in Cancer Samples Promises New Treatments. *BMJ* **2010,** *340*, c2149.
7. FrizzoBarker, J.; Chow-White, P. A.; Charters, A.; Ha, D. Genomic Big Data and Privacy: Challenges and Opportunities for Precision Medicine. Comput. Support. Cooperative Work (CSCW) **2016,** *25*(2–3), 115–136.doi:10.1007/s10606-016-92487
8. Robertson, J. A. The $1000 Genome: Ethical and Legal Issues in Whole Genome Sequencing of Individuals. *Am. J. Bioethics* **2003,** *3*(3), 35–42. doi:10.1162/152651603322874762
9. Guerrini, R.; Noebels, J. How Can Advances in Epilepsy Genetics Lead to Better Treatments and Cures? *Adv. Exp. Med. Biol.* **2014,** 309–317.doi:10.1007/978-94-017-8914-1_25
10. Wang, S.; Jiang, X.; Singh, S.; Marmor, R.; Bonomi, L. et al. Genome Privacy: Challenges, Technical Approaches to Mitigate Risk, and Ethical Considerations in the United States. *Ann. NY Acad. Sci.* **2016,** *1387*(1), 73–83.doi:10.1111/nyas.13259
11. Nayeri, N. D.; Aghajani, M. Patients' Privacy and Satisfaction in the Emergency Department: A Descriptive Analytical Study. *Nurs. Ethics* **2010,** *17*(2), 167–177. doi:10.1177/0969733009355377
12. Naveed, M.; Erman, A.; Jacques, F. et al., Privacy in the Genomic Era. *ACM Comput. Surveys* **2015,** *48*, 1–44.
13. Greenbaum, D.; Sboner, A.; Gerstein, M. Genomics and Privacy: Implications of the New Reality of Closed Data for the Field. *PLoS Comput. Biol.* **2011,** *7*(12), e1002278. https://doi.org/10.1371/journal.pcbi.1002278
14. Raghupathi, W.; Raghupathi, V. Big Data Analytics in Healthcare: Promise and Potential. *Health Info. Sci. Syst.*, 2014.
15. Cao, L.; Zhang, H.; Zhao, Y.; Luo, D.; Zhang, C. Combined Mining: Discovering Informative Knowledge in Complex Data. *IEEE Trans. Syst. Man Cybernet. B June* **2011,** *41*(3), 699–712.
16. Akgün, M.; Bayrak, A. O.; Ozer, B. et al. Privacy Preserving Processing of Genomic Data: A Survey. *J. Biomed. Inform.* **2015,** *56*, 103–111.
17. M. Shamila, Amit Kumar Tyagi, Genetic Data Analysis, Book: Handbook of Research on Disease Prediction Through Data Analytics and Machine Learning, 2021, Pages: 15, DOI: 10.4018/978-1-7998-2742-9.ch017.
18. Gillala Rekha, Amit Kumar Tyagi, and V. Krishna Reddy, "A Wide Scale Classification of Class Imbalance Problem and its Solutions: A Systematic Literature Review", Journal of Computer Science, Vol.15, No. 7, 2019, ISSN Print: 1549–3636, pp. 886–929.
19. King, T. E.; Jobling, M. A. What's in a Name? Y Chromosomes, Surnames and the Genetic Genealogy Revolution. *Trends Genet.* **2009,** *25*(8), 351–360.doi:10.1016/j.tig.2009.06.003
20. Kumari, S., Muthulakshmi, P., Agarwal, D. (2022). Deployment of Machine Learning Based Internet of Things Networks for Tele-Medical and Remote Healthcare. In: Suma, V., Fernando, X., Du, KL., Wang, H. (eds) Evolutionary Computing and Mobile Sustainable Networks. Lecture Notes on Data Engineering and Communications Technologies, vol 116. Springer, Singapore. https://doi.org/10.1007/978-981-16-9605-3_21
21. Erman, A.; Louis, R. J.; Paul, J. et al. *Privacy-Preserving Computation of Disease Risk by Using Genomic, Clinical, and Environmental Data*; Healthtech13: Washington, 2013.
22. Raisaro, J. L.; Tramèr, F.; Ji, Z.; Careyat, K. al., Addressing Beacon Re-identification Attacks: Quantification and Mitigation of Privacy Risks. *J. Am. Med. Info. Assoc.* **2017,** *24*(4), 799–805. doi:10.1093/jamia/ocw167

23. Eichmüller, S. B.; Osen, W.; Mandelboim, O.; Seliger, B. Immune Modulatory microRNAs Involved in Tumor Attack and Tumor Immune Escape. *JNCI* **2017,** *109*(10). doi:10.1093/jnci/djx034
24. Backes, M.; Berrang, P.; Humbert, M.; Manoharan, P. Membership Privacy in MicroRNA-based Studies. *Proceedings of the 2016 ACM SIGSAC Conference on Computer and Communications Security—CCS'16*, 2016. doi:10.1145/2976749.2978355
25. Ayday, E.; Humbert, M. Inference Attacks against Kin Genomic Privacy. *IEEE Sec. Priv.* **2017,** *15*(5), 29–37. doi:10.1109/msp.2017.3681052
26. Abouelmehdi, K.; Beni-Hessane, A.; Khaloufi, H. Big Healthcare Data: Preserving Security and Privacy, *J. Big Data* **2018,** *5*, 1.https://doi.org/10.1186/s40537-017-0110-7
27. Ayday, E., Raisaro, J.L., Hengartner, U., Molyneaux, A., Hubaux, JP. (2014). Privacy-Preserving Processing of Raw Genomic Data. In: Garcia-Alfaro, J., Lioudakis, G., Cuppens-Boulahia, N., Foley, S., Fitzgerald, W. (eds) Data Privacy Management and Autonomous Spontaneous Security. DPM SETOP 2013 2013. Lecture Notes in Computer Science(), vol 8247. Springer, Berlin, Heidelberg. https://doi.org/10.1007/978-3-642-54568-9_9.
28. Knoppers, B. M. Framework for Responsible Sharing of Genomic and Health-Related Data, *HUGO J.* **2014,** *8*(1). doi:10.1186/s11568-014-0003-1
29. Harmanci, A.; Gerstein, M. Quantification of Private Information Leakage from Phenotype-Genotype Data: Linking Attacks. *Nat Methods* **2016,** *13*, 251–256. doi:10.1038/nmeth.3746
30. Ramaswamy, S.; Rastogi, R.; Shim, K. Efficient Algorithms for Mining Outliers from Large Data Sets. *Proceedings of the 2000 ACM SIGMOD International Conference on Management of Data–SIGMOD '00*, 2000. doi:10.1145/342009.335437
31. Amit Kumar Tyagi, Poonam Chahal, "Artificial Intelligence and Machine Learning Algorithms", Book: Challenges and Applications for Implementing Machine Learning in Computer Vision, IGI Global, 2020.DOI: 10.4018/978-1-7998-0182-5.ch008.
32. Mathais, H.; Kevin, H.; Ayday, E. et al., De-anonymizing Genomic Databases Using Phenotypic Traits. *Proc. Priv. Enhanc. Technol.* **2015,** *2015* (2), 99–114.
33. Kim, M.; Lauter, K. Private Genome Analysis through Homomorphic Encryption. *BMC Med. Inform. Decis. Mak.* **2015,***15*. doi:10.1186/1472-6947-15-S5-S3
34. Kaul, C.; Prasad, B.M.K. Analysis of the Cyber Attacks over the Past Decade. *IJIET* Dec **2015,** *6*(2).
35. Tyagi, Amit Kumar; Nair, Meghna Manoj; Niladhuri, Sreenath; Abraham, Ajith, "Security, Privacy Research issues in Various Computing Platforms: A Survey and the Road Ahead", Journal of Information Assurance & Security. 2020, Vol. 15 Issue 1, p1–16. 16p.
36. Shamila M, Vinuthna, K. & Tyagi, Amit. (2019). A Review on Several Critical Issues and Challenges in IoT based e-Healthcare System. 1036–1043. 10.1109/ICCS45141.2019.9065831.
37. Tyagi, A. K.; Gupta, M. Aswathy S. U., Ved, C. Healthcare Solutions for Smart Era: An Useful Explanation from User's Perspective. In *Recent Trends in Blockchain for Information Systems Security and Privacy*. CRC Press, 2021.
38. Madhav A. V. S.; Tyagi A. K. The World with Future Technologies (Post-COVID-19): Open Issues, Challenges, and the Road Ahead. In: Tyagi A.K., Abraham A., Kaklauskas A. (eds) Intelligent Interactive Multimedia Systems for e-Healthcare Applications. Springer, Singapore. 2022. https://doi.org/10.1007/978-981-16-6542-4_22

39. Nair M. M.; Kumari S.; Tyagi A. K.; Sravanthi K. Deep Learning for Medical Image Recognition: Open Issues and a Way to Forward. In: Goyal D., Gupta A.K., Piuri V., Ganzha M., Paprzycki M. (eds) Proceedings of the Second International Conference on Information Management and Machine Intelligence. Lecture Notes in Networks and Systems, vol 166. Springer, Singapore. 2021. https://doi.org/10.1007/978-981-15-9689-6_38
40. Tyagi, A. K.; Aswathy, S. U., Aghila, G., Sreenath, N. AARIN: Affordable, Accurate, Reliable and INnovative Mechanism to Protect a Medical Cyber-Physical System using Blockchain Technology. *IJIN* **2021,** *2*, 175–183.
41. Tyagi, A. K.; Nair, M. M.; Deep Learning for Clinical and Health Informatics. In *Computational Analysis and Deep Learning for Medical Care: Principles, Methods, and Applications*, 2021, DOI: https://doi.org/10.1002/9781119785750.ch5
42. Kumari S.; Vani V.; Malik S.; Tyagi A. K.; Reddy S. Analysis of Text Mining Tools in Disease Prediction. In: Abraham A., Hanne T., Castillo O., Gandhi N., Nogueira Rios T., Hong TP. (eds) Hybrid Intelligent Systems. HIS 2020. Advances in Intelligent Systems and Computing, vol 1375. Springer, Cham. 2021. https://doi.org/10.1007/978-3-030-73050-5_55.

Index

A

AI-based robotics in e-healthcare
applications, 249
biomedical devices in hospitals, 257–260
case studies (medical robots in healthcare), 261
bots for stem cell growth, 263–264
bots in disease detection and treatment, 263
brain biopsy robot, 264
Chabot's, 263
radio surgery for tumor, 264–266
rehabilitation robots, 262
robotic nurse, 262
sanitizing and disinfecting robots, 263
spine assist robot, 266–267
surgical assistant, 261–262
deepbot in rehabilitation robots, 260–261
robots for tracking, monitoring patients, 251–252
deep learning-based robot for patient monitoring, 254–256
deepbot, communication module in, 256–257
patient monitoring system, 252–254
Automated health monitoring system, 271
background, 274–282
fourth revolution in healthcare technology, 282–283
advanced automated in, 284
artificial intelligence, 285–286
big data, 285
cloud computing, 286
cyber physical systems, 283–284
intelligent sensing, 285
internet of health things, 284
robotics, 286
internet of things (IOT)
applications, 273–274
challenges in, 273
cloud layer, 290–291
end device layer, 287–289
in field of healthcare, 272
fog layer, 289–290
improving healthcare, 287
methodology and implementation, 292–293
pill dispensing system, 291–292
system analysis, 293–294

B

Big data visualization, 4
Biomedical data analysis, 297
background work, 305–306
brain networks, 309
activity process, 310–311
neuroimaging data, 309–310
challenges, 316–318
future research directions, 316–318
genomic association studies, 313–314
image processing, 301
genome and genomics, 302
next generation sequencing data, 301
proteomics and metabolomics, 302–305
imaging, 300–301
imaging today and tomorrow
scope of, 307–309
informatics, 299
molecular genetics, 311–312
motivation, 306–307
open issues, 316–318
and population genetics, 311–312
tools available for analyzing, 314–315
analytic *vs.* previous approaches, 315
biological network-based integration, 315–316
types, 312
SLY, 312–313
Blockchain for wearable internet of things, 223
challenges, 231
communication, 232

healthcare infrastructures, 232–236
sensors for, 236–237
wearable devices, 231
internet of things (IOT), 224–225
wearable devices, 225–227
wireless communication standards, 227–228
operating systems, 228–230
digital forensics, role, 230–231
security, 238–239, 245–246
challenges, 238
different levels, 239–241
technology, 241–242
Bitcoin, 242–243
blockchain, 243–244
working, 244
Brain networks, 309
activity process, 310–311
neuroimaging data, 309–310

C

Case studies, 23–24
medical robots in healthcare, 261
brain biopsy robot, 264
Chabot's, 263
disease detection and treatment, bots in, 263
radio surgery for tumor, 264–266
rehabilitation robots, 262
robotic nurse, 262
sanitizing and disinfecting robots, 263
spine assist robot, 266–267
stem cell growth, bots, 263–264
surgical assistant, 261–262
Cloud applications, 373
challenges, 383–384
motivation, 379
open discussion, 387–389
popular issues face, 379–383
related work, 376–379
suggested solutions, 386–387
Computer vision and artificial intelligence, 69
background, 73–75
methodology, 75–78
slam algorithm, 78–81
3D Convolutional neural network
lung nodule detection, 85
experimental results, 94–99
related work, 90–91
system architecture, 91–92
system architecture
data acquisition, 92
lung nodule detection, 94
preprocessing, 93–94

D

Data mining techniques, 105
experimental results, 114–127
literature review, 107–109
preprocessing, 109–110
proposed methodology, 110
logistic regression, 113
naïve bays, 110–111
random forest, 112–113
scaled conjugate gradient, 113–114
Data visualization, 4–5
health care
area chart, 29–30
bar chart, 24
circle chart, 24–25
collection, 22–23
data wrapper, 22
dimension of data collected, 23
emergency wait time, 20–21
forms of visuals, 6–8
fusion charts, 22
high charts, 22
hospital readmission rates, 16–17
importing, 23
line chart, 29
measurements, 14
packed bubbles, 26
patient safety, 19–20
patient satisfaction, 18–19
patient wait time, 17
QlikView, 22
scatter plot, 28
SEO (search engine optimization), 8–9
sisense, 9–10, 22
tableau, 21
test and deploy, 23
treatment costs, 15–16
tree map, 27
visualize and monitor in, 23
transforming, heath care, 10
custom data visualization, 13

infographics and mini-infographics, 11–12
interactive widgets, 13
leaders, 12–13
motion graphics, 13–14

F

Fetal electrocardiography (FCG), 201–205
online available FECG database, 205–206
performance assessment tools, 206–208
signals, algorithms, 213–218

G

Genomics and genetic data, 395
analysis, 405–407
applications, importance, 402–403
revolutions, genomic databases or genetic analysis, 403–405
challenges and opportunities, 414
issues, 414
affordable personal genomes, 418–419
common challenges, 416
keeping data safe and useful, 417–418
open issues, 414–416
opportunities, 416
motivation, 402
privacy, 407–409
related work, 399–400
genomic data sharing, improvement, 401
19th century to 21st century, 400
threat model, 409–410
attacks, 410–411
tools, models or algorithms, 411–413
protecting privacy and confidentiality, 413–414

H

Healthcare 4.0, 325
artificial lung on, 340–341
futuristic approach, 327–330
centered patient medical care, 331
instruments, 335–340
precautionary and persistent medical care, 330–331
precision and personalized medical-care, 331
lung research, 341–343
respiratory system-oriented challenges, 331–332
process for test, 332–335
techno-commercial business model, 343–345
Healthcare technology
fourth revolution in, 282–283
advanced automated in, 284
applications, 273–274
artificial intelligence, 285–286
big data, 285
cloud computing, 286
cyber physical systems, 283–284
intelligent sensing, 285
internet of health things, 284
robotics, 286

I

Image fusion, 56–57
correlation measure, 62
entropy, 61
factor, 61–62
peak signal-to-noise ratio (PSNR), 62–63
performance metrics for, 61
Image processing, 301
genome and genomics, 302
next generation sequencing data, 301
proteomics and metabolomics, 302–305
Internet of things (IOT), 224–225
applications, 273–274
challenges in, 273
cloud applications, 373
challenges, 383–384
motivation, 379
open discussion, 387–389
popular issues face, 379–383
related work, 376–379
suggested solutions, 386–387
cloud layer, 290–291
end device layer, 287–289
in field of healthcare, 272
fog layer, 289–290
improving healthcare, 287
methodology and implementation, 292–293
pill dispensing system, 291–292
system analysis, 293–294
wearable devices, 225–227
wireless communication standards, 227–228

K

Knee osteoarthritis
  detection and classification, 151
    literature survey, 155–157
    microtexture descriptor, 160–161
    proposed work, 157–159
    result and discussion, 162–163
    structure, 153–154

L

Laplacian pyramid transform (LPT), 58
Lifestyle revolution, 349
  angles of lifestyle, 354–355
  big data and lifestyle, 361–368
  Covid-19 and changes in Indian, 361
  history, importance of, 353–354
  literature review, 352–353
  MDPR lifestyle
    pets, 357, 358
  research problem statement, 351–352
    insights, 352
  yoga- a way to healthy life, 358–360
    promotion, 360
Lung nodule detection, 85
  experimental results, 94–99
  related work, 90–91
  system architecture, 91–92

M

Machine learning (ML), 199
  bar graph, 45
  coronary illness, 34
  dataset
    scaling and imputing, 45
  design and implementation
    data flow diagram, 43
    decision tree, 39
    logistic relapse, 40–41
    naïve bayes (NB), 39
    support vector machine (SVM), 41–43
    UML diagram, 44
  developments in, 208–209
    algorithms performance assessment tools, 212–213
    popular algorithms, 209–212
  fetal electrocardiography (FCG), 201–205
    online available FECG database, 205–206
    performance assessment tools, 206–208
    signals, algorithms, 213–218
  methodology
    coronary illness forecast, 35
    past exploration contemplates, 36
  performance evaluation
    accuracy, 47
    F1 score, 48
    precision, 47
    recall, 47
    ROC (receiver operator characteristic), 48–50
    specificity, 47
  results and conclusion
    data preparation, 44–45
    exploratory analysis, 46–47
  system architecture
    accuracy measure, 38
    attribute selection, 38
    classifiers, 38
    dataset, 37
    disease prediction, 38
    preprocessing framework, 37–38
    rescaling data, 36
    standardizing data, 36
Mathematical model of Covid-19 diagnosis prediction, 131
  background, 136
    computed tomography scan diagnosis, 137
    infection prediction, 137–138
    x-ray diagnosis using deep learning, 136–137
  discussion, 144–146
  experimental analysis, 142
    evaluation metrics, 142–144
  methodology, 138
    dataset, 139–142
  related work, 134–136
Medical image processing, 53
  image fusion, 56–57
    correlation measure, 62
    entropy, 61
    factor, 61–62
    peak signal-to-noise ratio (PSNR), 62–63
    performance metrics for, 61
    root mean square error (RMSE), 63
    standard deviation, 61

structural similarity index measurement (SSIM), 62
proposed methodology, 57
discrete wavelet transform, 59
Laplacian pyramid transform (LPT), 58
multiresolution singular value decomposition, 60
wavelet transform, 58–59
working procedure, 60
sample results of, 63–66
Medical robots, 261
brain biopsy robot, 264
Chabot's, 263
disease detection and treatment, bots in, 263
radio surgery for tumor, 264–266
rehabilitation robots, 262
robotic nurse, 262
sanitizing and disinfecting robots, 263
spine assist robot, 266–267
stem cell growth, bots, 263–264
surgical assistant, 261–262

P

Peak signal-to-noise ratio (PSNR), 62–63
Predicting pandemic diseases, 167
machine learning (ML), 170–171
approach, 175–179
epidemics and methodology, 171–174
open discussion, 193
prediction and discussion, 190–193
proposed theoretical models, 179–190
sensor cloud, 169–170

R

Robots
tracking, monitoring patients, 251–252
deep learning-based robot for patient monitoring, 254–256
deepbot, communication module in, 256–257
patient monitoring system, 252–254
ROC (receiver operator characteristic), 48–50
Root mean square error (RMSE), 63

S

Structural similarity index measurement (SSIM), 62
Support vector machine (SVM), 41–43
System architecture
accuracy measure, 38
attribute selection, 38
classifiers, 38
dataset, 37
disease prediction, 38
preprocessing framework, 37–38
rescaling data, 36
standardizing data, 36

T

Tracking, monitoring patients, 251–252
deep learning-based robot for patient monitoring, 254–256
deepbot, communication module in, 256–257
patient monitoring system, 252–254

X

X-ray diagnosis using deep learning, 136–137